D0389548

A
MIND
SO RARE

A MIND SO RARE

The Evolution of Human Consciousness

MERLIN DONALD

W. W. NORTON & COMPANY
New York / London

Printed in the United States of America
First Edition

The text of this book is composed in Adobe Garamond
with the display set in Amanda Inline
Composition by Gina Webster
Manufacturing by Maple-Vail Book Manufacturing Group
Book design by Judith Stagnitto Abbate/Abbate Design
Production manager: Leelo Märjamaa-Reintal

Library of Congress Cataloging-in-Publication Data

Donald, Merlin, 1939–
A mind so rare : the evolution of human consciousness /
by Merlin Donald.
p. cm.
Includes bibliographical references and index.
ISBN 0-393-04950-7
1. Consciousness. I. Title.
BF311.D57 2001
153—dc21 00-053721

W. W. Norton & Company, Inc., 500 Fifth Avenue, New York, N.Y. 10110
www.wwnorton.com

W. W. Norton & Company Ltd., Castle House, 75/76 Wells Street,
London W1T 3QT

1 2 3 4 5 6 7 8 9 0

Let the Laces Decide

At the edge of my visual fishbowl
a melted computer
whines and whispers.

An intrusive neon bulb
shimmers indistinctly above
an arrow of desk aimed at my heel.

My shoes resting on the table
droop their knotted laces
into the momentary center of my world.

Room and body radiate away in warm brown rings,
books conspire in coves,
slightly out of focus, murmurming in groups,

perched like gannets on a cliff,
ready to soar out
and pick over my mind.

What do they want with me?
Should I shoot them down in midflight,
or should I lay down my arms

and welcome the invasion
of other, possibly dangerous
minds? I waffle.

Perhaps I should hire a diviner
to read the knotted entrails of my shoes,
and let the laces decide.

Contents

Prologue

There is no whole self. Any of life's present situations is seamless and sufficient. Are you, as you ponder these disquietudes, anything more than an indifference gliding over the argument I make, or an appraisal of the opinions I expound?

—JORGE LUIS BORGES

Science can often lead us in directions that are very counterintuitive. For example, the Ancients understandably expected that the commonsense "atoms" of Democritus would prove to be solid—that is, that they would epitomize the most obvious property of solid matter. Similarly, they expected that transformations of matter into gas or into liquid should presumably reflect, on the atomic level, the surface appearance of such transformations. Thus they expected that atoms should melt, flow, and rigidify and that when water evaporated or wood burned, the corresponding atoms should also vaporize or burst into flames.

Of course, atoms do no such things. They are made up mostly of empty space. They are not the irreducible particles the Ancients expected. Rather, they are elaborate little universes in themselves, made up of even more microscopic, or subatomic, particles, some of which at the end of the day don't look at all material. In fact, the line between matter and energy gets blurry at this level, leading to another counterintuitive conclusion: that matter and energy are interchangeable. Many people still have great difficulty accepting that matter itself is ultimately insubstantial and that the world's appearance of solidity is more a reflection of how our sensory systems are constructed, and how brains do their work, than of anything else.

The point is, theories that may appear absolutely crazy to common sense, or much too speculative, may turn out to be true. To discover such explanatory theories, scientists have always been willing to take risks—carefully argued risks, to be sure—but risks nonetheless. As the Nobel laureate Niels Bohr is supposed to have said to a colleague, "We all agree that your theory is crazy; where we disagree is whether it is crazy enough."

The same principle applies to the mind. Many of the so-called theories of psychology are really too commonsensical to be called theories. For instance, the Freudian notion of unconscious motivation, the Behavioristic idea that reward and punishment govern behavior, and the theory that old memories never really decay are just elaborations of common sense. They achieve little more than to hammer home what Granny probably suspected all along. We should really be asking deeper questions, such as how the brain could represent something as complex and subtle as a "reward" in the first place and how mere neurons could record a disposition as rich as a lasting grudge or a deep hostility toward authority. Or, more perplexing, we should wonder how a brain, built from proteins that last, at the most, a few days, can keep a memory fresh for eighty years. Even memories of unimaginable complexity, such as a double-bind relationship with one's mother, can persist for a lifetime in a person whose body has not retained a single atom of its younger physical self. How is this possible? Commonsense theories will never tell us because they have avoided such questions in the first place.

Commonsense theories of mind are, in a sense, failures of will. They are reassuring, but they demonstrate our avoidance of the fact of our own strangeness to ourselves. We don't like to confront what many scientists expect must be the case: that a deeper theory of how the human mind works may be just as alien to our common sense as our theories of the physical universe have turned out to be. But before we have theories of the mind that work, we may have to abandon some of our more comfortable homilies and categories. This sentiment has been stated often, but in spite of this, we have rarely taken the speculative risks with the mind that we have with matter. Where we have done so, we have sometimes had to encounter disturbing ideas, such as the notions that conscious awareness is merely an illusion and that our brains are ultimately as empty, and insubstantial, as the atoms from which they are constructed. Many people, including scientists, are terrified that the ethical, social, and political stakes are just too high to venture down such roads. But fear is no basis for rejecting reasoned ideas or important questions. Even though they encroach on forbidden territory, they deserve our consideration.

One such question centers on our individuality. It is common sense to assume that the individual mind starts in the brain, with a set of innate capacities, sufficient to cope with the challenges of life. On this assumption, a baby comes fully equipped to learn about the world, acquiring its own special experiences and memories with the tools it has been given. Certainly, this idea seems to apply very well to most other species, and surely it applies to us as well.

But this is where our common sense can mislead us. Unlike that of other species, the human mind has a collective counterpart: culture. We stubbornly adhere to the idea that we are distinct individuals, yet we are also highly cultural beings. Indeed, humanity might be defined as the only species on earth that combines individual with collective cognitive processes and in which the individual can identify with, and become part of, a group process. We can see this in our corporations and other institutions. The life of the human imagination oscillates between these two polar extremes, individual and corporate. In our most inward-looking moods, we have the impression that our minds are the only enduring reality for us and that the world "outside" is either an illusion or, at best, a source of experience. In our most outward-directed moods, such as in moments of war hysteria, while joined closely to a corporate process, we meld with the group, "losing ourselves" in whatever the group dictates.

Importantly, the cognitive tools that we use to do much of our thinking seem to be dependent on our cultural institutions, and all our symbolic tools are imported from outside—that is, from culture. This raises questions about the sources of human awareness and the role of consciousness in a being that is capable of such intense collective identification. This book proposes that the human mind is unlike any other on this planet, not because of its biology, which is not qualitatively unique, but because of its ability to generate and assimilate culture. The human mind is thus a "hybrid" product of biology and culture. It is important to realize that I am referring to the mind itself, not merely particular experiences. The human mind cannot come into existence on its own. It is wedded to a collective process, and the very sources of its experience are filtered through culture. The generation of culture is thus a key question in human evolution.

The key to understanding the human intellect is not so much the design of the individual brain as the synergy of many brains. We have evolved an adaptation for living in culture, and our exceptional powers as a species derive from the curious fact that we have broken out of one of the most critical limitations of traditional nervous systems—their loneliness, or solipsism. From our earliest birth as a species, humanity has relied upon creating "distributed" systems of thought and memory, in which intellectual work is shared across many nervous systems. I articulated this idea in an earlier book on human cognitive evolution, in which I said that the progress of the human mind could be adequately described only in terms of its cultural achievements, which in turn reflected an emerging cognitive-cultural symbiosis. I do not want to suggest, however, that there is a "group mind." Our minds are still very much sealed into their biological containers. But they can do remarkably little on their own. They depend on culture for virtually everything that is unique to the human world, including our basic communicative and thought skills.

The word "culture" usually connotes something other than its cognitive aspect. It usually refers to a set of shared habits, languages, or customs that define a population of people. It may be these things, but on a deeper level, any given culture is a gigantic cognitive web, defining and constraining the parameters of memory, knowledge, and thought in its members, both as individuals and as a group. Cultures can differ hugely in their power in this regard. Our cultures are thus our best friends and worst enemies, because we rely on them so completely for many of our most crucial mental powers while at the same time they threaten our intellectual autonomy. They can rob us of the freedom to think certain kinds of thoughts, and even in the most supposedly liberal of cultures, few of our ideas and experiences can really be called our own, so thoroughly have they been washed and filtered through the fine cloth of the culture itself.

All very well, but how did our archaic ancestors escape the confines of their isolated nervous systems? Where could this ethereal cultural network, which lifted humanity out of the intellectual prison it previously inhabited, have come from? The answers I propose will surprise many. In this book I suggest that culture itself, as well as its two principal by-products, languages and symbols, are consequences of a radical change in the nature of consciousness. We are the species that invented culture, but the roots of that remarkable public invention lie, of all places, in our most private place, the conscious mind. Moreover, that place, as we shall see, is not at all what we usually think it is.

A
MIND
SO RARE

1

Consciousness in Evolution

Successive decompositions of the more complex phenomena of intelligence into simpler ones, and these again into still simpler ones, have at length brought us down to the simplest; which we find to be nothing else than a change in the state of consciousness. This is the ultimate element out of which alone are built the most involved cognitions.

—HERBERT SPENCER

Many books have been written on the subject of consciousness, and at least as many on human evolution. But they do not tend to cross-reference one another very often. The few evolutionary thinkers who have shown an interest in consciousness seem to believe that it is Public Enemy Number One, a devilish distraction with no functional importance, which deserves to be banished from the Academy. In 1991 the philosopher Daniel Dennett wrote an influential work called *Consciousness Explained* (which might more aptly have been entitled *Consciousness Explained Away*), in which he argued that consciousness has virtually no influence on mental life. In itself, that book could have led in several directions, but a few years later, in another book, he had an astonishing, St. Paul–like conversion to hard-line neo-Darwinism. On reflection, I shouldn't have been surprised by this. He had merely pushed his own brand of determinism to its logical conclusion.

The neo-Darwinian Hardliners Dennett has embraced are a staunch band of revolutionaries (or inadvertent reactionaries, depending on your point of view) who believe in applying the strong form of Darwin's ideas to human nature. They share an uncompromising belief in the irrelevance of the conscious mind and the illusory nature of free will. Theirs is a self-complete world view, rather like that of neoconservative economists, inasmuch as they tolerate little dissent, except over minor matters of detail. One of the most articulate of these thinkers is MIT psychologist Steven Pinker, who has appropriated large chunks of modern psychology for his agenda. He has argued that human nature, including the intellect itself, is fixed in genetic concrete. Thinking, social behavior, emotions, and language are predetermined in their development, form, and usage, down to a startling level of detail. For Pinker, we don't just have emotions. We have them for very specific reasons, and they play themselves out in predictable scenarios that are culturally universal. We don't just see. We parse the visual world in a very particular manner. He also claims that we have a specialized social intelligence, with built-in cognitive tools, such as cheater detectors, built into our brains, which can quickly recognize people who do not give a good return on emotional investments. These capacities are wired in, complex, and completely unconscious and automatic. They have to be, to work so efficiently.

To Hardliners, the innateness principle extends even to the highest games of the intellect. We supposedly carry around a whole armamentarium of unconscious cognitive weapons that come into play, automatically and unreflectively, whenever we engage with society. These weapons include analytic thought and language, which originated in what they call the "cognitive arms race" that started with our ancestors' first great brain expansion, two million years ago. Hardliners don't deny consciousness, but they trivialize it. They reduce its role to that of a facilitator, like those that run executive training seminars. Facilitators don't say anything substantive themselves. They just grease the machine and keep things going smoothly. But the real work is done by others. If we are to take this idea seriously, it follows that our human mentality has been shaped in great detail by unconscious forces, and we are powerless to do anything about it. According to their scheme of things, all that prating religious nonsense about moral self-discipline, denying the flesh, turning the other cheek, gaining control over one's base desires, not coveting thy neighbor's wife, and so on is bound to fail because it contradicts our biology.

Of course, this is exactly what Darwin feared might happen to his ideas. Victorian gentleman that he was, he delayed the publication of his first book on evolution for two decades, and his ideas on human origins were held back for another fifteen years, because he feared where they might lead. But most modern neo-Darwinians do not worry about any of this. They have seized what they perceive as the moral high ground. It is irrelevant, they claim, whether Darwin would have approved their agenda. He was a man of the nineteenth century, and here we are in the twenty-first. In the past many scientists were frightened of the places to which their observations could lead. Galileo, like Darwin, was one of a long line. But ultimately the truth wins out. We cannot put off facing reality forever, even if we fear our own creations.

DEMONS AND MEMES

Where its existence is reluctantly acknowledged at all by Hardliners, consciousness is treated as a quirky vestigial artifact, a freak show curiosity in our ongoing cognitive circus. They have argued that the conscious mind gives us the pleasant illusion of control, while in reality it can do nothing but stare helplessly and stupidly (since it is also inherently shallow) at the game of life as it passes by, because all our important mental games are played entirely unconsciously. For Hardliners, consciousness is not even necessary for abstract thought. This may sound like an exceedingly odd idea, but it comes from a highly respectable source, computational science, which has shown that computers can think and solve problems without being conscious. If computers can, why can't we? Computers are full of unconscious mental operators that are sometimes known in Artificial Intelligence (AI) circles as demons, or agents. They are the real players in computation. By extension, Hardliners think that we have demons in our brains for all kinds of specialized operations, such as computing the structure of three-dimensional space, calculating social distance, and deciphering the grammars of language. We also have demons that produce specific actions. Certain demons, deep under the surface of our minds, allow us to crank out impeccably correct sentences without any conscious planning on our part. By definition, these putative cognitive demons are supposed to operate entirely outside consciousness.

This is all highly mechanistic, Predestinarian dogma, the same basic set of ideas that an earlier school, Instinct Psychology, flogged one hundred years ago. Not surprisingly, the more ambitious neo-Darwinians are also taking dead aim at culture, just as their predecessors did. In fact, for them, culture itself is just another Darwinian killing field. The very stuff of culture—rituals, fads, styles, customs, symbols, myths, ideas, and the like—are said to be engaged in a perpetual game of *Mortal Kombat* for survival. They fight to the death in a public cultural arena that parallels the struggles that take place in the competitive ecologies studied by evolutionary biologists. But all their significant fights take place in the unconscious mind. For these neo-Darwinians, myths and archetypes originate in deeply entrenched social algorithms hammered into the unconscious of our race over many millennia.

These algorithms are generated, unconsciously of course, by a language apparatus that has also been bred in the bone. Thus, they claim, our cultures are ultimately products of natural selection, just like our genes, and generated unconsciously, like all the rest of the incredible complexity of the universe.

Richard Dawkins has coined a term for the irreducible entities that wage this cultural battle: memes. Memes are supposed to be the cultural equivalents of genes. They are seen as the elementary, reproducible components of culture, the way genes are the elementary, reproducible components of life. They transcend the individual, emerging from the swamps of the unconscious, battling for survival in cultures that are, just like Mother Nature, red in tooth and claw. Memes include beliefs (such as dying for one's country), images (such as the crescent and the cross), and customs (such as Mardi Gras). These cultural memories are parasitic on individual minds and can affect a person's likelihood of survival. They also vary in their own survival value. May the best meme win, as it were. This may sound like the script of a bad horror movie or a *Jurassic Park* of the mind, but the bottom line is that in all this, the conscious mind does not count, not even a bit. Awareness is captured and enslaved by memes, and those memes can determine what we do, whom we love, and even how we see ourselves. The conscious mind is thus triply handicapped. It is battered about by parasitizing memes, chained down by its own genetic constraints, and racked by a series of unforgiving environments. It must suffer this indignity passively, because it cannot play the game. From this, it follows that the cultural environment, like its physical twin, is seen as completely unconscious and mechanistic in its origins. Memes are said to rise and fall, like nematodes and dinosaurs, according to the edicts of the Universal Acid of Darwin's central idea, natural selection. But consciousness has nothing to do with it. The conscious mind is just a harmless voyeur, prone to delusions of grandeur about its ability to influence events. Welcome to the Brave New Mind.

QUESTIONS OF DEFINITION

From the vantage point of neuroscience and cognitive science, large parts of this deterministic theory of mind do not make any sense. In general, it has been built by making great inferential leaps from genes to culture and ignoring a good deal of what we know about the brain. Laboratory scientists are expected to clean up the ensuing mess. But in the laboratory and in the clinic we don't have the luxury of nailing sweeping theoretical manifestos to the door of the Academy. We are exposed on a daily basis to the messy realities of mind and brain, and we can see that consciousness is the single biggest determinant of what the brain does. Far from an incidental player, or an epiphenomenon, the overwhelming presence of consciousness pervades every study of perception, action, or cognition. Precisely because it interferes, many researchers in the physiology of sensation prefer to work with anesthetized animals, so that attention and awareness cannot upset the orderliness of their experimental results. The empirical truth is that brain and awareness interpenetrate each other's domains like the cogs and wheels of a clock, with some (not all) of their temporal relationships timed to a fraction of a second. They also interact in very tangible and measurable ways inside the physical space of the brain, and our understanding of sensation cannot advance any further until consciousness is included in the equation.

In the clinic the presence of consciousness is even more evident. Oliver Sacks has written several extended accounts of patients who suffer from a selective loss of conscious awareness as the result of brain injury. What a devastating loss that can be! One of Sacks's most famous descriptions was titled *The Man Who Mistook His Wife for a Hat*. This unfortunate man had a profound disturbance of awareness (it isn't easy to mistake your wife for a hat), in which his knowledge of the world was disconnected from his conscious vision. As a result, even though he could see objects and faces clearly enough to draw them accurately, the world looked unfamiliar. He could not even recognize himself when he looked in the mirror (for that matter, he probably wouldn't have known what the mirror was for). He no longer knew his wife by sight. But when he heard her voice, he recognized her instantly, and all his knowledge about her came flooding back into his awareness.

Sacks's patient suffered from a neurological disorder called agnosia. Patients affected by agnosia typically retain their storehouse of knowledge about the world. But this knowledge is blocked selectively from awareness and can be studied only implicitly, through their unconscious emotional reactions and brain waves. Such cases allow us to conclude that perceptual awareness is mediated by certain paths in the brain, and not others. This does not necessarily localize that kind of awareness to one specific part of the brain, but it does show that consciousness depends on the existence of certain precise linkages in brain circuitry. Other similar phenomena reinforce the idea that consciousness is contingent upon specific circuits. Mel Goodale, a Canadian neuropsychologist, and his colleague, A. D. Milner, described a woman who lost her awareness of visual orientation and could not deliberately perform any action that demanded such awareness. For example, she couldn't intentionally place a card in a slot at the correct angle when instructed to do so. But astonishingly, she could adjust her hand *unconsciously* to use that same slot if her attention was directed elsewhere, for example, when trying to use an automatic banking machine. Hardliners may explain this by insisting that there is a visual region in the brain that controls the motion of the hands unconsciously. In their view, the demons again win out, and awareness is an unnecessary perk. I don't deny that there are visual pathways that can circumvent awareness. But this kind of unconscious visual control is very limited in its speed, range, and especially its flexibility. The patient's inability to guide her hands visually *in consciousness* left her devastated and unable to carry out even the most menial tasks. When the chips were down, her unconscious demons couldn't help her. They couldn't achieve much without the beam of consciousness to guide them.

These and countless other examples clearly illustrate the vast difference between what we can achieve with and without conscious deliberation. The evidence for the importance of consciousness seems undeniable. At the very least we must draw upon our limited conscious capacity to carry out virtually any complex mental activity, including most forms of learning. Several generations of scientists have tried to prove that we can learn without consciousness, and this is possible, but only to a very limited degree. For example, we can learn not to fall out of bed. In this case, learning must occur unconsciously since we usually wake up only after we have fallen out. We must have registered the negative reinforcer (the shock and pain of hitting a hard floor) while still asleep and somehow modified the way we move in bed the next time we are sleeping, to avoid falling out again. But this kind of unconscious learning occurs only in light sleep and is extremely limited in its application. Most learning requires the concentration of all our conscious resources on the task at hand.

Another area of debate has been the question of subliminal, or unconscious, perception. On the one hand, there are those who do not accept the evidence for subliminal perception. On the other, there are those who believe that it exists, like Tony Marcel, a British psychologist who has had to go to exotic lengths to come up with any data that can pass scientific muster. Arien Mack and the late Irvin Rock have recently argued that we cannot learn to perceive even the simplest object without dedicating conscious attention to that end. Even Marcel will concede that most of the time, perception demands considerable conscious involvement on the part of the observer. Moreover, if unconscious recognition is a marginal phenomenon, unconscious thought is even more marginal. Conscious engagement seems vital to any complex or fast-moving cognitive activity. The claim that children usually learn language unconsciously (an idea well loved by Chomskyans and neo-Darwinians alike) is patently absurd. Infants cannot even learn to perceive vowels and consonants without directing their conscious attention toward the relevant aspects of the world. Once these sound patterns have been learned, their perceptions of them will become automatic, but the learning of new sounds and new words is never easy. Infants must deal with these challenges consciously. Indeed, this is the methodological assumption underlying most modern infancy research on language. Peter Juszyck has used the child's shifting patterns of attention to demonstrate that infants learn much more about language and at a much younger age than was previously believed. By tracking how they allocate their conscious attention, Juszyck found that infants scaffold their language learning on a chain of learned attentional linkages that are acquired very early. While they may not yet have the representational tools to reflect on what they are doing, they are already fully conscious, deliberate, self-assembling knowers.

The normal reaction of Hardliners to any defense of the importance of consciousness is that we don't have our definitions straight. Hardliners claim that we mean different things by consciousness and cannot identify conscious capacity with consciousness in the philosophical sense of explicitly representing what we experience. They argue that a cat may have conscious capacity, in the sense of self-directed attention, but that it is not fully conscious, in the sense of being self-aware. This complaint has some validity. Consciousness can be defined in several different ways and is not an all-or-none phenomenon. However, conscious capacity is ultimately the foundation of self-awareness, and there is no escaping its crucial role in cognition.

BUILDING OUR OWN DEMONS

Despite our disagreement over the role of consciousness, I agree with many of the conclusions of neo-Darwinians. I agree that evolution is a vital consideration in theoretical psychology. I also agree about the unconscious origins of many aspects of the human world, including our sensory capacities and most of our emotional repertoire. We have inherited many unconscious demons, and these forces play an important role in cognition, but only to a degree. Like many enthusiasts fixed on a single idea, neo-Darwinian Hardliners have vastly overstepped the valid reach of their evidence. Their dismissive attitude toward consciousness is their great Achilles' heel, a flaw so fundamental as to jeopardize their larger agenda, especially when they try to apply their ideas to human culture, because they have denied the efficacy of the very feature of mind that generates it, indeed the only feature that allows us to construct a credible scenario for our own evolution.

In this book I shall argue that our capacity for consciousness is our distinguishing trait. It can account for the evolution of many of the most revolutionary features of human thought, including language. The reader might now perhaps intuit the immensity of the gap between the kind of evolutionary theory I propose and those put forth by Hardliners. They are evidently repelled by the notion of consciousness, whereas I revere it. I confess freely to this. I lionize it. In my world, consciousness is king. It defines human nature. The elaborate apparatus of consciousness in the brain is front and center in my thinking, because it has evolved, in very special ways, in humanity. The greater part of our mental evolution was based on an enlargement of our capacity for conscious processing. The emergence of languages, symbols, and many social institutions, including organized religion, has all been driven by a felt need to extend the reach of human consciousness.

Moreover, the existence of conscious capacity has changed the rules of the cognitive game because it is a self-regulatory system, fundamental to all voluntary mental operations. It bestows an unusual degree of mental autonomy on its possessor. Whereas most other species depend upon their built-in demons to do their mental work for them, *we can build our own demons.* Conscious capacity is the basis of all human cultural skills, including language. After all, languages are nothing less than assemblages of cognitive demons, arranged in elaborate operational hierarchies and stored in brain tissue. To a growing nervous system, such skills as playing the piano and speaking English are novel patterns constrained by thousands of rules, linkages, and contingencies. These need to be assembled in memory, and they are never assembled unconsciously. The conscious mind supervises their assembly and ultimately controls the hierarchies of demons that make culture possible. Memes are simply the painstakingly generated products of these processes. If we want to understand what makes human nature special, this is the place where we must start: neither with the end result (memes) nor with the antecedent (the genetically installed demons in the minds of most other species, some of which we have inherited), but with the creative engine itself, the generator of novel demons, the center of human genius: our capacity for consciousness.

HOW THIS BOOK IS ORGANIZED

This book proposes a theory of consciousness that stays carefully on the functional level and does not try to "explain" how awareness could have emerged from a material thing such as a brain. I believe that we might someday understand how this came to be. However, in my opinion, our present intellectual and scientific resources are not sufficient to give us even the beginnings of such a theory. We are only starting to comprehend just how difficult it will be to solve this problem conclusively. A mature theory of consciousness will have to describe not only the causal chains linking the activity of the brain to the detailed properties of subjective awareness but also the transformative powers associated with conscious capacity and the feeling tones that pervade conscious experience, including such things as emotions, moods, urges, and subtle feelings like doubt, envy, and ambivalence. We are not close to being able to develop such a theory. Although we made great progress in sensory physiology during the past thirty years, there is nothing in the exquisite machinery of sensation, not even in our best-understood system, vision, to help us in our quest. Nothing in the columns and bands of the visual cortex, or in the infinitely complex microcosms of the retina, gives us the slightest hint of how brains might become aware of such things. Some recent proposals about the origins of consciousness, notably those of Roger Penrose and Francis Crick, are stimulating and extremely constructive, but they seem unlikely to lead to any fundamental breakthroughs on the special nature of human consciousness in the immediate future, for reasons that will come to light later, especially in Chapter 5.

What, then, is the use of a theory such as this that does not even attempt a full explanation? The simplest answer is that it might allow us to gain perspective on the problem. Consciousness is a big problem, and its contours are still uncertain. When you want to establish the existence of something large, such as the San Andreas Fault, you need to stand back and take a picture from a distance. A microscope can tell you about the soil in and around the fault or the nature of the rocks in the area. But you cannot establish that the fault exists in the first place, or the nature of the phenomenon that you need to explain, with a microscope. You need a telescope, or a Landsat photo, to get the larger picture. So it is with consciousness. To describe the nature of this complex territory, we need to take an omnibus approach, as inclusive as possible. Someday, when the larger geography of consciousness has been agreed upon, the data gathered with our laboratory microscopes might make more sense, because they will fit into a secure theoretical framework. But first we need to get the big questions straight. Only then can we expect to construct the kinds of detailed linkages we need for a mature theory.

The big questions about consciousness still revolve around its definition. I argue here that we cannot be satisfied with a narrow definition of consciousness. It is, by its very nature, an inclusive concept. Chapters 2 and 3 review some of the issues involved in defining consciousness, and I discuss some of the many attempts that have been made to simplify the concept and reduce it to a single dimension. Chapter 2 reviews the Hardliner approach to consciousness, which emerged simultaneously in both psychology and philosophy, reinforced by work in linguistics and computational theory. Hardliners are a varied group, and the only thing that they have in common is a desire to strip consciousness of its complexity and tie it down to a simple operational definition. Different Hardliner schools have proposed distinctive ways of narrowing the definition of consciousness. Some have tried to reduce it to sensation, while others have preferred to make it entirely contingent on language. Others have identified it with short-term working memory. The resulting theories are as cacophonous as one might expect from an orchestra with no conductor and no agreement on what piece of music to play.

I have difficulty with virtually all Hardliner proposals, and my objections are further outlined in Chapter 3, which provides a different perspective and points in a very general way to the direction my own theory is taking. Here I outline some of my core proposals. I argue that consciousness is far more extensive in time and space than most Hardliners believe. It is not sensation, which runs only in the foreground of human awareness. Nor is it language, which is, at best, the bastard child and obedient slave of awareness, ever at its beck and call. Rather consciousness is a multilayered, multifocal capacity and a deep, enduring cognitive system with roots far back in evolution. Awareness occupies a position of governance in human life, and it is special, for a variety of reasons. In fact, our extraordinary capacity for conscious processing might be singled out as one of our defining traits as a species.

Chapter 4 discusses the evolutionary history of consciousness and the species that might be included in the Consciousness Club. I suggest that the Consciousness Club should include many nonhuman species, which have some of the conscious capacities of humans, judged from their brain anatomy and behavior. Chapter 5 covers the evolution of the executive brain system, a complex anatomical system we have inherited, whose history defines the basic levels of awareness as they evolved. I propose that there are three levels of basic conscious capacity. The first, which enables level-1 awareness, is basic perceptual unity, or binding, the mechanisms of which seem to have emerged in the common ancestors of birds and mammals. The second, which enables level-2 awareness, is short-term working memory, which assumes the existence of a binding mechanism and extends the reach of awareness over time. This is especially characteristic of mammals but may also exist in a few nonmammalian species. The third, which enables level-3 awareness, is what I call intermediate-term governance. It is found in some of the social mammals, including primates and ourselves. Level-3 awareness carries the time parameters of working memory further along and introduces an evaluative, or metacognitive, dimension to conscious processing, which allows the mind to supervise its own operations, to a degree.

Chapter 6 returns to the problem of our distinctively human consciousness, which seems to be contingent on four things: an expanded executive brain system, extreme cerebral plasticity, a greatly expanded working memory capacity, and especially a process of brain-culture symbiosis that I have labeled "deep enculturation." In this chapter I discuss Constructivism, an approach to human cognition that originated in French philosophy, with Condillac. Constructivism holds that the mind self-assembles, according to the dictates of experience, guided by a set of innate propensities, which correspond roughly to the basic components of conscious capacity. Here I also review the important case of Helen Keller and try to draw from it, and other similar cases, lessons about the absolutely critical importance of culture in the self-assembly of a human mind. I conclude that by combining the Constructivist approach to self-assembly with a Darwinian account of conscious capacity and merging these with a culture-centered account of modern cognitive structure, we have a very powerful theoretical foundation on which the distinctiveness of human consciousness can be fully accounted for, at least in terms of stretching out its larger dimensions.

In the final two chapters, 7 and 8, I work out some of the implications of this approach, in light of my previous theoretical work on human cognitive evolution. I argue that our conscious capacity provides the biological basis for the generation of culture, including symbolic thought and language. Conversely, culture also provides the only explanatory mechanism that can unlock the distinctive nature of modern human awareness. Without deep enculturation, we are relatively helpless to exploit the potential latent in our enormous brains because the specifics of our modern cognitive structure are not built in. Our brains coevolved with culture and are specifically adapted for living in culture—that is, for assimilating the algorithms and knowledge networks of culture. In a sense, our brain design "assumes" the existence of a cultural storage mechanism that can ensure its full development. This is the only feasible way to build a continuity theory of language evolution and maintain a smooth linkage with our deep evolutionary past. Cultural mindsharing is our unique trait, linked as it is to our conscious capacity. Human culture started with an archaic, purely nonlinguistic adaptation, and we never had to evolve an innate brain device for language per se or for many other of our unique talents, such as mathematics, athletics, music, and literacy. On the contrary, these capacities emerged as by-products of our brain's evolving symbiosis with mindsharing cultures. Language emerges only at the group level and is a cultural product, distributed across many minds.

This is why we have evolved such a novel evolutionary strategy, which relies on off-loading crucial replicative information into our cultural memory systems. The algorithms that define the modern human mind may have originally been generated by collectivities of conscious brains living in culture, but these accumulated storehouses have now assumed a certain autonomy and have become an essential part of the mechanism by which we replicate, and continue to extend, the domains of our awareness. We have evolved into "hybrid" minds, quite unlike any others, and the reason for our uniqueness does not lie in our brains, which are unexceptional in their basic design. It lies in the fact that we have evolved such a deep dependency on our collective storage systems, which hold the key to self-assembly. The ultimate irony of human existence is that we are supreme individualists, whose individualism depends almost entirely on culture for its realization. It came at the price of giving up the isolationism, or cognitive solipsism, of all other species and entering into a collectivity of mind.

2

The Paradox of Consciousness

> Were you to live three thousand years, or even thirty thousand, remember that the sole life which a man can lose is that which he is living at the moment; and furthermore, that he can have no other life except the one he loses. This means that the longest life and the shortest amount to the same thing. For the passing minute is every man's equal possession, but what has once gone by is not ours. When the longest- and the shortest-lived of us come to die, their loss is precisely equal. For the sole thing of which any man can be deprived is the present; since this is all he owns, and nobody can lose what is not his.
>
> —MARCUS AURELIUS

The subjective world of the mind can be viewed only from consciousness. We suspect that things might lurk in our unconscious minds—urges, impressions, memories—that lack the familiar resonances of the conscious world. But we shall never know if this is true since by common definition we can only *know* things by making them conscious.

This chapter, for instance, this paragraph, this sentence exist for readers only inasmuch as they are made conscious. For their brief moment in the sun, while they remain conscious, the written messages are vivid, warm, familiar, tangible; they gain presence as well as meaning. When they fade, first into a place in the mind that is not too far away and then completely out of awareness into a kind of oblivion, their continued existence depends, for a while (or so we currently think), on electrochemical ripples in distributed nets of neurons in our brains and then, over the longer term, on traces left in tangles of swollen synaptic knobs and neuropeptides scattered throughout the nervous system. Once there, the traces of what once was conscious remain, relatively but not completely quiescent, stored in a fuzzy nonliteral form that we do not yet understand.

These traces may stay out of consciousness for many years, sometimes for the better part of a lifetime, until by some other, still unknown process we are able to call them back into awareness, back to the subjective side, where we can again reflect on them consciously. Only at that moment do the ideas latent in the brain's tangled networks return, as it were, to the land of the living, to the conscious mental landscape where we live.

A LIMITED INSTRUMENT

We have the strong impression that we inhabit a world of conscious mental events, and this presents us with a dilemma. On the one hand, our current objective science tells us that we are born with a very limited mental instrument, a tiny cognitive filter through which all our conscious experiences must pass. On the other hand, our Humanistic traditions have given us grand intellectual ambitions that weigh heavily on consciousness and appear far beyond its reach.

Particularly in Western culture, we have placed consciousness and conscious experience on a pedestal. Many of our religious and intellectual traditions emphasize the development of awareness in the individual mind. Our political traditions view society as a democratic assemblage of well-informed free citizens. Our legal traditions assign personal responsibility and guilt. These assume a citizenry that is conscious of what is important. Could words such as these be used to describe people whose minds were deemed to be largely unconscious, less than fully conscious, or worse, only apparently (but never really) conscious? They could not, of course. Our implicit value system assumes a rather elevated level of awareness and self-control, and our legal vocabulary is filled with concepts that reflect this assumption: "murder with intent to kill," "free elections," "intellectual property," "informed consent," and so on. Consciousness is our central tenet, our most precious, hard-won possession.

This is reflected, above all, in our educational system. We teach our children that to be truly free, one must be able to think freely. Moreover, to be free in this sense, we must be aware. The cultivation of awareness has been one of the major objectives of Western education since the Enlightenment. Schools and universities have emphasized the importance of a lifelong quest for learning. This idea traces back to the Socratic notion of the examined life—that is, a life made conscious by the habit of constant reflection. Ideally we are supposed to sustain an unremitting assault on our own ignorance, on any residual unconsciousness, until we die. This is intended to make us more aware of our options; to encourage us to seek further enlightenment through the fine arts and literature, which are supposed to make us more conscious of what is important; and to prepare us to be responsible citizens.

However, from the current perspective of cognitive science, this consciousness-raising ideal is built on sand. The conscious human mind, the instrument that mediates this finely tuned awareness, seems hopelessly unsuited for such challenges. During the past forty years, in countless laboratories around the world, human consciousness has been put under the microscope and exposed mercilessly for the poor thing it is: a transitory and fleeting phenomenon. The ephemeral nature of consciousness is especially obvious in experiments on the temporal minima of memory—that is, the length of time we can hold on to a clear sensory image of something. Even under the best circumstances, we cannot keep more than a few seconds of perceptual experience in short-term memory. The window of consciousness, defined in this way, is barely ten or fifteen seconds wide. Under some conditions, the width of our conscious window on the world may be no more than two seconds wide.

That is a shocking limitation. Where is the permanence or wisdom in a consciousness that cannot hold on to anything for more than a few seconds? To make matters worse, the brevity of this window is not restricted to sensory experience. It also limits our ability to think. Conscious thought is subject to the same temporal limitations as any other conscious mental activity. This presents us with an even more curious dilemma. We cannot usually complete a complex chain of thinking within a single subjective moment. To think really complex thoughts *in consciousness,* we have to string together a series of brief moments into some kind of systematic thought sequence.

The catch is that if control of that entire thought sequence is to be made conscious, it must emanate from the same brief time window. But how could it? How could we string our thoughts together consciously over a long sequence if the dwell time of our conscious searchlight is so brief? How could a person stitch together a meaningfully conscious existence from an endless series of two to fifteen-second samples of experience? Sisyphus, the legendary king of Corinth, had it easy by comparison. His labor was only physical—he had to push the same giant boulder up a hill in Hades over and over again—but at least he could reflect lucidly on his simple dilemma. His punishment seems mild when compared with the cognitive travails of the average human being, driven to synthesize a lifetime of memories while coping with the demands of day-to-day survival, and all this massive effort supposedly achieved within a sliding time window of fifteen seconds or less.

It may seem an absurd design for a brain, but this appears to be our all-too-human fate. Our internal video editor will never let us stop, even for a moment. It has no freeze-frame option, no slow advance, or slow reverse. It is unidirectional, unfolds in real time, and cannot hold more than a short, highly selective sample of the world—usually the one currently projected onto our brains by the environment. Consciousness is like the proverbial conveyor belt that will not stop for anything. Like Charlie Chaplin's hapless assembly line worker tied to an inexorable contraption, we are bound to an unstoppable stream of consciousness. All our adventures of the mind are ultimately limited by this fact.

We have a few moderately clever tricks to help us get around these limitations. For instance, we can try to grasp the larger picture by extracting generalities from the stream of consciousness, typically by creating verbal summaries for future reference. Still, even if this could be achieved perfectly, by constructing a kind of ongoing Proustian narrative in our own memories, how could we ever consciously appreciate the thousands of narratives built up over a lifetime, given that they would have to be held, and simultaneously viewed, in some sort of temporary memory store for that purpose? Because of our brain design, we have no such memory store. Therefore, we cannot view them all simultaneously; we can look at them only serially, one at a time. We seem doomed to a fate of endless repetition, reciting our conscious mantras over and over like monks doing their matins or lauds or Homeric bards practicing the *Odyssey*. We cannot be conscious of everything important all the time. We might keep returning to our personal narratives, but the business of living tends to intrude, and we really have very little truly conscious time, in the reflective sense of the term, once we have accounted for basic survival. Thus our ongoing narratives, our conversations with ourselves and others, are shattered and interrupted again and again.

Given these limitations, it is hard to imagine how any advanced form of thinking about the past or future could take place within such a tiny frame as human consciousness. But it is a fact that such thinking has taken place in human history, many times, in many places. If it was not achieved within the consciousnesses of individuals, where did it take place? Is our history wrong? Or is our science of mind perversely wrong?

MINIMALIST PEOPLE

This picture of the conscious mind has a long history, as exemplified by the opening quotation from the Stoic emperor Marcus Aurelius. But it has been reinforced by the application of laboratory methods to the study of the human mind. Some may object in principle to doing this. A common complaint is that experimental psychologists, like me, tend to examine people in stripped-down Minimalist settings. And Minimalist settings produce Minimalist people.

This criticism is substantially true. The psychological laboratory is very different from the comfortable psychoanalytic couch that Hollywood still presents to the public as the ideal of psychological examination. Victorian-minded psychoanalysts treated their patients as demigods (keeping in mind that their favorite patients were usually famous artists, wealthy professionals, or, on a bad day, minor nobility). They usually saw their patients in quiet, softly lit rooms filled with art and anthropological curios. The patients' minds were given great importance and were portrayed as battlefields of dark, arcane forces. This provided, among other things, for great theater. It has been said that psychoanalysis had the singular virtue of making dull people appear much more interesting than they really were.

Laboratory experiments on mental activity are, by comparison, lean and rather mean. They take an uncompromising, gritty approach. The aesthetics of the laboratory, if the word can be applied to such a setting, is closer to that of punk rock than to the aristocratic salons of psychoanalysis. There is no ego stroking here, only a stainless steel chair and concrete block walls. Expect to see Mad Max in full battle gear, armed to the teeth with rocket launchers and grenades, in the waiting room. If psychoanalysis can be accused of romanticizing the psyche, the opposite criticism can be made of the laboratory. Here cruel Irony rules. Here rough Pragmatism is king.

Experimentalists usually put their subjects (a revealing term that retains a whiff of storm trooper condescension) in austere institutional rooms, alongside banks of computers and other machinery. The machinery controls everything the subjects see and hear and dictates exactly what they can or cannot do. Subjects are reduced to playing a role in a temporary drama that frames their every action. They may be required to sit rigidly upright, wear earphones, or stare at computer monitors and respond to the stimuli delivered to them. Above all, they must obey, in great detail, whatever instructions they are given. The drama takes place in an experimental setting that has been rigorously designed in accordance with a theoretical construct called a paradigm. Here the paradigm rules. It is an odd ruler inasmuch as it enslaves both lord and serf—that is, it engages both experimenter and subject in complementary roles. Both are momentary components in a man-machine system that is larger than either of them.

This is surely not the most propitious setting for searching self-disclosures. But this is precisely the point. Self-disclosure is not on the menu; objectification is. In the laboratory we objectify humans in the same way that we do rats and insects. I do not say this accusingly. Science has no choice. Its methods are dictated by its specific questions. Either you want an answer to your question, or you do not. Sometimes the answer to a question cannot be obtained without unpleasantness. There is nothing unique about this. Science has always depended upon a certain ruthlessness. We would have learned nothing about human anatomy without first going through that revolting Renaissance business of illegally dissecting stolen corpses (even now dissecting a corpse is no picnic, so to speak). Of course seventeenth-century citizens objected vehemently to the theft and dissection of their dead relatives! Even today, knowing what we know about the success of medical research, many people would object to having Auntie Em dissected just to satisfy someone's curiosity. But the price that our ancestors paid for their strong prohibition against dissection was ignorance about the workings of the body. Somebody had to demythologize the human body before scientists were allowed to do terrible, profane things to it, on a routine basis, to find out how it worked.

In the same way, we have had to demythologize the human mind in order to subject it to scientific study. This is a dirty business, not unlike robbing graves. But instead of desecrating their bodies, we have stripped people of their illusions. For many, this is an even harder pill to swallow; hence the visceral rejection of Behaviorism by Humanists. But there you are: either you want an answer, or you do not. Either you want to see the world clearly, or you do not. For experimental psychology, the choice was made long ago. We rejected tradition, and having seduced subjects into our Orwellian laboratories, we made countless observations and took careful measurements. The analysis of experiments became a huge statistical adventure, a feeding frenzy for obsessives. Given our relatively advanced technology and our passion for detailed analysis, an analogy with the effect of the microscope on biology is not out of line. We spend inordinate amounts of time exploring mountains of empirical data. The most extreme case of over-the-top obsessive-compulsive data analysis that I know was a study where each hour of computer-monitored videotape took sixty hours to score. And that was just the scoring; the statistical analysis of the scored data took even more time. B. F. Skinner's old-fashioned pigeons, laboring long hours at their pecking keys, would have been envious. Prima donnas (or rather, prima Columbiformes) that they were, they never received such detailed attention.

Our examination of the microdetails of behavior over the last three decades has yielded at least one very consistent impression: Under these conditions, human consciousness measures out as a fleeting and narrow window. When people are removed from their familiar settings—work, school, home, city, street—and subjected to bare-knuckled scrutiny, their conscious capacity reveals its severe limitations. This includes everyone. There are no geniuses in the psychological laboratory. In real life you may be the local Einstein or the village fool, but it does not matter. The laboratory setting is the greatest leveler of people yet discovered. You might feel that this is inherently unfair, a nasty bit of Postmodernist theater designed specifically to reduce us all to nothing, an insensitive abasement of the human mind, and in a sense, you would be right. You might also object that human beings are not butterflies, not to be pinned in such a way. Again I agree, at least in spirit. Ideally we all deserve better than that. But in the final analysis, these Romantic objections do not really wash. They express rage at a loss of dignity that humans have suffered at their own hands. They reflect lost pride. But they cannot dismiss the stark reality of the laboratory findings. The results are very real. They cannot be ignored.

THE ULTIMATE FORM OF DECONSTRUCTION

Any mental operation can be broken down into smaller atoms. Even a creative act, such as writing a poem or creating a computer program, can be segmented and analyzed. The larger-scale components of most tasks can be split fairly easily, as is often done in mass production and efficient management. Thus a manager can divide any manufacturing or accounting operation into a series of smaller tasks, which can in turn be broken down into smaller operations, sometimes called microcomponents. These are closer to the true atoms of thought and experience than large-scale operations.

But microcomponents are less easy to study than large-scale mental operations because they happen very fast, and are especially hard to observe because they usually occur inside a single mind. Nevertheless, they are measurable. For instance, if an employee is asked to multiply several large numbers, the solution to the task can be broken down into a chain of large-scale operations, such as "Multiply the first digits in the first two rows." These can be split into smaller operations, such as "Locate digit x in the table," or into microoperations, such as "Recognize digit x." These in turn can be reduced to sequences of even more elementary operations that might be described thus: "Move the eyes four minutes of arc to the left and fixate" or "Resolve relative location of visual input."

Each component of a mental task can thus be isolated in the laboratory and brought under scrutiny. When this is done, it is often possible to assign fairly precise numbers to each microcomponent's duration. On average, it takes a very small amount of time—usually one-tenth of a second or less—to perform each of the microcomponents in a task hierarchy. Individual differences in this regard tend to be minimal, and the vast majority of people perform the required operations in a fixed order, within a fairly predictable time period. Keep in mind that each of these microcomponents involves a cognitive operation of some complexity and that the underlying neural operations are necessarily faster and even more complex. In terms of brain activity, even the lowest-level operations in a task hierarchy tend to involve large-scale neural events affecting hierarchies of neural nets. Behavior and cognition result from the synchronized actions of millions of such hierarchies.

There is a universal rule here that seems to apply to nearly any aspect of behavior. Large-scale cognitive operations can be broken down into smaller ones and ultimately into discrete microcomponents strung together into a neural hierarchy. This applies to everything we do, even to our most elevated thoughts and experiences. It is really this discovery, rather than the specifics of any particular experiment or theory, that stands out as cognitive psychology's most important contribution in the past century. This breakthrough was comparable in importance to the formulation of the neuron doctrine in neurophysiology, which was based on the finding, proved conclusively only about sixty years ago, that there are actually tiny spaces between neurons. Neurons are therefore treated as discrete cellular entities, rather than as parts of an unbroken reticule or network, and the nervous system can be fragmented into millions of component parts. This fragmentation principle also applies to cognition. Just as brains are made of discrete neurons connected together, so complex thoughts and perceptions are made up of discrete microcomponents strung together in intricate sequences.

This methodology of behavioral deconstruction is very, in fact one might say relentlessly, powerful, much more so than the methodology of Humanistic deconstruction. It is applied to the mental work of individuals, small groups, organizations, and corporations. It is applied routinely in government, management, research, systems design, and education. Moreover, it works. The problem is, we do not much like the outcome. The romance we once had with ourselves, the magic of being human, is hopelessly defeated by assuming this rigorous stance. After many years of carrying out such experiments myself, one image still haunts me: the helplessness of people once they are brought under the laboratory microscope. Even the experimenters appear helpless, reduced to a managerial role in paradigms of their own design. So are all the other managers and specialists who design working environments and the bureaucrats who design large-scale social service systems. They all subscribe to the same design philosophy and are caught up in something larger and far more revolutionary than they can control. Human beings have brought reflection to bear on the thought process itself. The human mind, the source of all our deliberations, is finally managing itself as a mere component of a collective enterprise. In effect, we are industrializing our own cognitive process.

Humanity seemed such a formidable species when we could romanticize our collective achievements and when we demythologized only the rest of the biological world, keeping our own minds sacred. But once our taboos were broken, we fell from grace with ourselves, leaving our conscious minds dissected on the laboratory bench. The disillusioning result is that we have apparently uncovered the true extent of our limitations.

THE TUNNEL OF CONSCIOUSNESS

If we look closely at a conscious mind going about its business, consciousness looks less like a brilliant searchlight than a tunnel through which everything must pass, and a small, dark tunnel at that, affording a rather poor view. We live at its mercy most of the time, groping our way from one slow-moving, fuzzy moment to the next. We might manage to achieve occasional epiphanies, those precious moments of clarity that illuminate our lives. Nevertheless, on the whole, we depend on external cues to keep us going. We call this predictable envelope our routine. This is a kind of algorithmic cocoon that reduces uncertainty and stress by programming our lives. But it does more than reduce stress; it also provides our awareness with a stable framework. Without habit to carry us along, consciousness can become disconnected, vague, and lacking in focus.

This can be seen most clearly when people are locked into sensory isolation. In sensory deprivation studies conducted at McGill University during the late 1950s students were paid very generously to do absolutely nothing, in dark, totally soundproofed rooms. Some of them were floated on water beds, with cuffs covering their hands and arms to diminish their senses of touch and gravity. Except for occasional breaks for eating and self-care, they were isolated from the external world and forced to fall back exclusively on their own internal resources; that is, on their fantasies, images, and memories. Were they happy with this condition? Quite the contrary, the result was more often madness than joy. No volunteer lasted more than seventy-two hours in such isolation without showing signs of temporary psychosis. Many reported hallucinations. None could sleep normally. We may intuitively understand their vulnerability in this regard. Most of us can recognize that we need constant stimulation. We usually arrange our lives so that our environment is properly structured to keep us in line and on course, to move us along in what we perceive to be an orderly manner from one moment to the next. This creates a sense of continuity and security. But our autonomy and independence are only illusions. Consciousness cannot apparently be left simply to drift off on its own, into inner space, without any external correctives on its activity. If it is liberated from the external world, its very survival is in jeopardy. This holds true whether we are presidents or peasants, cooks or CEOs. There are no exceptions.

The term we sometimes use to describe the inherent temporal limitations of consciousness is "capacity-limited." In the jargon of science, our conscious capacity is limited so that we can consciously attend to only a few things at a time. If we are confronted with too many things simultaneously, we cannot attend to all of them, no matter how hard we try, and as a consequence, we start making errors of omission. Attending to one thing blocks out other things we might want to take in. But we cannot take everything in. Only a few things can be housed simultaneously inside our consciousness. In this regard, the conscious capacity of human beings, by laboratory estimates, seems no larger than that of many animals.

As we have seen, the time window of our awareness is very small, fifteen seconds at most, but there are also serious *spatial* limits to consciousness. Its measurable breadth is small. During any simultaneous sample of experience, the average person can grasp only a few things at a time. To be precise, we can attend to just six or seven different things in a single mental glimpse. This has been known since the time of William James and repeatedly demonstrated in various clever ways. James's original demonstration was the most gripping, if not the most convincing from a methodological viewpoint. He scooped up a handful of marbles, threw them into a box on the floor of his office, and tried to guess, in a single glance, as they landed, how many he had thrown. He discovered that he could actually see the number of marbles he had thrown, without counting them, as long as there were five or six. If the number was much larger than this, he had to count. Thus he concluded that a single perception could only take in between five and seven distinct things. This rule has been verified many times. George Miller's famous paper on this topic, "The Magical Number Seven, Plus or Minus Two," was still required reading when I was a graduate student. The rule of seven plus or minus two applies not only to vision but also to touch and sound, to abstract and concrete ideas, to short-term memory, and even to images held in the imagination. We cannot simultaneously take in or attend to more than this. Thus, conscious perception is very limited, in space as well as time.

Even these facts do not do full justice to our limitations. Not only are we limited in time and space, but we are forever prevented from becoming directly aware of most of our thought processes. We cannot introspect on most of the mental steps that make up the larger sequences of our behavior.

For instance, if we ask people whether they knew what they were about to say just before they intended to speak, they will usually report that they did not. They will say, "The words just came to me." They have no idea where the words came from. Words just fly out, and speakers generally hear what they have just said at about the same time everyone else does. A group of turn-of-the-twentieth-century German psychologists known as the Würzburg school explored some of these introspective issues in great detail, and their conclusions must have been very depressing to anyone who valued the primacy of consciousness. They concluded that most thought processes, including those that seem most conscious, such as thinking and reasoning, are largely inaccessible to conscious reflection!

Can it possibly get any worse? Indeed, it can and it does. Consciousness is also notoriously vulnerable to interference. Left on our own, our minds tend to wander all over the map, and we are inevitably drawn to attractive nuisances. Our vulnerability is especially evident when we are stressed or exposed to major distractions. This is due to our very strong startle response or orienting reflex, an automatic attention system that interrupts our awareness whenever there is a major change in the inflow of experience. This system interrupt response is part of a very ancient brain reflex that evolved long before mammals came along. There are very good evolutionary reasons for this. In a world of predators and other dangers, it would not be very adaptive to get so lost in thought that we failed to notice the arrival of a serious threat. Humans can gain some mastery over their orienting reflexes, and indeed, some meditators claim to have achieved complete control over them. But this is questionable, and at best this sort of mastery can be demonstrated in very few people, and only under special circumstances. By and large, we are all highly vulnerable to distraction.

This has consequences. We cannot retain complete conscious control of what we attend to, especially in a stimulating environment. This is easy to demonstrate in the laboratory, where under controlled conditions people can be distracted and driven to error with ridiculous ease. The conscious mind is a delicate entity that can be captured and held hostage by virtually any strong stimulus. For this reason, we find it difficult to maintain a sharp focus in the presence of strong competing attractions. Such interference effects can be quantified, traced to specific stages of the thought process, and measured to show how many tenths or hundredths of a second were lost at each stage. From this type of analysis, we know that the whole cognitive hierarchy can collapse under the effects of stress. Not only is the highest level of control—the top management of the mind, as we have called it—disordered by interference, but the whole hierarchy of microcomponents can also malfunction, down to the lowest level. The functional unity of behavior itself can dissolve, even with mild stress, and with very high degrees of stress, our whole cognitive house of cards will collapse. Our commonsense terms for such a breakdown include "battle fatigue," "nervous breakdown," and "mass hysteria." Under such circumstances, awareness is volatile and fragmented. Entire episodes in a person's life may fail to be registered in permanent memory.

It is sometimes objected that the apparent limits on human consciousness might be an artifact of the particular laboratory paradigms used. There is some evidence in favor of this. Certain animals look much smarter in the wild than they do in the laboratory. In a stripped-down conditioning box, a normally bright dog may need thousands of trials just to learn a simple discrimination. However, that same animal—a hunting dog, for example—can find his way through a forest, leap over obstacles, cut across fields and climb mountains, purposively toss aside various distracting events (such as confrontations with other animals) recognize and corner his quarry, and, if he likes his job, remember not to eat what he finds. On the same afternoon, he might also have to deal with other challenges, such as the need to assert dominance over other dogs, to get to the quarry first, to understand and follow the rules of the hunt (in the canine manner, of course), and to obey various vocal and trumpeted commands coming from the hunting party. Could this be the same dog that looked so dumb in the conditioning box? Laboratory results sometimes clash horribly with the rich facts of ethology.

This principle also applies to humans. People are original, daring, shrewd creatures in the wild; that is, in their natural environments, whether city or rain forest. Humans take magnificent risks. They negotiate intricate social contracts. They also wage wars, invent languages, build cities, and construct literary cultures. But in the laboratory, under the light of our deconstructive methodology, the same individuals that are collectively capable of such great achievements suddenly appear very limited, especially when forced to do things entirely within the span of their conscious awareness. How could this be?

I do not want to give away the store by answering that question up front. Dealing with it is largely what this book is about. As we shall see, the answer is rather complex, but I cannot comfort the troubled reader at this point by concluding in advance that our science is all wrong. I am certainly not willing to say that there is anything wrong with the thousands of controlled experiments published on the subject of our limited conscious capacity. There is not, as far as I can see, any fundamental flaw in the methodology of these studies. The evidence on limited capacity is depressingly conclusive and has been replicated many times in different places. Moreover, the replications of these experiments have been often carried out by scientists who would have dearly loved to contradict their colleagues. Any experimental psychologist could attain considerable fame by disproving the whole body of work on capacity limits. So conspiracy theorists can rest easy. There is no conspiracy to devalue human beings.

Despite our limitations, we try to sustain the continuity of our lives by becoming cognitive acrobats, juggling several things at once, always on the brink of dropping everything. We are a society of anarchic mental jongleurs who fake reason and order, even awareness. The miracle is that human life seems as coherent as it does.

THE PARADOX OF PARADOXES

Thus the paradox of paradoxes: Consciousness, our intellectual home, the cradle of our humanity, appears to be the most limited part of the mind. Given its great prestige in our culture, consciousness seems a poorer thing than we might have expected, powerless to intervene in any significant way in human affairs. It is hard to accept that such a limited mental instrument could have created the subtle, complex fabric of knowledge we have woven or the elaborate cognitive worlds we inhabit.

The most radical resolution of this paradox is widely accepted in cognitive science, which states that most of what we achieve takes place unconsciously! This conclusion is not intended in Freud's relatively soft use of the term "unconscious." Freud may have emphasized our unconscious sources of motivation, but he preserved his old-fashioned belief in the supremacy of our conscious, rational Ego in certain domains. In that sense, Freud was a pussycat. But the Minimalists are not. They propose a much more radical dressing down of the human spirit. The Ego that Freud thought was the conscious part of our psyche is perceived by them to be mostly unconscious in its operation. According to this doctrine, we act, think, speak, and imagine mostly in the unconscious. Most of our knowledge is stored away in memories that exist outside consciousness. They are retrieved into it only when immediately necessary and are made more vivid in our working memory. But even this process is generally unreliable. According to the Epiphenomenalist doctrine, we are only vaguely, inaccurately, and occasionally conscious of what we are doing. Moreover, even in these few instances, we are probably deluded about the degree of control we consciously exert.

In other words, the glory of our humanity is seen as our millstone, our ball and chain. It is extremely significant that experimental psychologists apply the notion of limited capacity only to conscious behavior and never to the unconscious. The unconscious, or automatic processing systems of the brain, as they are known, can operate in parallel, without interfering with one another. In a way, the most important conclusion to be drawn from the evidence on capacity limits is that they may not apply to automatic or unconscious mental activity.

The unconscious systems that dominate the brain allow us to develop what cognitive neuroscientist Bernard Baars has called the unconscious context of knowledge, a huge reservoir of unconscious or automatic cognitive processes that provide a background setting within which we can find meaning in experience. By relying on these deep automaticities, we can achieve great things intellectually. We can even carry on several parallel lines of cognition at the same time, provided they are kept out of consciousness. Musicians know this. When professional pianists play, they cannot afford to become overly conscious of their fingering or the specific notes of the passage they are playing, particularly the more rapid ones. That kind of self-consciousness is paralyzing. They have to automatize those difficult passages, or they will make major mistakes. The same rule applies to speaking.

But unconscious operations do not satisfy our very basic need to try to grasp the events in our lives consciously and to assert some degree of control over what we are and what we do. How ironic that the very design of our brain seems to conspire against us! No matter how wise or learned we become, in terms of the knowledge we have stored away, the video editor of our conscious experience will not let us stop, even for a moment, to reflect. Also, just as we think that we are grasping some important point, just when we seem to be experiencing some personal peak of awareness, the whole thing begins to slide away. Those moments of clarity simply cannot be frozen in consciousness forever. Regardless of how many items we may recall from our inexhaustible long-term memory banks, we can hold them in consciousness for only a few seconds, and then they too slip away, pulled into the unending stream. This unstoppable serial process of becoming aware of something, then of something else, and then of something else again, and so on without end seems bent on preventing us from fixing our experience into a wider span.

Then what are we to conclude? How have we accomplished all that we have achieved? Unconsciously? Are we nothing but clever automatons with absurd after-the-fact pretensions to awareness, freedom, and conscious dominion over what we do? Is the only resolution to this paradox to demote consciousness and dump it on the scrap heap? This is not just another esoteric debate without consequences. A demotion of consciousness to such status would involve a full acknowledgment of the implications of automaton theory down on the ground; that is, in daily life. Such a move would actually threaten the conceptual underpinnings of our civilization. It would suggest that we are, for the most part, preposterous illusory creatures who inhabit the cognitive underground along with the rest of creation while harboring delusions of intellectual grandeur.

If we are these unconscious, illusory creatures, why do we keep trying so hard? If our awareness is so limited, why is it so common for human beings to keep wanting to improve the unimprovable? We have been driven, generation after generation, to seek enlightenment, and we continue to strive to raise or expand our own consciousnesses, to look for some means—chemical, disciplinary, meditative, whatever—to extend awareness. Indeed, there have been periods in human history (emphatically not this one!) when the need for enlightened awareness has been one of the principal drivers of society, including the economy. Even after one accounts for their enormous economic and political importance, the cathedrals of medieval Europe were primarily an architectural means for achieving a simultaneous, conscious experience of light, beauty, and infinity. They were the one place in that world where one could focus *consciously* on what were believed to be the loftiest and most important issues in life, a place for personal integration and reflection. They may have been many other things as well, but in their highly contrived and symbolic design, they were virtual environments, conceived toward a primarily cognitive end, to focus and nourish the conscious mind, toward gaining an experience known as illumination. Even in our extraordinarily secular civilization, people have sought to use new technologies to attain spiritual and intellectual relief from the cognitive chaos of lives lived too rapidly, without time for conscious reflection or vision. Every new piece of chemical or electronic technology has been harnessed, at one time or another, for the purpose of raising, expanding, or developing awareness, with the notable exception of television and computers. But perhaps we should give them more time.

Our pursuit of expanded awareness reflects a deep fear of being less than human. The notion that we might be only a step removed from those imaginary zombies that populate horror films, staggering through life with merely the thinnest veneer of awareness, is repulsive to most of us. This revulsion was evident in the violent reaction of Victorian society against Freud, who elevated the role of the unconscious. Ironically, our response to his much more radical successors, the cognitive scientists, has been one that Freud would have found amusing because it is so Freudian: massive denial.

Understandably, most psychologists deny the darker implications of their discipline. The only people I know who are not threatened by the notion that we are cognitive zombies are certain philosophers and cognitive scientists who actually seem to revel in the idea that our mental processes are mostly unconscious. They are the true Hardliners in this debate. Like the demons that drove the Gadarene swine into the Sea of Galilee, they are legion. And at the moment they are on a bit of a roll.

HARDLINERS

The Hardliners are a disparate group. They agree on little among themselves. They will be (somewhat unfairly) grouped together here only because they share one important feature: They reject the pragmatic, fuzzy definition of consciousness that is commonly used in cognitive research. Without exception, they attempt a more rigorous definition of consciousness and carry it to its logical conclusion. Their motive seems virtuous enough. They want to introduce more logical rigor into the study of consciousness. Some Hardliners have narrowed the definition of consciousness even more, almost to a vanishing point. Some have questioned whether consciousness has any function in cognition. Others have dismissed it as a superficial by-product—that is, an epiphenomenon of no substance. Still others, notably Patricia Churchland and Georges Rey, have challenged the value of the term altogether. They see it as a commonsense word for which we have no need.

In this, most Hardliners go much further than most experimentalists. The experimental tradition has generally accepted the idea that some sort of central processor, roughly synonymous with consciousness, plays a role in cognition. Note the vagueness of the terminology: "some sort of central processor," "roughly synonymous," and so on. Empirical scientists are pragmatists and have always tolerated fuzzy categories, especially when the theory in question has not yet really crystallized. They tend to home in on the truth more like hound dogs than philosophers, sniffing out solutions, sometimes highly improbable ones, by trial and error, intuition, and indulging in "stamp collecting"; that is, compiling lists of observations in a search for patterns. This requires flexibility, opportunism, and a quick response to sudden ambush by unpredictable results. But it does not require ideology or the a priori imposition of constraints.

Hardliners, on the other hand, led by a vanguard of rather voluble philosophers, have gone that extra mile. Not satisfied with merely saddling consciousness with limitations, they are determined to assign it a passive role. Many would prefer to toss the whole idea into the trash can. They believe not merely that consciousness is limited, as experimentalists have been saying for years, but that it *plays no significant role in human cognition.*

They believe that we think, speak, and remember entirely outside its influence. Moreover, the use of the term "consciousness" is viewed as pernicious because (note the theological undertones here) it leads us into error.

The error that Hardliners so abominate is implicit Dualism, the complete separation of the domain of mind from that of matter and natural science. Hardliners proclaim that we must purge, once and for all, the Cartesian Theater assumption that underlies so much experimental work because it leads us unwittingly into the clutches of Dualists. Dualists capitalize on our intuitive belief in the existence of a little-man-in-the-head, or homunculus, who is the apparent ultimate destination of everything the mind knows and the initiator of whatever it actively does. This homunculus supposedly sits in a little-room-in-the-head (aka the central processor), located in the center of the mind, watching the show, making decisions, and so on. The homunculus presumably experiences a Jamesian stream of consciousness. But this assumption is (to Hardliners) obviously false. There is no little man sitting in the head, and our homuncular assumption is the main source of our confusion. As long as psychologists continue to use the term "consciousness" in this Cartesian context, Hardliners argue, they will inevitably lapse into Dualism, the very error that entrapped even so eminent a thinker as René Descartes. Hardliners have been conceded this point altogether too easily, in my view, because most experimentalists are too busy, too tired, or perhaps too suggestible to resist. Sad to say, there has been a virtual stampede of researchers vying with one another to confess, beg forgiveness, and recant publicly their real or imagined Cartesian beliefs.

To one who is afraid of closing off potentially important routes to knowledge, this putative victory over the Cartesian Theater seems a Pyrrhic one. If we strip consciousness thus, it may salve our bruised consciences and allow us to proclaim ourselves cleansed of the Dualist heresy. But the central scientific problem remains. If there is no homunculus, how can our experience of conscious unity be explained? Where is our experience of agency rooted? Hardliners try to answer these questions by replacing the traditional homunculus of Descartes with a committee of equally inscrutable quasi homunculi. There is no one home in the Cartesian Theater, they say. Instead there is an unregulated free market of the mind, a state of cognitive anarchy that computationalist Marvin Minsky has called a Society of Mind. As Dennett put it, there is a Pandemonium of mind, in which there is no central control, only a chaotic sequence of unconsciously generated images and representations, out of which only a fortunate few surface into what we experience as consciousness.

But wait a minute. This ploy is transparently obvious. Hardliners have simply imbued those unconscious, anarchic forces of the mind with homuncular properties of their own. Dennett's unconscious cognitive demons, as well as Minsky's unconscious agents, compete, know, formulate thoughts, and even win control of consciousness. But, Hardliners are anxious to add, these agents are not subject to the same objections leveled against the homunculus because, *mirabile dictu,* they are not conscious. Under the surface each of them is just a specialized neural machine, as empty as a tomb. They know, remember, think, and so on without being conscious. How can they do this, we may be forgiven for asking, if those actions seem by their very nature conscious? Well, their real agenda becomes visible at this point. Their answer is: because we know that computers can do these things unconsciously. We know this? Or we assume this? This is dogmatic stuff, the core of classic AI theory. They are slipping one past us. Hold on to your wallets.

The conscious mind itself is supposedly thus emptied of superstitious entities and ghosts and reduced to its essence: the pointless spinnings of what William Calvin has called a Darwin Machine, a kind of fast-time evolutionary computer that crunches numbers through time till the cows come home, producing a user illusion in any nervous system that just happens to be lying around, soaking up the environmental action. The receptivity, the very presence itself of that user illusion, *constitutes* consciousness. Oh, they add, this is a very unusual user illusion, because there is *no user*. NO USER? Yes, Virginia, this is a very unusual user illusion!

This tired old AI view, a powerful school of thought for more than two decades, proposed that the mind is made up of software agents (presumably they all wear long black overcoats and dark glasses), which battle for survival in a kind of Darwinian cognitive underground. More anthropomorphism. More dead algorithmic clickety-clackers (my word for the blind, dumb, obedient digital logic machines that I used to build as a graduate student) into whom the Strong AI theologians have breathed anthropic souls. The only thing missing is the river Styx and a few hooded boatmen. This barrage of metaphors transports us to an unconscious computational netherworld deep under the surface of the mind, and old-fashioned consciousness simply slips away from our grasp, fading into the mist. Minsky's agents and Dennett's demons are of course nothing but those same ancient psychoanalytic demons revisiting us in sci-fi dress. The unconscious is no longer a decadent Parisian salon filled with predatory pimps and harlots trying our virtue. It is now the deck of the starship *Enterprise,* filled with humanlike computers and computerlike androids, none of which are what they appear. Catchy stuff, but they are singing the same old tune, about a human psyche overpowered by forces that it cannot control, that it cannot even see.

Hardliners are not anxious to own up to the practical consequences of their deliberations, which, if taken seriously, would be far more serious than the limitations reviewed in the previous section. The practical consequences of this deterministic crusade are terrible indeed. The denial of the homunculus implies that there is no center to the human cognitive universe. There is no sound biological (or ideological) basis for selfhood, willpower, freedom, or responsibility. Dennett and the philosopher Richard Rorty, among others, have tried to head off this rather Draconian conclusion, making a distinction between the ritual exorcism of the homunculus and the apparent destruction of personal identity that it implies. The latter can be kept, they argue, as long as we understand that selfhood is a representational invention, a cultural add-on. But this exercise in evasion fails; we are not deceived. The notion of conscious life as a vacuum leaves us with an idea of the self that is arbitrary, relative, and, much worse, totally empty because it is not really a conscious self, at least not in any important way.

For cognitive scientists, the consequences are difficult to accept. The central process, the old workhorse at the very heart of the entire experimental literature, is being gutted before our very eyes. Perfectly workable notions about the organization of the brain's supervisory systems, cogent ideas about attention, working memory, controlled processing, automatization, explicit memory, and a major chunk of the clinical neurological literature would need a complete overhaul to accommodate this new vocabulary. The Hardliners are not letting on, mind you.

RADICAL PRESUMPTION

The hardest of Hardliner theorists define consciousness as the direct awareness of sensation and nothing else. This definition has a long history and has reappeared in several recent books, encompassing views as diverse as those of psychologist Nicholas Humphrey and philosopher Robert Kirk. To quote Humphrey, "To be conscious is essentially to have sensations. . . . There are no non-sensory, amodal conscious states. If and when we claim that another living organism is conscious we are implying that it too is the subject of sensations." This also applies to nonliving entities: "A mechanical robot for example would not be conscious unless it were specifically designed to have sensation as well as perception (whatever that design implies)." Note that Humphrey does not deny robots the capacity of perception; that is, the ability to detect properties of the environment. But, he notes, unless they also have subjective sensations, they cannot be called conscious.

In his book *Raw Feeling*, Robert Kirk made the same point in a different way. He used the expression "raw feeling" to describe what happens whenever we use any of our senses, including internally generated feelings, such as dreams, hallucinations, and mental imagery. One who experiences raw feeling must be said to have at least minimal consciousness. The question of privileged access is important to Kirk: individuals alone have direct access to their feelings, and others cannot gain that kind of privileged access. Raw feelings cannot be defined in relation to anything else; in fact, they do not involve anything beyond physical events in brains.

In Kirk's view, consciousness is better regarded as a passive by-product of activity in neural nets, especially those that make up the sensory cortex. This position implicitly concedes consciousness to many species since many of the behavioral and physical correlates of human consciousness occur in nearly all mammals and birds and probably other genera as well. He is alluding to the principle of emergent complexity, a principle that has been widely accepted in biology for some time, as reflected in the work of writers such as Edmund Sinnott in the sixties and Ernst Mayr in the eighties. In this view, there is no need to invoke a special principle to explain the emergence of conscious mind, any more than such a principle is needed to explain life itself. When complex molecules reach a degree of complexity, they are alive, their status being an emergent property of their organization. Similarly, we do not need a separate principle to explain why living neural nets can become conscious and eventually self-aware. Once neural nets reach a certain level of complex structure, they just are or, at least under the right circumstances, they can become, conscious, to varying degrees. This position allows for gradations of consciousness in different species. Thus fish might have a primitive sort, cats a more advanced form, and humans an even higher level of awareness.

I agree with much of this. The principle of emergent complexity is a very useful idea as long as it is articulated in sufficient detail, but it runs into some trouble in this particular application, where consciousness is virtually identified with sensation. Its application exclusively to sensation is really quite arbitrary and internally incoherent. If neural nets just emit consciousness, why not motor nets? Or odorless, colorless, symbolic nets? Why only sensory nets? This is not only an arbitrary stance but also open to attack from within its own ranks. If we accept that consciousness is nothing but sensation, then how do we verify its existence in anyone other than ourselves? The verification of theories of consciousness is a thorny issue for anyone who accepts the nonfunctionality assumption, the idea that consciousness, as a passive reflection of neural activity, does not actually *do* anything.

But Hardliners invariably believe this. It follows that if consciousness is neither functional nor accessible to objective study, we could not in principle tell the difference between a robot that behaves just as we do but is not conscious and one that is truly conscious, as we presume ourselves to be. It is conceivable that we could live in a world where every other apparently conscious person was a fake, a clever simulacrum, and never realize that they all lacked consciousness. We could not tell the difference between such a world and the real one, and thus the problem of consciousness is inherently unsolvable. If we strip away the philosophical pyrotechnics, this is the essence of the unsolvability thesis, the position taken by many Hardliners.

Presumably we should all go home at this point and turn to something more productive. Fortunately, physical scientists did not abandon their quest when it was pointed out long ago that they could not tell the difference between the real physical world and an illusory one, cleverly contrived by the gods to fool them. They still could not tell the difference, at least in theory, despite all the victories of natural science, if a virtual world were convincing enough. But this precious insight has had precisely zero implications for science or, for that matter, for anything else. The same criticism applies here. The so-called unsolvability thesis is an empty solution to a phony problem, a word game for the chattering classes. It rests on the questionable assumption that a world of intelligent minds, intelligent in the same sense that humans are intelligent, but completely lacking in consciousness is a real possibility.

There is absolutely no reason to accept this idea a priori, let alone build an entire philosophy around it. No autonomously intelligent robots have ever been manufactured, and until they have been, we cannot assume that they can be produced, even in principle, without possessing consciousness. The closest things we have to simulations of real living minds are a few existing neural net simulations that might be called autonomous, but only in a very limited sense of the term. Even the most impressive neural net simulations I have seen do not come anywhere close to the intelligence of, say, a honeybee. If you doubt this, try to imagine an unguided, unpiloted helicopter as autonomously smart as a honeybee. Engineers are still so far from being able to simulate the complexities of mammalian brains that it is difficult to speculate about what properties neural net simulations might need to acquire in order to achieve this. At least we will need a new mathematics that can simulate the complex analogue computational style of real neural nets. In the long run, when we know more about such things, we might be able to answer the question of whether we can simulate complex cognition without building in, by design or otherwise, the functional equivalent of consciousness. My bet is that we cannot, for reasons that will become clear later.

The other Hardliner assumption—the nonfunctionality assumption that is often used to shore up the unsolvability thesis—is also entirely arbitrary. Who says that consciousness serves no function? On what evidence? The answer to this question is entirely dependent on the particular definition of consciousness we wish to adopt. Hardliners have taken a position and decided the issue axiomatically, rather than on evidence. But these axioms are not verifiable in themselves. They are only the starting points of an inquiry, and if one assumes from the start that consciousness is passive, then it is hardly surprising that it is found to be passive.

Some of the Hardliners' proposals have more radical implications, and at times the debate has been so intense as to become almost unstable. Some Hardliners assert that we can never become directly conscious of abstract ideas despite the distinct feelings of control that many people report while thinking and their insistence that their thoughts are indeed conscious. To this, they reply that direct awareness of thought is only an illusion, and abstract thoughts are really only familiar to consciousness by means of their sensory attributes. That is to say, ideas can become conscious only inasmuch as they elicit concrete feelings and images in the observer.

According to this view, I can only become aware of an idea such as "Fascists are unpleasantly pushy people" if I reflect on the imagery associated with such a thought. This might consist of the sound of a voice actually saying, "Fascists are unpleasantly pushy people," or it might flow from my own mental images—perhaps brown armbands and arms raised in salute, my own inner voice, a memory of a dictator's voice, or even the emotion of fear and anxiety associated with these ideas or images. But the key point is that my awareness must be built out of these kinds of data, not out of meanings. Meanings without a sensory or perceptual aspect have no direct entry into consciousness.

A corollary of this position is that in its very origins, consciousness is nothing but an internal mirror of the external world. This ties us tightly to the external environment. If consciousness is nothing but sensation, then our awareness can have no significant independence from the external world that generates either sensation itself or the images that it spins off in memory. When defined in this way, the mirror of consciousness might include the neuronal flotsam and jetsam that result from our reprocessing of sensory experience; that is, the brain's distorted and transformed internally generated imagery. But it cannot and does not include abstract ideas. Conscious awareness is restricted to the surface of ideas and cannot penetrate the underlying depths.

The radical presumption of these Hardliners becomes clear: *Consciousness is inherently meaning-less; that is, truly empty of meaning.* This notion may appear exceedingly arbitrary to the uninitiated, but it is attractive to some because it greatly simplifies the debate by eliminating the sticky issue of precisely how we might become directly aware of abstract meanings and by narrowing the whole argument to a regional squabble over the organization of sensory systems.

EVEN MORE RADICAL PRESUMPTION

This sensory reductionism is unsatisfactory to another school of Hardliners, who hold to a representational approach that, in its own way, is just as uncompromising. This school believes that language is the sole legitimate basis of consciousness. Instead of Bishop Berkeley's *esse est percipi* (to be is to be perceived), they might substitute *esse est loqui* or perhaps *esse est verbum feci* (to be is to be spoken, or made a word). Consciousness is linked to symbolic thought, entirely dependent on language, and unique to humans. Adherents to this approach do not accept that simple sensory awareness is ever sufficient evidence for true consciousness. For them, language alone is the key to consciousness. Some have taken this idea to extremes. Julian Jaynes once proposed that consciousness, in the sense of self-consciousness, was strictly a cultural invention, and a very recent one at that. Thus he restricted consciousness not only to those with language but to those with certain ideas and thought habits that are to be found only at specific times and places in human history.

Jaynes's position is often seen as idiosyncratic, indeed eccentric, and not representative, yet it is not that far from many mainstream theories, such as Dennett's. For most of those who adopt such a viewpoint, only people who can capture their mental contents in language could be described as truly conscious. Presumably children, or the nonsigning deaf, or a variety of other people with disabilities could not become fully conscious unless they acquired sufficient proficiency in language. Some people have actually claimed this in writing. This includes Richard Rorty, who wrote in his book *Contingency, Irony, and Solidarity*: "We have no prelinguistic consciousness to which language needs to be adequate." This confirms the experience of Helen Keller, who, in her autobiography, testified that before having language, she was not fully conscious. However, as we shall see, this was a naive claim on her part. When we look at her own testimony about her life before she had language, we are led to believe that she was also conscious at that time. More about this later. We are obviously introducing quite a different meaning of the term "consciousness" when we identify it with language and symbolic representation.

Another philosopher, Hans-Georg Gadamer, has argued that all human experience is essentially linguistic. In fact, in most Postmodern scholarship, conscious awareness of reality is often assumed to be entirely a product of language and sometimes, even more radically, of crafted texts. In this view, conscious awareness is something that is fashioned entirely in culture and has no objective reality to use as a referent. Dennett has referred to this aspect of consciousness as a collection of memes. More to the point, consciousness is seen as having nothing whatever to do with sensation. Human awareness is seen as having no rivals, no equivalent in other beings, and mere sensate organisms do not qualify as having consciousness at all.

Curiously, and characteristically in this furiously reductionistic century, many members of this language-as-consciousness consortium agree with sensory Hardliners on one thing: they support the downgrading of consciousness to the status of epiphenomenon. Polar opposites though they may be on so many other issues, they agree that consciousness is a secondary by-product of the brain's activity, a superficial manifestation of mental activity that plays no functional role in cognition, a mere consumer of experience that gets the news too late to do anything about it.

If consciousness is always too late and too passive, surely it is irrelevant whether the news it receives comes from sensation or from language. For both schools of thought, the action is down in the depths of the mind, in the places where the brain actually knows things, thinks about them, and remembers them. It is pointedly not in consciousness, which is a useless evolutionary side effect. I have already pointed out how the first kind of Hardliners came to this belief (they invoked the principle of emergent complexity and put consciousness in the position of passively reflecting the state of a given neural net). However, it may not be obvious why the second kind of Hardliners, those who believe in language-as-consciousness, would agree with them on this issue.

The reasons for this strange partnership trace back to the linguist Noam Chomsky, who has fought for three decades to defend the notion that language has unique properties that require a built-in brain device for its generation. He, and his MIT colleague Jerry Fodor, articulated a well-known theoretical defense of linguistic modularity as a general principle. Two ideas are central to this defense, and both were developed in the psychological laboratory, where they held great currency in 1983, when Fodor wrote his book *Modularity of Mind.* The first idea is that there are only two major classes of cognitive entities in the brain. One class consists of mental modules, the unconscious specialists of the mind. These include, for example, all the unconscious computations that underlie vision, which we can never consciously experience. Our awareness doesn't know how the brain extracts visual images; it simply receives them from the unconscious modules that produce them.

The other class of mental entity is the central processor, which, in contrast with modules, is a conscious generalist. The central processor is coextensive with consciousness and is domain-general in its influence because it ranges widely over our mental landscape.

The central processor is said to gather visual impressions simultaneously with those of other senses, while also drawing from memory and monitoring the body and its actions. Despite this awareness, however, it cannot actually *do* anything. It sits in magnificent isolation from the real world, relying on platoons of menial specialists to provide it with material for experience and other specialists to implement action. Consciousness is thus placed in a powerless minority position. Modularized brain systems must be encapsulated—that is, protected from interference by conscious awareness—and it is easy to understand why. If we could decide to turn off our color vision, for instance, or our hearing, the results might be fatal.

In this view, modules are designed to optimize their special competence and generally function near the theoretical limits of their potential efficiency. The central processor is the opposite. It is rather slow-moving because it lacks a specialized competence. But it has one indispensable feature: general access to all the outputs produced by the specialized modules of the mind. Only it can integrate the experiences generated by the various modules, uniting vision and hearing, touch and inner feeling into a unified stream of awareness. As such, the central processor is located at the geographical top of the cognitive hierarchy. But it is denied true governance.

Fodor's second principle is the critical one for our purposes: The details of whatever modules do are not accessible to consciousness. Only the outputs of modules are. In his theory, consciousness is condemned to perpetual shallowness. It cannot look inside the mental modules or change what they are doing; rather it has to accept the outputs as they are. Thus, if I view one of Escher's impossible figures, in which the image violates the rules of three-dimensional perception, no amount of knowledge and no attempt at conscious intervention will make the illusion disappear. Illusions are involuntary and cannot be modified or dissolved voluntarily because they occur within the sensory module producing them. I will continue to see the illusion, despite what I might know to be the case, because my visual brain does most of its business without the benefit of consciousness.

The most controversial part of Fodor's proposal is his agreement with the Chomskyan notion that language is one of the modules of mind rather than part of the central processor. This seems unlikely at first; language has general access and cuts across all our sensory modalities. We certainly are free to talk about whatever we see, hear, or feel, and in that sense, language must be domain-general and accessible to consciousness. But in other important ways, language acts more like a module. There is an involuntary dimension to the perception of language and even to its acquisition. For instance, while listening to a speaker in our native tongue, we cannot stop ourselves from hearing the speaker's words *as words*. We cannot simply decide to hear them as ordinary sounds and switch off their meanings. On the contrary, we are compelled to hear them as language. We cannot consciously intervene in this process.

In fact, the apparatus of speech seems to operate outside the reach of consciousness, and this principle extends to grammar as well. We are not aware of the grammatical processing of speech when we listen to it or when we produce it. Grammar usually does its work unconsciously, and we are generally not aware of its action, either when we construct sentences or when we parse the sentences of others. Moreover, when children acquire language, they master grammar very rapidly, without specific instruction on many points, and at a faster rate than the growth of their other skills or general knowledge. This suggests that grammar is embedded in a special module, designed to optimize language acquisition, rather than identified with general awareness.

But if language, including grammar, is indeed modular, then it follows from Fodor's own principles that anything understood in terms of language must be essentially unconscious in origin. To say the least, this has major implications for those who wish to believe that consciousness is entirely mediated by language. Fodor realized this and left himself an escape hatch. He relegated the most abstract aspects of language to cognitive limbo, allocating them to both the central processor and the language module. However, most neo-Chomskyan theorists of language do not like this messy solution and have not availed themselves of it.

Let me repeat that statement for emphasis: Fodor's proposal implies that *language is not conscious in its generation* because it originates in a mental module. It also follows that we only become conscious of incoming language after the fact; that is, after our modules have decoded it. This includes even our own inner voices. But how can this be? How can language be simultaneously a precondition of consciousness yet unconscious in its own generation? Indeed, this is a bit of a problem, and here is where some further logical gymnastics come in handy. The solution offered is roughly as follows. The contents of awareness are manufactured exclusively and indirectly by inaccessible modules in the brain, whose activity must always remain unconscious. They are responsible for determining every aspect of conscious experience, whether sensory or linguistic. If we combine this idea with the notion that language provides the material of fully fledged consciousness, it follows that awareness is epiphenomenal and cannot play an active role in anything important in cognition.

In other words, for these theorists, consciousness is an epiphenomenon, and most of our cognitions, including those mediated by language, are unconscious in origin. As soon as they concede that language should be filed away in a Fodorian module, the Romantics among the language-as-consciousness gang are in deep trouble. If language is unconsciously generated, and if the conscious mind gets to view the action only after it has already unfolded in the specialized modules of the brain (and even then only selectively and occasionally), it follows that we (our conscious selves) do not really play much of a role in generating language. So when I talk to you, and you talk to me, my unconscious is addressing your unconscious. Our unconscious modules are engaged deeply in conversation. Our conscious selves are occasionally allowed to peek over the transom but never permitted to join in. This has oddly Jungian resonances, and given the widespread contempt for psychoanalysis in these circles, it makes for a delicious irony indeed.

DENNETT'S DANGEROUS IDEA

The *capo di tutti capi* of the Hardliner school is Daniel Dennett. By his logic, the illusion that we are consciously in charge is itself an after-the-fact perception, a by-product of brain processes that we cannot consciously control. This makes some intuitive sense. As I write this sentence, the words tumble out, and I type them onto the screen, but I really do not have any awareness of where the words are coming from. Something in my brain must be generating the sentences, but I am not aware of the teeming minions down in the boiler rooms of my mind, assembling sentences and ordering my fingers to move to the correct keys. By and large, I seem to become aware of my words only *after* they have been formulated or even typed. Consciousness trails too far behind. It does not seem to have played a role in the specific formulation of my text.

The same applies to spoken language. I babble away at a fantastic rate. Like most people, I can generate about fifteen distinct speech sounds, or phonemes, in just one second. Also, like most people, I am blissfully unaware of the unconscious modules that are serving up my ideas, choosing my words, assembling my sentences, and even arranging rhetorical flairs and voice modulations for the benefit of my listeners.

The orderliness of my speech, says Dennett, can be attributed to various demons that are beavering away in my unconscious. The traditional little-man-in-the-head, or homunculus, simply isn't there and should properly be replaced by his conceptual Pandemonium. My anarchic inner world is teeming with ideas and images, all of which are simultaneously fighting for supremacy. But I can never see or hear it. All the real cognitive action is below the conscious surface. I am aware of my own utterances only after I hear my own voice (as a child), or my inner voice (as an adult), or see what I have gestured, signaled, or written. The conscious "I" did not really generate them. I have no idea where these things came from because I cannot even access the parts of my mind that produced them and can become aware only of their most superficial attributes.

For most of us, this may be a radical enough statement for a day's work, but not for Dennett; his deconstruction does not stop here. He carries it much further than other mortals. Tough hombre that he is, he faces up to the inevitable conclusion of this line of thinking, unlike many of his camp followers, who tend to bail out at much lower altitudes. Consciousness, in his epiphenomenal view, does not even possess the reality of a helpless voyeur passively watching the world go by, because even those thoughts that it mistakenly believes are its own are an elaborate Darwinian illusion. Admitting the existence of a Central Meaner, as he puts it, is tantamount to admitting that there is a homunculus viewing reality in an imaginary Cartesian Theater. Such a notion is regarded by Dennett as fundamentally flawed, a lapse into simple belief that willfully denies the brutal implications of any mechanistic explanation of mind. It mistakenly concedes a degree of biological reality to the self. It is always foolishly believing itself to be made of solid stuff, an entity in the world in the same sense as, say, a living thing, such as an animal. But it does not, and it is not. It is a mere figment of the imagination. Dennett wants us to swallow our cod-liver oil. The self is a convenient fiction, invented and transmitted, like all fiction, exclusively in culture.

The radicality of this point is easily missed in the dazzle of Dennett's deconstructions. He is not merely reiterating the laboratory evidence, reviewed in the last section, that showed that consciousness is tremendously limited. No, he wants to go beyond that difficult, but at least defensible, point. He is actually denying the biological reality of the self. Selves, he says, hence self-consciousnesses, are cultural inventions. There is no proprium, as Gordon Allport once called it, no innate realm of selfhood.

Dennett's initial aim was to formulate a frankly mechanistic theory of consciousness. His conclusion is that this is unnecessary and impossible because the concept of consciousness is a distracting commonsense notion that has led us astray. This applies even to sensation. His main argument is centered on a series of empirical demonstrations that supposedly show that consciousness is not part of the inner cabinet of the parliament of forces that runs the mind. The essence of his message is that the initiation and execution of mental activity are always outside conscious control. We may have a strong illusion that there is a place in the psychological firmament where it all comes together. We feel that knowledge is assessed, memories are retrieved, choices are made, and actions are initiated in consciousness. But this is an illusion, no more.

Where does he get the evidence for this? For the most part, he draws freely from experimental psychology and cognitive science. He is very selective, an understandable strategy for anyone approaching the treacherous shoals of experimental psychology from the outside. His most powerful demonstrations are classic experiments that, when taken in isolation, seem innocent enough. He cites some of the standard demonstrations of unconscious influences on cognition, which show what he calls the unconscious editorial activity of the brain. For example, the so-called McGurk effect demonstrates that the actual sounds of speech can be unconsciously corrected by the brain to fit the visual image of the speaker's lip-synching. Because of this, the actual sound track of a film may sometimes be overridden by what the viewer sees. The viewer's brain evidently modifies what the ear was given by way of data, and it achieves this without the conscious knowledge of the listener. Similar editorial effects abound in vision. The simple fact that the world is stable, despite the constant jiggling of the eyeball in its socket, implies such a process. Somehow the brain establishes a stable world despite the chaotic sequences of images presented to the retina, and it does this unconsciously and automatically.

One of Dennett's favorite examples is something called the Phi phenomenon, an illusion that forms the basis for motion pictures. When several images of roughly the same thing are presented to the eye rapidly, one after the other, the object seems to move. Viewed at just the right speed, about twenty-four frames a second, such a sequence will evoke a smoothly moving image. The simplest demonstration of Phi is to alternate two light sources spaced a few feet apart: on-off-on-off, etc. At a certain speed of alternation the viewer no longer sees two lights turning on and off but instead reports seeing a single light moving back and forth. The observer's brain has unconsciously resolved the alternating images into a single moving picture.

A variant of this effect, called color Phi, was discovered by a former colleague of mine at Queen's University, Michael von Grünau, and Paul Kolers, of the University of Toronto. They found that when the first stimulus alternates with a light of a different color—red followed by green—the result is very odd: The eye tries to fuse the images into a single moving light, but the color switches from red to green at midpoint between the two images! This happens even on a single exposure. Even if the two lights are flashed in very rapid sequence just once, the first appears to move toward the second and to change color in midstream. But how could this be, since the eye cannot possibly know in advance that it is about to see a color change? The first image should retain whatever color it was until the last moment and only then be replaced by the other color. But this does not happen. This suggests to Dennett that there are unconscious editors at work here, creating a solution to a perceptual conundrum. It is hard to dispute his conclusion. Whatever causes these effects at the neural level, consciousness has nothing to do with them. Consciousness only catches up later, after all the real work has been done.

Once we are in this selective state of mind, there are many other phenomena that can be added to Dennett's list. The random dot illusions of Bela Julesz illustrate a different kind of unconscious editing. Julesz invented a brilliant trick: One eye is presented with a computer-generated display of random dots, but unknown to the viewer, the other eye receives an identical display, except for a systematic displacement of some of the dots. The pattern of displacement—that is, the discrepancy between what the left and the right retinas receive—forms a figure, such as a square or triangle. The figure cannot be seen by either eye alone since each retina sees only random dots. There is no image of the figure in the brain either, except in the region that compares the two eyes. This comparison is done by the binocular neurons of the cortex, which receive projections from both retinas. Comparison (probably by some form of subtraction) of the two sources produces an anomaly, an image that does not exist on either retina but only somewhere in the visual brain. Unlike normal images, these Cyclopean images, as Julesz called them, take quite a while, sometimes minutes, to emerge in awareness. The experience is uncanny: After the stereograms have been viewed for a while, the hidden image gradually emerges and becomes quite vivid. The perceived Cyclopean shape may hover in the foreground, like a sharp-edged cloud suspended over a sandy beach. The point is that this fusion process cannot be achieved through conscious effort. Viewers must wait until their visual demons do the work for them. This is a particularly clear case of the brain's carrying out its visual computations outside awareness. Undoubtedly consciousness plays second fiddle here.

I can add one more example. One of the most compelling illusions I have ever seen is one that many psychologists are not familiar with. Known simply as Bidwell's phenomenon, it was discovered in the nineteenth century. In the sixties, when I worked for a drug trials project in Montreal, we used Bidwell's phenomenon to track the effects of drugs on the nervous system. The observer's eye is exposed to red light for a brief exposure of roughly one-sixtieth of a second, followed immediately by a bright white field of light. Strangely, under the right conditions of intensity, the observer sees only a green light, never red. The green image is what we call an afterimage of the red (afterimages often occur at night just after we turn off the lights). Colored afterimages are usually seen in the complementary, or opposite, color of the stimulus. In the nervous system the color opposites are blue for yellow and red for green. Thus, if the lamp we turn off is yellow, we see a bluish glow in the darkness; if the lamp is red, we see a greenish afterimage; and so on. The eye remembers the opposite of the color last seen, resulting entirely from the way the visual system is wired. Normally the afterimage is preceded by an experience of the real image that triggered it.

The strange thing about Bidwell's phenomenon is that the actual color of the stimulus, red, is never consciously registered by the subject. How, then, could its complementary aftereffect be consciously registered? Clearly, at some time the original color must have been strongly present in the visual system to generate such a strong, complementary aftereffect. If so, why would it never have entered consciousness? One theory holds that in this effect, since the white field is much brighter than the red, it causes the afterimage to catch up and erase the original red image. This is a particularly dramatic example of unconscious editing.

Dennett extended his argument by citing evidence that consciousness is very poor at tracking the timing of events in the world. He relied especially on the work of Benjamin Libet, a neurosurgeon who revealed a considerable delay between the sensory registration of an event in the brain and the patient's awareness of that event. While testing the brain's electrical responses, he stimulated a nerve in his patient's arm and tracked the arrival of the nerve volley at the cortex. He found that the nerve volley arrived very fast, in less than a tenth of a second. He then tried to determine whether the patient's awareness of the stimulus coincided with the arrival of the signal in his cortex. It did not. Libet could block awareness long after the nerve volley had already arrived in the cortex by delivering an electrical pulse directly to it. He could wait as long as a third of a second after stimulating the arm and still block the arm sensation from reaching consciousness, even though the sensory cortex had already responded electrically. That delay is a very long time, as the nervous system goes, and Libet concluded that even simple sensory registration takes considerable time to reach consciousness after it has arrived at the cortex. Awareness is significantly out of synch with the real world. It trails by about a quarter of a second.

Another of Libet's demonstrations shows this in a different way. He studied the relationships among (1) the exact moment when his subject was aware of making a decision to move a muscle (the subject watched a clock hand and reported its position); (2) the timing of the brain's electrical command to the muscles; and (3) the time the muscle actually started to contract. Intuitively, we would say that such a sequence should unfold in a fixed

1-2-3 order; that is, the subject should first make the conscious decision to move, then give the movement command in the cortex, and finally initiate the actual movement. However, Libet found that the order was in fact 2-1-3. The brain started to prepare for a motor command as much as a half second before the subject was even aware of having made any decision to move. In other words, consciousness lagged behind the actions that it presumed to initiate. The subjects were convinced that they had consciously initiated the sequence. But their brain waves contradicted this impression.

In short, Libet claimed that the conscious self gets the news too late to affect what the body has already decided to do. Thus it cannot prepare intentions or make decisions. They are made unconsciously, in the brain, before awareness develops. We may think we intend to act, but in fact, we only think that we intend to act. The real work has already been done, down in the depths of the brain, out of the reach of consciousness, long before it can intervene.

Where does the illusion of conscious control originate then? Dennett's proposal can be summarized briefly as follows. Cognition is basically unconscious, running on a brain that is organized like an enormous parallel computer. To achieve consciousness, says Dennett, we must "become the objects of our own perceptual systems." This is facilitated by human culture, which arranges things so that we are actually re-presented to ourselves. Because we are embedded in elaborate symbolic cultures, we transform ourselves into what Dennett calls Joycean machines; that is, virtual serial machines running on a parallel architecture. The Cartesian self is thus merely an illusion produced in our brains by representational feedback originating in culture. To use the computer metaphors Dennett is so attached to, selfhood and consciousness itself are software rather than hardware, in fact, worse: They are "meme-ware," if such a word exists. As for the biological self? It does not exist. There is no homunculus, except as a representation in the mind.

CONCLUSION

So we have come full circle since the pre-Socratic philosophers. After two and a half millennia of endless fussing and analysis, we must conclude, as they did, that All is Illusion. Except that, by denying the existence of the Central Meaner itself, Hardliners have made this conclusion much more devastating than the pre-Socratics ever did. The pre-Socratics held that external reality was an illusion, but human awareness was not part of that reality. It always stood apart and provided the arena in which our intellectual excursions could take place. But now our own awareness itself has been folded into that chaotic external reality. To say that we are reduced in status by this would be an understatement. It would be more accurate to say that we are annihilated. Human awareness, by this doctrine, is nothing but an aspect of the soulless ether, not even an organizing force, just a clutch of algorithms turning tricks in an endless vacuum. And that is it. Things pass through our conscious minds, but this impression is entirely illusory; there really is nobody home.

I confess that I bail out at this point. Maybe I am a more practical man than I thought I was. Maybe I lack the infinite faith in words and semantic hairsplitting that seems necessary to qualify as a Minimalist or Hardliner. Maybe I am just irritated by the academic habit of indulging in exotic forms of intellectual *seppuku,* a sort of methodological *rite de passage* whereby one conducts sadomasochistic experiments with various forms of *angoisse* (one of which is the overuse of foreign words) and occasionally, if necessary, defines oneself out of existence (I notice that there is usually a temporary lull in this exercise in self-annihilation around tenure time, however).

Of course, Hardliners and Minimalists are not ordinary self-effacing academics. With their customary flair, they have made their point with truly Rabelaisian excess. Their message is obvious enough: Consciousness is an illusion, and we do not exist in any meaningful sense. But, they apologize at great length, this daunting fact Does Not Matter. Life will go on as always, meaningless algorithm after meaningless algorithm, and we can all return to our lives as if Nothing Has Happened.

This is rather like telling you that your real parents were not the ones you grew to know and love, but rather Jack the Ripper and Elsa, She-Wolf of the SS. But not to worry.

3

The Governor of Mental Life

> The only states of consciousness that we naturally deal with are found in personal consciousnesses, minds, selves, concrete particular I's and you's. Each of these minds keeps its own thought to itself. There is no giving or bartering between them. No thought ever comes into direct *sight* of a thought in another personal consciousness than its own. Absolute insulation, irreducible pluralism, is the law. It seems as if the elementary psychic fact were not *thought* or *this thought* or *that thought,* but *my thought,* every thought being *owned.*
>
> —WILLIAM JAMES

For those who care about these things, there is a terrible, compelling logic to the Hardliners' case, one that gives James's words a somewhat bitter ironic resonance. (*Owned*? By whom? A phantom in cyberspace?) But before we banish consciousness from our Pantheon of cognitive heroes, we should consider the flaws in the Hardliners' arguments. Moreover, we should try to build an alternative theory. There may be a way to salvage what is solid in their ideas, while constructing a broader theoretical framework for consciousness.

As we shall see in this chapter, there is much more to human awareness than is dreamt of in the Hardliners' philosophy. Their database is not wrong in itself. Rather it is incomplete. Their theoretical universe has been constructed from a narrow set of paradigms that are commonly used in laboratory studies of human performance. These include tests of perception, attention, voluntary movement, and short-term working memory, and in these domains their ideas seem valid. But human consciousness is far wider and deeper than any of these paradigms, and we have reason to doubt whether we can generalize their ideas about awareness to the real mental worlds in which we live. When viewed in this broader perspective, human consciousness appears neither passive, nor eternally shallow, nor impotent, as they believe. On the contrary, in most situations, it emerges as the very governor of mental life. This does not necessarily leave us in a theoretical limbo. Instead it suggests some changes that need to be made to our standard theory to accommodate a more inclusive and ecologically valid definition of consciousness.

THE TIME FRAME OF AWARENESS

The most serious flaw in the Hardliners' case may be traced to an apparently innocuous factor, the time scale employed in most laboratory paradigms. It is simply too short to yield an adequate image of human consciousness. Laboratory studies are tightly designed, but they are obsessed with the here and now, especially with sensation and memory in the short term. They are also atomistic and reductive, following the deconstructive methodology developed during the Cognitive Revolution in the fifties and sixties. This methodology creates an exclusive focus on relatively peripheral or front-end phenomena, such as short-term memory, visual imagery, perceptual illusions, and the allocation of attention, which must be crammed, because of methodological necessity, into a time window of fifteen seconds or less. With this built-in, albeit unintended, bias, such experiments look only at the lower limits of conscious experience, reducing awareness to the status of a badly edited video, filmed with faulty cameras (our brains).

But this portrayal is wrong, utterly wrong. It seriously misrepresents the object of our study. Human consciousness casts a shadow far beyond the narrow time corridor of our standard laboratory paradigms. Consciousness is not sensation. Sensation may run in the foreground of human awareness, but it is the wider background, the larger landscape of awareness that really matters, in the human case. This wider awareness insinuates itself into the slower-moving metacognitive spheres of social cognition, assuming control over the processing of much larger realities than those framed by the immediate moment. I am speaking here of the intermediate time frame within which most conscious human action takes place. This time frame is a much larger window of experience than short-term memory. It lasts for minutes and hours, rather than the seconds and milliseconds of laboratory paradigms, and it is the phenomenon we should be studying because it is the heart of consciousness.

TABLE 3.1

Starting framework for a unified conversational episode in the intermediate zone

Topic: Spielberg's film *Saving Private Ryan*, just viewed by all eight participants.

Duration: Three hours

Starting lineup

Speaker A:	English/French, expert film buff, hated the film.
Speaker B:	French/English, moderate expertise, mixed love-hate reaction.
Speaker C:	English/Spanish/French, moderate expertise, complex reaction.
Speaker D:	Spanish/French, expert film historian, loved the film.
Speakers E, F:	Czech/English, little or no expertise, uncertain reactions.
Speakers G, H:	English only, moderate expertise, mixed reactions.

Opening move: Speaker A makes a strong case against the film. She raises four themes that will recur in various ways throughout the conversation:

1. Historical inaccuracy
2. The comparative virtues of other great war films
3. The destructive effects of the star system in Hollywood
4. The question of what war is, really, and how it should be depicted on film

Ancillary considerations

- The unique background assumptions, emotions, and social sensitivity of each speaker affect the way each point is made and taken. This applies to quips, puns, barbs, glances, and body language, as well as to substantive arguments. Since the participants are mostly new to one another, these attributes are learned on the fly and must be made constantly available to each speaker, presumably held in some form of working memory.
- Speaker D speaks no English, and all points made in English need to be translated to him and are answered in French, which four participants do not speak. Thus there is a greater danger of miscommunication in his case, and most points have to be simplified for his consumption.
- Subthemes tend to emerge between the Spanish and French subgroups, in parallel with the larger group themes that are shared by all. These themes eventually enter the mainstream exchange.

A simple example, outlined briefly in Table 3.1, might help illustrate what I mean by this and why I am so concerned about the choice of time frame. Say we are sitting with friends and acquaintances in a café, and we have all just seen a controversial new movie. To make matters a bit more complicated, three languages are being spoken at the table, and not everyone speaks all of them (a fairly common scene in the part of the world where I live). Of the eight participants, two are very naive about cinema and lack strong opinions, while two others are highly expert, strongly opinionated, and in constant disagreement. The others, including me (Speaker C), are somewhere in the middle, fairly knowledgeable but less articulate than the self-appointed experts. I find myself informally chairing the debate and mediating between our warring experts, while trying to sway the softer-minded majority toward my own covertly held, but firm, views. Although our discussion may have been triggered by the film we have just seen, the conversation wanders all over the map, from political theory to aesthetics, from cinematic history to disagreements over what the director was trying to say.

Such an exchange is a supremely conscious occasion for everyone, because it demands extensive controlled processing, a term that experimental psychologists use almost interchangeably with conscious processing. Controlled processes subsume numerous suboperations, including attentional selection, the maintenance of vigilance, the allocation of priorities, and the upkeep of working memory. By definition, these are part of our conscious capacity. In order to follow the conversation, we must keep the ideas of the various participants in separate areas of memory, properly labeled and updated. The conversation is loaded with information, feeling tones, and clashes. Even the humor and the personality conflicts must be noted and remembered. We place a huge load on conscious capacity when we are faced with such a challenge, especially such an unpredictable one. In technical terms, we draw on our finite conscious capacity. A complex conversation can push that capacity to its limits because it generates novel, rich, and meaningful memory material, highly changeable from moment to moment, as the conversation shifts from topic to topic, language to language, and speaker to speaker. It is an incredible achievement that despite this complexity, a conversation usually coheres in memory as a single unified episode.

Moreover, a single conversational episode can extend over a very long period of time, often exceeding an hour or more. During that period conversations are converted, in memory, into gigantic clusters of events and hierarchies of ideas that have momentums of their own. Usually no single person controls a conversation, and no one can predict its direction with complete accuracy. As a result, to remain a player, each participant must track it more or less continuously. This places a great demand on one of the main functions of conscious control in each participant, the updating of working memory. To lag behind, even for a minute, is to lose the thread. This requires continuous self-monitoring, not so much over the short term of milliseconds and seconds as over the intermediate term of minutes and hours. Conversation is very demanding in terms of continuously self-reflective, metacognitive governance, a process that oversees the stream of conscious experience for extended periods, keeping the ongoing temporal and spatial linkages of the whole episode alive and unified into a meaningful memory structure.

The width and depth of working memory in such situations are much larger than those suggested by traditional laboratory techniques. Defined in terms of the number of simultaneously active or primed traces it imposes on the brain, when compared with any laboratory task I have ever seen, the load generated by such a conversation is tremendous. To return to our hypothetical movie discussion, each of us must remember: (1) which language, which opinions, and which level of expertise apply to each participant, and these estimates are often constructed on the fly, as the conversation progresses, especially if some of them are new to the group; (2) what each person, including oneself, has actually said, and in what order; (3) where the larger conversation itself is heading—that is, which camp is winning, and which losing ground; (4) whether one's own opinions might be changing and what strategic effect such a shift in position might have on the overall balance of power; (5) whether the opinions of the other players might be changing, and in which direction; and so on. All this requires attentional management as well, because the balance of power and meaning are rapidly evolving, and shifts in background and foreground are constantly taking place. A point that seems relevant right now, in the midst of someone else's monologue, might not be relevant a minute from now, when the opportunity arises to make it, but we may want to return to it, so we keep the argument alive, on a shelf somewhere in the background, waiting to pounce when the time comes. The smallest hints of weakness, or subtlest shifts in opinion, may be accompanied by furious self-defense, counterattacks, and changing alliances, which exist only in the intermediate term. This generates several simultaneous streams of working memory, all active at once, which must be kept separate and not confused with one another.

The conversational episode is always situated, and bracketed, within a larger physical and social reality. All participants must keep this clear. The entire episode must be neatly placed within that frame, or they will lose contact with reality. Conversation or not, there is a continuing need for each person to maintain normal contact with the surrounding physical and social context. If the conversation were to take over awareness completely—that is, occupy all available capacity, to the elimination of other considerations—the result would be a disastrous loss of contact with reality, akin to what some schizophrenics might experience. The participants are thus burdened with yet another metacognitive, or evaluative, obligation. They must keep track of the other current realities imposed by the situation; that is, remember where they are, behave appropriately in the restaurant (as opposed to the way they might behave in a bar or bedroom), deal with waiters and other patrons of the café, settle the bill in an equitable and civilized manner, walk home, and so on, all the while continuing the conversation. Not all this is consciously achieved, of course. Much of this behavior may be managed by one's calling up unconscious scripts for behaving in a restaurant, and so on. But these must be consciously managed. The conscious mind must periodically check the surrounding social realities and remain sufficiently attuned to notice significant signals (competing adjacent conversations, glowering waiters, flirtatious customers) when they arise.

All participants must also remember huge amounts of knowledge on various topics and store away new facts, images, and impressions in memory, for later use. This requires enormous active on-line memory organization. All this detail must be stowed away in various corners of the mind without breaking the larger framework of meaning that has been constructed. In comparison with the apes, we are so good at doing this that even the most ordinary people might appear as geniuses. Certainly, when compared with ape communication, a conversation of the most mundane sort emerges as a gigantic achievement of memory management. Yet for us, it is commonplace. Most people in most cultures routinely participate in this kind of verbal jousting in some arena or another, whether they are discussing games, gossip, farming, music, religion, politics, dynastic succession, taxes, or whatever.

Although we might draw many conclusions from this example, one in particular stands out. The time parameters of such an extended conversation are completely over the top, by laboratory standards. They violate all the basic Hardliner and Minimalist parameters. Our laboratory estimates of human conscious capacity seem to have been horribly off the mark, at least when it comes to language. As a conversation extends further and further in time, its complexity in working memory grows, while all this unfolds under continuous conscious governance. Moreover, the importance of sensory awareness in all this is minimal. We may track the sounds, gestures, eye contacts, and moods of our colleagues, but these are secondary to our awareness of the conversation itself. Sensation is reduced to the role of a television set in a room full of chattering people, and most of the time the chatter dominates.

What kind of memory record are we discussing in this case? We have no good theoretical models of such a powerful storage system. It falls outside the reach of our common memory theories, most of which have been based on how people memorize lists of words. The intermediate-term working memory store has properties of both short-and long-term memory but is typical of neither, as usually defined. Short-term memory is identified with consciousness by many psychologists, and the long-term memory store is regarded as unconscious. Its memories are effectively "made conscious" by being retrieved into short-term memory. Thus the intermediate time frame seems to be neither fish nor fowl in its content, neither fully conscious nor truly unconscious. Ulric Neisser, one of the godfathers of the Cognitive Revolution, came to this conclusion some time ago, arguing that our laboratory paradigms have very limited real-world generality, but his words fell mostly on the deaf ears of those with strong vested interests in the status quo ante.

Anders Ericsson and Walter Kintsch have reviewed a large body of research in the field of expert systems, a computational field of research that focuses on the special cognitive skills of chess players, mathematicians, military strategists, and other experts. They suggested the term "long-term working memory" to describe the working memory framework of expertise. Experts must expand the range of their working memories while applying their expertise. For example, during a game of chess, grand masters will typically draw on a plethora of stock board arrangements, and this triggers their knowledge of specific strategies that might be applied on a specific move. After years of training, their speed in calling up such strategies, presumably from long-term memory, is almost as fast as the normal reaction speed with which we react to things that are held in short-term memory. In fact, their reaction times often come within a quarter of a second of the speed of our usual short-term recall. That implies the existence of a different sort of working memory, one that dwells in the background of performance but cannot be classified as truly short term because it lacks the vivid, literal (sensory) nature of short-term memory. This formulation is quite attractive because it allows us to maintain the traditional Hebbian dichotomy between active (short-term) and inactive (long-term) storage.

However, it might be time to abandon that dichotomy. The memory system that Ericsson and Kintsch have described (and that roughly corresponds to what I have called intermediate-term awareness) is not a series of piecemeal items pulled out of inactive long-term storage. It is a fully active, albeit nonsensory, store and therefore less vivid than traditional short-term storage. But it is a true working memory because it is constantly being updated and revised. Is such a working memory subset fully conscious? I would argue that it is a matter of gradations of vividness and not a simple dichotomous matter of either/or. Given its rapid reaction speed, it is undoubtedly engaged in the immediate task at hand. Certainly, it matches our widest, most inclusive definition of consciousness. These are the circuits that we try to tweak when we meditate, take drugs, or otherwise play games with our awareness.

The conversational example described above is challenging for any standard theory of working memory, conceived as a short-term store. Each ongoing conversational game demands a tangled narrative memory record that is kept both primed and current over a long period of time. This implies that it is a true working memory, extending over much longer periods of time than conventionally thought. Like working memory in the shorter term, it is capable of rapid recall, instantaneous growth, and reorganization and is very rich in current detail. While in use, it remains vivid and active, playing an immediate governing role in ongoing behavior. This contrasts with the patchy, rather loose, slow-to-surface long-term memory records we are likely to retain later. It also contrasts greatly with the concrete, literal nature of normal short-term recall. Sometimes virtually nothing of a conversation such as this will become fixed in either short-term or long-term memory as usually defined (especially if we have arguments like this every night). Yet *at the time it is held*, the conversation itself always coheres and stretches on for hours, demanding virtually all the conscious capacity of its players during that time.

Whatever happened to our infamous fifteen-second conscious span in this painfully ordinary piece of behavior? What happened to our video game model of awareness? They seem inadequate, dead in the water, off topic, even trivial. The fifteen-second, seven-items-plus-or-minus-two world of short-term memory may be appropriate for assessing the performance of on-line video games and pilot-training simulators, but it simply does not account for the kinds of conscious governance we are discussing. The experiments so often cited by Hardliners are undoubtedly replicable and valid, within their limited purview, but they are largely irrelevant in virtually any human context, other than the man-machine interactions they were developed to explain in the first place.

This is not a minor criticism. Many Hardliners have tried to avoid extremes of interpretation and cling closely to their data, but even then they have frequently dug themselves in too deep by overinterpreting some small subset of evidence. A good example is Benjamin Libet's demonstration described in Chapter 2, which showed that there is a long delay between cortical activation and consciousness and which has too often been accepted as evidence that consciousness lags hopelessly behind events in the real world. However, this generalization is not justified or at least not very significant, even if we accept that his controversial cortical timing data are reproducible (and it is perhaps worth pointing out that they have never been reproduced). As soon as we shift our temporal perspective to the intermediate time frame I have described, which extends over minutes and hours, why should a quarter-second time lag at the sensory interface bother us? Moreover, why should a fifteen-second limit on our capacity for literal recall matter at all? These are merely the lower limits of the system and apply only when conscious awareness is forced to track the sensory environment on a fast, moment-to-moment basis. Human consciousness is not usually very good at this. But lower limits notwithstanding, there is a critical intermediate time zone within which consciousness is much more effective and can deal quite adequately with the things that matter to it.

Time frame is not the only consideration here. So is capacity itself. In conversation, working memory capacity seems vastly larger than in the laboratory. It is much nimbler at retrieving material from an immeasurably huge working store. We can recall very complex things at remarkable speed in the heat of conversational battle, and we often seem to maintain several concurrent working memory systems. For instance, in response to a remark, I might spontaneously allude to some statement you fleetingly dropped two hours earlier and accuse you of changing your position several times during the argument, with all this happening against the backdrop of a deeper emotional game, centered on a personality conflict between us, perhaps a fight for dominance. You might reply, instantly and unexpectedly, that I was unduly influenced by the alliance of our two resident movie experts on some point or that I had also changed my view, an inflammatory suggestion that you knew would be guaranteed to raise my hackles. The sparring would continue, our conscious minds bobbing and weaving, revealing their subtlety, ferocity, and deadly accuracy under such circumstances. This is our true gladiatorial arena. We are clever beasts in the realm of words and gestures. Once we acknowledge this, the zone of conscious engagement must be expanded to encompass what we observe. It must be granted a larger influence than it has ever had in our laboratories.

But is this expansion of awareness unique to language? If so, this might be taken as oblique support for a language-as-consciousness position. But no, it is not unique to language, not at all. Our conscious span is deeper than language and is equally important in nonverbal domains. Dance and music are good examples of nonverbal activities that demand qualitatively, if not quantitatively, similar intermediate-term conscious capacities. They both demand the division of attention between multiple foci, constant updating of working memory, and the processing of very large amounts of simultaneous information. Dancers must keep track of their own actions over a fairly long run, with the complication that they might also have to keep track of several other dancers at the same time, not to mention the music, the audience, and the stage itself. Musicians, especially soloists or members of small groups such as trios or quartets, have an immensely complicated task when they try to track their own performances in relation to those of their colleagues. Live jazz performers must "hear" not only their own performances but also those of every member of the group and estimate the overall sound that results. The same applies to various trades, in which we must keep track of tools, equipment, and unfinished products in relation to a desired result. This is also true of a supervisor in control of complex machinery, such as an industrial assembly line. In every case their work calls for a fast, multifocal, intermediate-term form of active supervision.

But the best examples of our intermediate-term governor at work come from an even more mundane example, sport. Take baseball and soccer, two games close to my heart. Both demand a huge investment of conscious capacity and require continuous responses to ever-changing novel situations, with very frequent updating by both players and spectators. Roughly the same time parameters apply to these games as to language. As shown in Table 3.2, games are extremely complex. Players must remember details of each specific plan and adjust every move to the idiosyncrasies of the emerging scenario, which evolves along tangents that are particular to the current state of the game. Players must accommodate themselves to the specific combinations of expertise, emotion, physical condition, and skill that are currently engaged and change their strategy as new players, or new waves of emotion, come in. Unpredictable accidents count for a lot in such games. It matters whether opponents are temporarily demoralized, or have momentum, or if the balance of power has been suddenly changed by injuries or penalties, or how well a particular player might be playing on a particular day, and so on. All this demands close attention and a razor-sharp memory for relevant events, and trouble befalls the player whose concentration lags behind, even for a fraction of a second.

The load placed on conscious capacity in this situation is similar in principle to that imposed by our conversational example. This extends to the very wide distribution of attention demanded by such games. At any moment a player must be aware of other players on both sides, and this is comparable to what is demanded in a complex conversation. Games, like conversations, demand spontaneous strategic innovation and shifting patterns of cooperation with other players. In this sense, certain games might sometimes be even more demanding than conversation, especially in terms of the speed with which coordinated attack or defense might be required. In our baseball example, strategic decisions must be made by pitchers, catchers, managers, base coaches, and batting coaches, on a play-by-play basis, all within the intermediate term. Runners, fielders, and hitters have to make decisions in the shorter term as well.

Such observations are rich in detail and easily verifiable. If they had been made on any other species than ourselves, we would have dignified them

TABLE 3.2

Starting framework of a single baseball game episode
Final game of the Stubble Cup, 1999

Total duration: Three hours

Starting lineup

	Flin-Flon Clodhoppers	**Come-by-Chance Squid-Jiggers**
Starting pitcher	A: Good velocity, poor control	A: Tricky delivery, long-lasting, tough
Bullpen	B: Good control and velocity, leftie	B: Fast but quirky, a bit unreliable
	C: Good control, poor endurance	C: Good for an inning or two at most
Catcher	Solid, not a good thrower	Sometimes unable to keep up with pitcher A
First base	Very reliable, good hitter	Reliable in the infield, mediocre hitter
Second base	Good base stealer, poor fielder	Reliable in the infield, mediocre hitter
Shortstop	Fast, good hitter	Fast, good hitter
Third base	Slugger, good infielder	Slugger, mediocre fielder and runner
Left field	Excellent fielder, hitter	Mediocre fielder, hitter
Center field	Good fielder, hitter	Good fielder, hitter
Right field	Excellent fielder, poor hitter	Good fielder, excellent clutch hitter

Special conditions: Come-by-Chance's pitcher B is tentative; Flin-Flon's star shortstop was on the disabled list for a week, and this is his first game since returning to the lineup.

Opening move: Flin-Flon loads the bases on a walk, a single, and another walk. Its star shortstop, recently disabled, is up to bat, with no one out.

Ancillary considerations: The field is in terrible condition; it has just rained for four days. Flin-Flon's catcher and coaches know that their shortstop is very fit, but they have deliberately created the impression that he was more seriously injured than he really was. Come-by-Chance has concealed the fact that their star pitcher, A, has an injured rotator cuff and may not last the whole game. Both coaches have discovered each other's deceptions, but the players don't know this. Halfway through the first inning Flin-Flon's pitcher is hit hard by a line drive and knocked unconscious. He has to be replaced by pitcher B for most of the game.

with the highly respectable term "ethological." In a human context, since they are so common and easy to obtain, they may seem anecdotal. But there is much more to this criticism of cognitive Minimalism than a few anecdotes. Ericsson and Kintsch's review includes a number of studies of nonverbal expertise, which reinforce our conversational example. When expertise is applied in a given case, say, playing chess, the attention and working memory capacity of the expert is fully occupied with the challenge at hand, which often imposes multiple foci. That kind of intense intellectual engagement makes enormous demands upon conscious capacity, over substantial periods of time, in a manner that requires temporal integration.

Once we concede that there is an intermediate time frame of awareness, multilayered in structure, and with multiple foci, our model of experience no longer takes the form of a disconnected video. It resembles something far more stable, extended in time, coherent, and complex, more like a slow-moving medieval mystery play than a flight simulator. It is infinitely subtle, filled with overlapping shades of meaning, unfolding slowly, over minutes and hours, rather than frantically unraveling over the very short time frame of milliseconds and seconds. Short-term sensory displays are still there, of course, running in the foreground, in vivid little sensory windows. But the real business of awareness is elsewhere, running in the background.

CONFUSION OVER AUTOMATICITY

Another critical flaw in the Hardliners' case is the curious notion that consciousness does nothing and has no causal role in action. The main evidence cited to support this comes from a variety of demonstrations that claim to show that cognitive operations, such as speaking, perceiving, and thinking, are automatized. But Hardliners are very confused about automaticity. In itself, evidence for automaticity can never be used as proof against the efficacy of consciousness in any domain because the automaticity of learned motor skills constitutes one of the principal *benefits* of conscious processing. The fact that we can automatize our enormous repertoire of mental operations, including routines as diverse as driving, playing the piano, and speaking, is testimony to the power of human consciousness to supervise and install complex skill hierarchies in our brains. According to a long line of scientific research on human skill, one of the primary functions of conscious processing is the systematic refinement and automatization of action. Automatization is the end result of a process of repeated sessions of rehearsal and evaluation, which rely heavily on conscious supervision. This extends to intellectual habits, such as reading and math. Much of the elaborate architecture of a mature mind is made of hierarchies of automatized skills that are constructed in, and constantly revised by, consciousness. Let me emphasize what that means: We install our own demons consciously. Conscious processing is needed to establish and maintain our own internal cognitive habits. It is also needed to alter them. And it is needed as well to use them in any complex situation. Automaticity is not the antithesis of consciousness. It is a necessary complement to it. Moreover, it is one of its by-products.

Conscious processing is needed for most kinds of learning. One of the most durable facts of cognitive psychology is the virtual absence of evidence for unconscious learning. Whatever few shreds of evidence there may be, they have never amounted to a demonstration of any kind of complex unconscious learning. The British psychologist Tony Marcel has argued for some unconscious registration and editing of incoming information, but at best he has made a tenuous case for a very limited form of learning. Even a process as elementary as perceptual learning seems to depend upon the selective allocation of attention; that is, on conscious capacity. Arien Mack and Irvin Rock have recently pointed out that it is impossible for the human mind even to perceive a simple object outside awareness. The unconscious mind may passively register certain very elementary impressions of the world, such as color, brightness, and movement, but it will neither identify an object nor locate it in the world without the active participation of attention, which, in this context, amounts to conscious awareness. Conscious capacity is also needed for acquiring and automatizing complex skills and representations, including, of course, more elaborate symbolic skills, such as mathematics, music, writing, speaking, and computer programming. These skills are laboriously acquired, over many years, and become completely automatized only in their maturity. They cannot be acquired without devoting conscious capacity at the time of acquisition. We can draw a strong conclusion from this fact. Conscious processes are largely responsible for setting up the automatized cognitive routines of the human mind, except for certain basic built-in reflexive and instinctual responses.

Thus the existence of hugely complex hierarchies of automatized skills, such as languages, in the human mind does not constitute evidence that these skills reside in innate modules or that consciousness is out of the loop. Rather it testifies to the enormously important constructive role that consciousness plays in adult cognition. When we speak or see the world through the intervention of automatized subroutines, we are playing a cognitive instrument that was originally created and fine-tuned in consciousness and continues to be maintained by conscious processing.

This frees us from the conceptual prison of the Hardliners' philosophy. Inasmuch as awareness pilots the process of cognitive development and continuously reviews, fine-tunes, and modifies the status of our automatized cognitive routines, there is a much greater degree of long-term conscious control over cognition than may be evident in the short-term perspective adopted by most experiments. Instead of locking into the immediate moment, conscious awareness has a lifelong constructive impact, through its guidance of the process of cognitive epigenesis. Moreover, as the concepts and ideas of a growing mind develop greater breadth, intermediate-term awareness gains enormously in power.

By ignoring these dimensions of conscious supervision, and by focusing exclusively on the low-end limits of sensorimotor performance, the Hardliners have distorted our view of consciousness. This could easily have been avoided. Their thesis was constructed by excluding a great deal of important and, one might even say, obvious evidence. The reasons for this are best left to historians, but I suspect that this exclusionary zeal was at least partly rooted in the deconstructive frenzy that overtook so many academic disciplines in the postwar years. Shred first, and ask questions later. Better yet, just shred. The questions can wait.

However, in science, as in politics, excessive shredding is risky. In the process of stripping their captive conceptual carcass—consciousness—to the bone, the Hardliners missed a key theoretical point that can profoundly change our model of consciousness. In humans, particularly, consciousness is occupied more with intermediate-term governance, longer-term planning and supervision, and the process of self-assembly than with immediate sensation and reactive movement. It is a constructive device, ultimately responsible for assembling not only our representations of reality but the high-end cognitive system itself.

THE CLINICAL VIEW: CONSCIOUSNESS AND SELF-GOVERNANCE

Clinical neuropsychology has always been precariously situated between art and science. Perhaps because of this, it has taken a pragmatic approach to consciousness (jugglers of dangerous things, such as knives, flaming torches, and the very real psyches of real people, tend to be very practical). Clinicians have insisted on a flexible definition of awareness, compatible with both science and common sense. In doing this, they have history on their side. Perhaps more important these days, they also have evolutionary theory on their side. Clinical judgment obviously has roots in our special human capacity for mindreading or what some, unfortunately, call having a "theory of mind." This capacity allows us to read not only our own minds but also those of others. We can do this effortlessly, without direct training, because of a specialized form of social intelligence that is more highly evolved in humans than in other primates. As something perfected in the evolutionary arena, it is a survival-related skill, and we must have it to survive in human society. This capacity is obviously the instrument of choice for taking the measure of one's own or another's awareness. It allows us to form judgments about others that are effective, in the sense of having been proved in the field. Such judgments are also the daily fare of neuropsychological clinicians.

Although mindreading ability does not depend entirely on language, our linguistic vernacular reflects the importance of this type of cognition. Language is full of words for subjective states, including those of others. If someone says, "I don't *intend* to go to the meeting today because I *know* that Dr. Johansson *wants* a confrontation, and I am *not willing* to be a part of it," that communication, with its embedded attributions of subjective states, is essentially unimprovable. It achieves its pragmatic goal perfectly. It may not always relate the truth about Johansson, but the truth of such statements is not the point. Rather it is their social utility that counts. Such statements might sometimes be true, or they might be bald-faced lies (I may simply want you to believe that Johansson wants a confrontation and may also be trying to assess your susceptibility to such a suggestion). This would depend on the social purpose of the speaker. The objective truth of such statements is not usually an appropriate measure of their success. We do this by evaluating whether such utterances help us achieve our instrumental, emotional, and communicative ends. But those subtle ends can come into existence only because we have an extraordinary capacity for mindreading in the first place.

Children begin to acquire this ability around age three, and by age five they are so acute in perceiving the intentions of others that they can play games of deception. In such games there may be layers within layers of meaning. Seven-year-old children have no difficulty pretending they are someone else, a person who always lies, for example. My wife and I used to play this game with our children. The only rule was: Whatever you and I say, we really mean the opposite. Thus, when you feel happy, you say, "I am so sad." You cry at birthday parties, laugh when your team loses, point in the wrong direction when asked where you live, say you don't care when you're really angry, and so on. Children have great fun with this because in their daily lives they routinely experiment with such deceptions.

This talent is also reflected in the subtleties of street language, which one could argue is a perfect instrument for describing and attributing human intentions. The proof of its success is the very ordinariness of romance-novel sentiments such as: "Her husband was jealous, and she knew this, and he knew she knew it, yet she persisted in her flirtations to punish him. And even though he knew that he was being punished and that her flirtations had no meaning to her except inasmuch as they hurt him, he could not suppress his jealousy, nor she her rage at his weakness." All this is effortlessly expressed, partly because we can formulate narratives, of course, but primarily because of our mindreading capacity. We are comfortable in the realm of intersubjectivity.

This skill can be carried to exotic heights, to the point where it can bracket the mind states of the self. By this I mean that we are able to distance ourselves from our own acts of mindreading, subjecting even those acts to the same kind of analysis as we might perform on the mind states of other people. In this, our mindreading power is turned on its own activities, and used to read its own internal mind states. This is surely the key to our unique ability to analyze our own thoughts. To perform such acts of self-dissection, we must read our own mental states as if they were those of someone else. Lewis Carroll described this capacity at work in the following words:

> "Then you should say what you mean," the March Hare went on. "I do," Alice hastily replied; "at least—at least I mean what I say—that's the same thing, you know." "Not the same thing a bit!" said the Hatter. "Why, you might as well say that 'I see what I eat' is the same thing as 'I eat what I see.'"

This is pushed further in Thomas Carlyle's famous aphorism: "Under all speech that is good for anything there lies a silence that is better. Silence is deep as Eternity; speech is shallow as Time." *A silence deeper than what is said*? An unspoken meaning intuited by an observer, hidden deep within the very same mind, that can embrace a universe of other minds, treating the self as other?

This is metacognition with a vengeance! The mind has risen above its immediate engagement with the world, evaluating its own knowledge states with reference to others, some potential, some real. This is a recursive process, in the sense that it involves a virtually endless hierarchy of perspectives, each bracketing the previous one. Thus I may feel that you feel that my feelings about your feelings are not the same as my feelings about your feelings, if you know what I mean, in an endless spiral of self-referential mindscapes. It is an open question just how much this process depends on having language. But there is no doubt that in the context of comparative psychology, this human capacity to alternate between various self-perspectives and various other-perspectives, a kind of zoom lens aimed at inner thoughts and impressions, constitutes a very remarkable evolutionary innovation. Mindreading skill is a conscious process, not in the representational sense that we explicitly notice and represent every impression but rather in the functional sense that real-time mindreading demands conscious capacity, usually occupying it to the full. Human social games are too intricate, unpredictable, and, above all, treacherous to play them on automatic pilot.

This provides interesting ammunition in defense of clinical observation.

Here we have a proven instrument for assessing both what other people are aware of and our own awareness. Surely we should harness it as an observational device. Why settle for an inferior tool when we wish to pursue the scientific study of awareness? The standard answer is that mindreading skill is a mere commonsense instrument and that science is a different kind of thought game, played by different rules. True enough. In natural science, subjective impressions can be seriously misleading, and even in the study of human experience itself, we must be cautious how we use such evidence. Above all, subjective common sense is often fatal to good theory. Natural mindreading skill may be an effective tool in the front lines of social behavior, but it is fallible once we step away from its utilitarian uses into the arena of scientific theory.

This rather unsubtle answer is correct, as far as it goes, which is not very far. Obviously, human mindreading skill did not evolve for constructing scientific theories, and subjective impressions should be regarded with great caution, of course, but as a descriptive tool for exploring the phenomenology of experience, the mindreading skill of human beings, especially that of experienced clinicians, is a measuring instrument without peer. If our primary interest is to document the phenomenology of consciousness, it would be incredibly foolish to ignore our primary source of knowledge on ideological grounds. Evaluating the mind states of others is after all the very domain for which this capacity evolved in the first place. We have to take care not to confuse observation with theory, but this should not be particularly difficult. Descriptions of subjective experience are routinely used in other research fields, such as vision and hearing, because they enrich other data. How else could Ernst Mach have discovered the illusory bands that emerge when we spin an irregularly painted black and white wheel than by recording his subjective impressions? He had the good sense to use his introspections and then to include them, appropriately bracketed, in his database, to be verified later, like any other piece of data.

Clinical observations of disordered consciousness are often (but not always) focused on the intermediate time range of experience. This is no accident, and it is driven by the most respectable of scientific motives, sheer necessity. This is the time scale that matters in adjusting to the real world and the zone within which many disorders of consciousness reveal themselves. Many patients who have apparently normal conscious self-regulation on the microscale of seconds and milliseconds and therefore have normal short-term memory and perceptual attention fail miserably on the larger time scale, especially in self-monitoring. The reader might as well know my punch line from the start: Once we confront this evidence, the span of consciousness is stretched, and the paradox of consciousness appears not to be such a devastating paradox after all.

Twenty-five years ago, when casting about for my own theory of attention and awareness, I was brought up short when I encountered an apparently mundane problem that called into question the rigid testing methods I had been taught. Our unit in the neurology service of the New Haven Veterans Administration Medical Center was given an assignment, and like most of the young researchers of that time, I was determined to apply only the most rigorous and reductionistic procedures to my investigations. We were monitoring recent stroke patients twenty-four hours a day as they recovered consciousness. The long-term objective of the project was to improve the prediction of secondary strokes, so that measures could be taken to prevent them. Our short-term assignment was to find a good device for measuring the patients' recovery of consciousness hour by hour.

To do this, we needed a straightforward test that would give us numbers on a chart. Stroke patients are often in great distress as they emerge from their initial comas, and when they first come into consciousness, they are often awake for only a few moments at a time. Thus any test of their transient recovery episodes must be brief, portable, and extremely flexible. I remember vividly the advice of an experienced neurologist when we started this project: "Don't bother with any test that I can't carry around in my pocket or that will take more than two minutes. More important," he said, "a test should not be upsetting to the patient." Apart from ethical concerns, a disturbing procedure would seriously interfere with accurate measurement. After a few days of trying various methods, it became clear that the problem was not going to be solved with any of our standard laboratory paradigms, which were awkward to use in such a pressured situation and lengthy, lumbering behemoths to administer. They were also highly intrusive, utterly insensitive to the patients' distress, and upsetting to both patients and ward staff, rendering their validity even more questionable.

To my dismay at the time, we had to fall back on a modification of traditional clinical common sense. We harnessed clinical judgment and converted it into numbers on a rating scale. The mindreading skill of experienced clinicians became our primary measurement tool. In retrospect, it is revealing that we had to resort to such a strategy. Methodological necessity had compelled us to acknowledge that clinical judgment was superior to experimental paradigms in four crucial ways: speed, flexibility, sensitivity, and subtlety. It was undoubtedly inferior in many other ways. Clinical judgment cannot track at high speeds or with counterbalanced conditions, as a laboratory paradigm can. It can only measure the slower-moving aspects of consciousness, those that are manifest in the intermediate time zone. But the slower-moving global nature of clinical assessment turned out to be a virtue in this case.

Flexibility and the capacity of experienced clinicians to monitor several dimensions simultaneously were particularly important since patients might wake up in distinctly different states of awareness, and their return to consciousness might be revealed, perhaps only fleetingly, by any one of various signs—patterns of movement, shifts in gaze or attitude, gestures, emotional reactions, words, and so on—which would have to be tracked down simultaneously. No experimental paradigm could handle this degree of complexity. In fact, most paradigms are designed precisely to get rid of complexity, to simplify interpretation. The feel of real-world consciousness as it returns is not simple, but good clinicians can follow all its signals, effortlessly and simultaneously. They are quick to recognize the slightest sign, however indirect, that might suggest that the patient is regaining consciousness. Clinicians are not alone in this; family members can track the recovery of a loved one with great sensitivity. Good clinicians excel mostly because of the range of their experience, not because their antennae are necessarily any better than anyone else's.

Skeptics might question the value of clinical ratings. Considering how difficult it is to define the term "consciousness," how could philosophically unsophisticated clinicians agree on how to do this, especially under such demanding circumstances? How could they really know what they were measuring? Well, to be blunt, the abstractions of philosophy and experimental technique have nothing to do with it. Monitoring someone's awareness and constructing a description of it are not a matter of theory, but of observing with the best possible instrument. Again, we can fall back on the issue of social consensus. Our common descriptive tools were developed by consensus, out of concrete social need. Otherwise, language simply wouldn't work. This applies especially to our descriptions of internal states, or the inner life of the mind, which are one of the primary applications of natural language. Expressions such as "I think," "they know," "you feel," "he is sorry," "they are angry," and "I am happy" are the common mindreading vernacular of day-to-day discourse in all human societies. People agree on the basic labels for mental state categories because they share a common semantic landscape in this regard. Our most common words include state classes such as hate, love, deceive, imagine, envy, gloat, agonize, and so on, which emerge directly from a shared experience base. Not surprisingly, clinicians experienced with neurological and psychiatric disorders tend to agree on what they mean by disorders of consciousness. Their ideas on this have survived numerous attacks by Positivists, Behaviorists, and Cognitivists of various stripes because they are useful in the front lines, where it counts. While still a student, I coauthored a study of clinical evaluation of mental states in a neuropsychiatric setting that showed that the level of agreement among psychiatric clinicians, measured by correlating their ratings of common cognitive and emotional symptoms, rose directly with the number of years of experience, independently of the kind of training they had received. It did not matter whether they were trained as psychologists, psychiatrists, general practitioners, or nurses, provided they had logged the same amount of hands-on experience.

What is the nature of this putative consensus then? The clinical world has acknowledged for a long time that disorders of consciousness are complex and multifaceted. Consciousness is never treated as a unitary or simple thing. Clinicians never hang their assessments of consciousness on either language or sensation, the two crutches used by Hardliners when defining consciousness. In the clinical world, tests of these abilities are aimed at uncovering other disorders, such as aphasia (language disorder) and agnosia (perceptual disorder). Even a symptom such as blindsight, a neurological syndrome in which a patient has no conscious experience of vision but retains the ability to use it in implicit or indirect ways, is usually classified as a visual disorder and not primarily as a disturbance of consciousness. Blindsight has been discussed extensively in the literature on consciousness because it shows that we can use some aspects of visual input unconsciously, without consciously "seeing" the input (in a classic example the clinician tosses a ball at a patient who claims he can't see but nevertheless is able to catch it). But blindsight is primarily a disorder of perception.

This leads to a useful distinction between two kinds of disordered consciousness. Awareness can be fragmented by brain injury in two ways. First, there can be a fragmentation of content, as in blindsight and many other sensory and semantic disorders. In such cases, the main problem is often the inaccessibility of one perceptual domain to an otherwise normal conscious mind. Second, there can be a selective breakdown of the conscious process itself, in which one or more of the underlying self-regulatory functions of consciousness break down, and the effect is generalized to all cognitive domains. It is the second type of disorder that is usually called a primary disorder of consciousness.

A clinical examination of consciousness is usually focused on a breakdown of conscious function itself, on a level that reaches across several perceptual domains. It is often detected during a standard mental status exam, which is designed to evaluate the metacognitive or executive functions of the mind. The latter can be defined as a process of very high-level self-supervision, acting primarily in the intermediate time frame within which people normally act and think. One of the most important aspects of that process is self-location; that is, the person's orientation in space and time. In a mental status exam, people are asked questions that test their ability to place the immediate moment in a larger context. Thus they are asked to describe where they are, where they live, what year, month, and day of the week it is, and so on. Patients who have difficulty with these questions are usually judged to have a primary disorder of consciousness. Why? Because clinical experience has shown that normal consciousness routinely tests and verifies the larger coordinates of space and time. A disoriented patient is not fully conscious because this crucial verification function is not working as it should. The patient's capacity to build a representation of the world beyond what his or her own immediate perceptions would normally allow is impaired.

Although we are not usually aware of monitoring our subjective experience of time and space in any explicit way, it is common for us (and our close relatives and acquaintances) to notice when we are not properly oriented. Why we should do this is a bit of a mystery. Why should we wake up from sleep or a coma and expect to be oriented in space and time? Why should we, with our supposedly decentralized, unconscious brain processes, care what day it is, whether our memory record has holes in it, or whether we can readily recognize where we are? But we do: Our memory systems instantly reconfigure themselves upon waking, so that we awaken with precisely the same space-time orientation we had just before going to sleep. Even more puzzling, we sometimes awaken, after a full night's sleep, in the middle of the same thought sequence we experienced before falling asleep. The continuity lost during sleep is returned the instant we regain consciousness.

Given the amount of ongoing activity in the brain during sleep, it is astonishing that conscious registration of space and time can be maintained over the dead-air time of sleep and effectively made continuous over that period, despite interruptions of eight or more hours. When we awaken disoriented in space and time, it is an upsetting and usually temporary state of affairs. Some forms of amnesia are terribly disturbing to patients precisely because the continuity of these vital memory records has been interrupted. Consciousness does not explicitly monitor these records, but in a neurologically normal person it usually detects any protracted loss of orientation in space and time and acts to correct it. When this function is disabled, the consequences are extremely serious, and patients can become very disturbed about their disorientation. Indeed, their disturbance is a good sign, inasmuch as it shows that their verification process is still working.

When disorientation is successfully detected, the brain immediately devotes resources to reestablishing its time-space orientation. This suggests that the brain has a basic level of cognitive integrity, and this challenges the computationalist claim that the brain lacks such an integrator and is made up of nameless hordes of autonomous computational agents and pseudo-homunculi that lack any overall coordination. Why on earth would such a decentralized brain even notice whether it had a good long-term fix on space and time? Why should it care? Wouldn't any old set of semiconscious drafts of reality do, provided that it did not produce obvious perceptual dissonances? The idea that human awareness gets its space-time fix exclusively from the language-related aspects of culture is no valid solution to the problem either. Space-time integration is not uniquely human, whereas language is. Many higher mammals, such as dogs, with no symbolic mental devices to mediate such integration, show precisely the same tendency to maintain their orientation in space and time, keep it during sleep, and, occasionally, lose it in old age, just as humans do. The self-orientation function of consciousness is older than humanity, more basic than either language or symbolic thought, and buried deep within the mammalian neural blueprint.

Another measurable function that clinicians rely on heavily when investigating disorders of awareness is self-identification. The nervous system expects to feel self-familiar (certain neurological cases are exceptions to this rule) and does not easily tolerate the absence of this feeling. The self-image effortlessly combines a concrete, detailed, up-to-date image of the body, with a rather abstract autobiographical memory record. This merger also happens in the intermediate time zone. People who do not recognize themselves, or cannot accept who and what they are, are just as subjectively disoriented, in their own way, as those who lose track of external time and space.

Physical self-familiarity is one of our cognitive touchstones, perhaps the basis of all higher forms of self-awareness. Our bodies set the stage not only for conscious experience, but for memory. This point has been made by the philosopher Francisco Varela and his colleagues Evan Thomson and Eleanor Rosch and by the linguists George Lakoff and Mark Johnson. But no one has done so more elegantly than neurologist Antonio Damasio, whose "somatic marker" theory tried to reestablish the central role of emotion and feeling tone in self-familiarity, as well as our sense of familiarity with others. He theorized that one of the crucial memory markers for fixing and identifying any given experience is the bodily feeling tone associated with it. This does not imply that we carry around conscious images of our embodied selves in our heads. On the contrary, those images usually reside well outside consciousness. But consciousness immediately switches in if it cannot verify some aspect of the self-image or notices lacunae, inconsistencies, or absences. This also applies to our perceptions of others. We are very good at detecting any apparent failure of identity in others, even for a brief moment. We expect people to be on track all day, every day, and the slightest deviation is detected quickly (Did you see that expression? Jack is really *not himself* today). This testifies to our remarkable sensitivity to finely nuanced shifts of mind states in people we know well.

While studying a syndrome called "anosognosia," Damasio made some profound observations on the relationship between self-consciousness and body feelings. Despite obvious, debilitating problems with body recognition, anosognosic patients don't acknowledge that anything is wrong. There is a bizarre dissociation of physical self from their behavior. The name of the syndrome implies precisely this since it means "without awareness of ignorance." Damasio went straight to the heart of this syndrome, to the problem of self-reference. He realized that these patients cannot self-monitor over the longer term because they cannot update their self-representations to match their disabled condition. In his words, "Not one says, in effect, 'God, how bizarre it is that I no longer feel any part of my body, and that all that is left of me is my mind.' Not one patient can tell *when* the trouble with self-familiarity started. And they will never know, unless they are told."

This stands in complete contrast with most other neurological patients, who invariably use terms of self-reference in describing their symptoms. Thus, in Damasio's words, "When a patient develops an inability to recognize familiar faces, or see color, or read, or when patients cease to recognize melodies, or understand speech, or produce speech, the description they offer of the phenomenon, with rare exceptions, is that something is happening to them." This is self-reference, to know that something is happening to oneself. This capacity for effortless self-reference is lost not so much when the patients lose language itself (although this can sometimes interfere obliquely with communicating a sense of self) as when they lose track of their bodies, especially of their emotions and feelings. In anosognosia the vital linkages between physical self-familiarity and the rest of the cognitive domain have been interrupted. The result is one of the most serious losses of metacognitive governance of any neurological syndrome and illustrates the complexity of conscious awareness, since the anosognosic patient clearly has some aspects of basic awareness intact yet suffers from a terribly disordered consciousness. Clearly, normal consciousness depends on the cooperative functioning of many different brain systems.

What about the role of language? Although it is usually assessed separately, language loss can sometimes raise questions about a patient's consciousness. Broca's aphasics, whose disorder is usually restricted to expression, do not appear to lose any component of basic consciousness, and this militates against any simplistic identification of consciousness with language expression. On the other hand, Wernicke's aphasics, who lose their understanding of language, sometimes find themselves in deeper trouble, and the possibility of a special kind of disordered consciousness has to be considered in their case.

Unlike Broca's aphasics, they seem to be isolated from conversation, even from inner speech, and lack insight into their condition. Because of their linguistic isolation, the measurement of conscious function in Wernicke's aphasics is much more difficult. They suffer from a primary disorder of representation and cannot formulate clear ideas about their illness. But the most basic aspects of awareness—their basic orientation in space, their understanding of objects in the environment, and their self-familiarity—are often intact.

The independence of conscious self-orientation from the language brain can be seen clearly in a case of reversible paroxysmal aphasia reported by neurologists André Roch-Lecours and Yves Joanette, which I have reviewed in detail elsewhere. This patient, known simply as Brother John, was a monk who worked as the editor of a small newsletter. He suffered from temporary language impairment, caused by highly localized epileptic seizures that lasted for up to ten or eleven hours. Significantly, during those periods of complete language shutdown, he remained completely conscious, and he was later able to remember the whole episode in detail. His language, including speech comprehension, speaking, reading, writing, and "inner speech," which is the habit of silently talking to oneself, was globally impaired. However, many of his cognitive functions were normal. This included metacognition, or self-evaluation. During his seizures he was able to recognize his dilemma over the intermediate term and self-supervise his activities. As a result, he could modify his behavior in a remarkable manner.

For example, he would carry around a portable radio and listen to it periodically, to test whether his comprehension of speech was returning. As soon as he was able to understand some fragments of the radio sound, he knew he was recovering (this implies that he could recognize whether a given sound stream was classifiable as language or not, much as an infant can, before it understands language). During his spells he could also operate familiar machines, such as an elevator or appliance, provided that they did not require him to read instructions. His space-time orientation was apparently good, in a practical sense. He could find his way around, not only on home territory but also in unfamiliar settings. He behaved appropriately in specialized spaces, such as restaurants, hotels, bathrooms, and train stations. He could signal to his friends that he was temporarily out of commission and adjust his social behavior accordingly, to minimize inconvenience and embarrassment.

Brother John's behavior shows that human consciousness is multilayered and complex. When he was temporarily without those aspects of awareness that depend upon language, he could not follow a conversation, read, or entertain such thoughts as "That headline is too aggressive." But many components of awareness, including some of his most important metacognitive skills, proved to be independent of his defective language. His recall for each episode (except for the initial few minutes) was excellent. His conscious perception, attention, short-term memory, and even ability to set and hold priorities all seemed to work well without the on-line participation of language. There may have been implicit effects of language on his coping behavior, but these could not explain his excellent episodic recall for each seizure period. His awareness was in some ways the inverse of the anosognosic patient. His self-familiarity was intact while his language system was temporarily shut down.

All this leads to one conclusion: Consciousness has independent review functions that check and maintain our orientation in time, as well as in physical and social space. It also verifies and positions the self accurately within an autobiographical memory context on several levels. Awareness is therefore closely tied to, and dependent upon, our working models of the world. The entity that clinicians call consciousness constructs and maintains the larger course that an organism takes over its lifetime. It is a surveillance system, a metacognitive governor that reviews the general state of the organism, maintains a fix on space and time, and formulates the mental models that give meaning and self-referential resonances to experience.

In executing this role, human consciousness reveals its adaptive value. It can extend its umbrella over a much longer time range and deal with a larger environment than the one afforded by a short-term, or perceptual, view of experience. This idea extends and enriches our vision of attention and awareness. A clinically diagnosed deficit in conscious function is rarely a simple question of a unitary deficit in short-term working memory, although there are such deficits, in a small number of cases. It is usually diagnosed as a serious intermediate-term dysfunction, involving metacognitive self-supervision. Loss of this conscious review function is akin to losing the pilot of mental life and the governing monitor of existence.

In this sense, awareness is more like a long-range guidance system and less like the proverbial one-armed attentional paper hanger that we usually study in the laboratory. Intermediate-term awareness is an active, influential, organizing force in mental life. We can see this with particular clarity in certain clinical case histories.

AN AGGRESSIVE, INTERVENTIONIST CONSCIOUSNESS: A CASE HISTORY

The conscious mind can be highly aggressive in asserting its guidance function. There is strong confirmation of this active role in the well-known case of a man known in the clinical literature by the pseudonym

Zasetsky. His history was documented by Alexander Luria, the Russian pioneer of neuropsychological case histories, in a book translated into English as *The Man with a Shattered World.* The case of Zasetsky shows with unique clarity how the conscious governor exerts a great influence on mental life, even under extreme duress.

Like thousands of other wartime victims in Russia, Zasetsky received a gunshot wound, as the result of which he lost major parts of the occipital and parietal cortex on the left side of his brain. He also lost vision in the right half of his visual field (vision is completely crossed over from each retina to the brain, so that injury to the left side of the visual brain destroys the view of the right side of the world for both eyes). He had occasional seizures, during which he experienced vivid visual hallucinations (he described them as dots that swarmed across his visual field), especially when his eyes were closed. Also, although he could still see fairly well on the left side of his visual field, this residual visual capacity was partially disordered.

Like most patients with this kind of brain trauma, he was faced with a devastating and permanent loss of control over language, memory, and thought. His voluntary recall of past experience was almost completely destroyed. His conscious experience was fragmented into a thousand pieces, and he was not even able to fit those few segments together. Zasetsky's conscious recall from memory was also affected; his subjective stream of consciousness was reduced to tiny snippets of clarity gained here and there, seemingly at random. His dilemma could be construed as an exaggeration of the tunnel of consciousness described in Chapter 2. He had an abysmally narrow range of awareness, an extremely limited short-term memory, and an inordinate reliance on unconscious or automatic cognition, with a concomitant loss of conscious control over his cognition.

However, all was not lost. He could hear normally, and his control of movement (including his pronunciation of speech sounds) was good. Moreover, his long-term memory for his earlier life was essentially normal. He could still remember the distant past and therefore knew who he was and where he was from. Luria speculated that this was because Zasetsky's frontal lobes were intact. Because this part of his brain, where the future is constructed, plans are made, and the long-term control of attention resides, had been spared any damage, he was able to realize what he had lost. His goals in life were still intact. His persistence, motivation, and determination to recover some of his mental clarity were there, as were his expectations. In other words, he still had his mental governor. But he faced a dilemma: While his capacity for metacognition was undamaged, he was greatly disabled in terms of the shorter-term aspects of consciousness, especially attention and working memory. To Hardliners, Zasetsky would be an anomaly, if not an impossibility. He lost the shorter-term aspects of his conscious process but retained the intermediate-term ones.

The disability suffered by most patients with this kind of injury makes it impossible for them to give detailed testimony about their illness, but Zasetsky was different. Although his mind lacked short-term continuity, and most of his cognitive capacities were running on automatic, through incredible persistence he managed to do something unique. Over a period of twenty-five years, although his attention span was confined to a time window of only a few seconds (he could understand only one or two words at a time), he fought his limitations and managed to write a three-thousand-page diary, giving a firsthand account of his experiences and describing what it feels like to have nothing but a fragmented and transitory consciousness. It is impossible to grasp the depth of the frustration this intelligent man must have felt as he embarked on his epic literary journey, a far more heroic journey, I think, than those of most autobiographical writers, who enjoy the services of a normal conscious governor.

Ironically, and incredibly, while writing his journal, he remained unable to recall directly from memory. In fact, he was unable to extend his awareness beyond what was immediately in front of him. But he kept trying to widen the window of his conscious mind. By his own report, this exercise was driven by only one objective: He wanted to make his condition consciously transparent to himself. His diary is the most extraordinary exercise in self-understanding, or rather in deliberately reconstructed self-consciousness, that I have ever encountered. It provides an object lesson for anyone who wishes to understand what conscious governance is really about.

Since his main injury was in that part of the brain where vision and hearing are unified in our perception of the world, Zasetsky's impressions were terribly chaotic, his perception of space was distorted, and there was very little he could do about his condition, which he described as follows: "Ever since I was wounded, I have had trouble sometimes sitting down on a chair or on a couch. I first look to see where the chair is, but when I try to sit down I suddenly make a grab for the chair since I'm afraid I'll land on the floor. Sometime this happens because the chair turns out to be further to one side than I thought."

Once-familiar visual routines were lost. If his mother asked him to chop wood or bring up milk from the storeroom, he would not know how to proceed. He had to relearn all the steps involved in the simplest of tasks, and he often had trouble remembering them. Immediately after his injury, he did not know how to hold a spoon or shake hands. He had forgotten how to use a hammer and nails or what to do with needle and thread. He could not remember how to search for things in his house, and he tended to get lost very easily. He would often lose track of where he had come from and where he was going. He had forgotten the streets of his hometown and lost his sense of direction. He could not read a map or a mechanical drawing, despite the fact that he had been a platoon commander and had received extensive training in those tasks.

His voluntary recall was greatly limited. He had difficulty remembering things he knew he should know, and this was especially upsetting to him. He had lost all his technical knowledge of mathematics and had to relearn the simplest arithmetical operations, such as addition and subtraction, but never again managed to multiply or divide. He lost his ability to play chess and dominoes, two games at which he had excelled. He sometimes regained a lost memory when it was triggered involuntarily by something in the environment or by a fleeting emotion. Interestingly, he could recognize this—that is, he would consciously realize that this memory was one for which he had been searching—but the rediscovered memories usually faded quickly from awareness, and then he would have difficulty getting them back. However, less explicit or less controlled forms of remembering were still intact. His fantasies, imagery, and imagination seemed normal, as long as he did not try to control them consciously. So, to outwit his limitations, he deliberately avoided conscious recall! His consciousness thus achieved an amazing and paradoxical thing. It voluntarily defeated its own tendency to recall from memory in order to achieve a longer-term goal.

His major complaint was a failure to remember words. He could not remember the names of things. He could speak fairly well when he was not under any stress or social pressure, but even under the best circumstances he had difficulty finding words. He could understand short sentences but found it hard to follow longer conversations and to remember and interpret gestures. He could not follow a movie because he could not retain such a long series of events in his memory. He also lost the ability to read. This is how he described the moment when he realized that he had become illiterate: "I went into the hall to look for a bathroom I'd been told was next door. I went up to the room and looked at the sign on the door. But no matter how long I stared at it and examined the letters, I couldn't read a thing. Some peculiar, foreign letters were printed there—what bothered me most was that they weren't Russian. . . . Was it possible that I couldn't read Russian any more, not even words like Lenin and Pravda? Something was wrong. It was ridiculous."

He registered for therapy, was highly motivated to recover reading, and made tremendous efforts where many might have given up. These eventually paid off, and he gradually learned to read and write again. However, he could not achieve this in the normal way and had to cobble together an extremely slow, tortured method of reading by capitalizing on his few remaining abilities. He described it as follows:

> I read printed matter letter by letter. When I first started to read again, I often couldn't recognize a letter at first and had to run through the alphabet until I found it. But later I did this less and less and tried to remember it myself—just waited until it came. . . . Only after I read a word and understand it can I go on to the next word, and then to the third. By the time I get to the third word I often forget what the first or sometimes even the second word meant. I also have to stop at the fourth letter of every word, because even though I can see it and know how it's pronounced, I've already forgotten the first three letters. While I'm looking at the fourth letter, I can still see the second and third, but not the first letter of the word, which is completely blurred.

It took him years to relearn to read and even longer to reacquire writing. He could not remember how to form letters no matter how often he tried, but discovered that he could write if he did not think about the letters, and wrote the words automatically; that is, without conscious monitoring. Once a word came back, he realized (on conscious reflection) that it tended to stay and that he could write it in the future. Since his recall of words was so poor, composing sentences was a major problem, but again he outwitted his disability by turning off his conscious efforts at recall. He knew that words came to him only passively. Yet he turned the tables by using his knowledge of his disability.

This strange strategy demanded great personal discipline. He had to write down the names of things and objects whenever the appropriate words came to him. This realization had to be instantaneous or he would have lost the words, so he collected them immediately in a notebook he carried around with him. He would then experiment, off-line, with various groupings of the words he had collected until the sentences seemed to make sense. Again he could do this only because he was consciously aware that he had a use for such words.

> I've repeated the same points over and over again in my story and may do it again, because I'm always forgetting what I've written and what I still want to say. . . . Sometimes I'll sit over a page for a week or two. . . . I have to think about it for a long time, slowly considering what I want to say and then comparing various kinds of writing so that I can figure out how to express myself. . . . For months

> on end, I'd spend day after day putting together a vocabulary from my scattered memory, collecting my thoughts and writing them down. . . . By working on that one story of mine every day—even small amounts at a time—I hoped I'd be able to tell people about this illness and overcome it.

On a good day he could write half a page of his reminiscences. He toiled in this way for a quarter of a century. At the time Luria published his journal, Zasetsky was still working on his project, still striving to regain his lost mind. This is one of the last entries before publication: "Time is flying. Over two decades have slipped by and I'm still caught in a vicious circle. The average person will never understand the extent of my illness, never know what it's like unless he experienced it himself."

There is no difficulty reconciling Zasetsky's aggressive, directive, interventionist efforts with the version of consciousness we see repeatedly described by clinicians. But there are all kinds of difficulties reconciling Zasetsky's journal with a passive, epiphenomenal, decentralized consciousness. Zasetsky's story conflicts with the idea that awareness is only a passive by-product of neural activity. His consciousness was very much a causative agent in the world, a strategic device without which online, real-time decisions could not be made, reconstructions of ideas and memories could not take place, and new plans could not be assembled. There is nowhere else in the world of the mind where such governance can happen.

Zasetsky's awareness also had properties that excluded language as either mediator or causative agent. If his consciousness was entirely language-bound, how could he have known that he should have known something he could no longer find words for? How could he know that he once had known a word but no longer did? How could he persist on a specific word search for days or weeks at a time? This must have been sustained at the level of his semantic understanding of the world; that is, in the prelinguistic engine deep within his mind.

This gives us a hint of what consciousness is really about, at least in the human species. It is much deeper than the sensory stream. It is about building and sustaining mental models of reality, constructing meaning, and exerting autonomous intermediate-term control over one's thought process, even without the extra clarity afforded by having the explicit consensual systems of language. The engine of the symbolic mind, the one that ultimately generates language to serve its own representational agenda, is much larger and more powerful than language, which is after all its own (generally inadequate) invention. Above all, this deliberative capacity is an active and causative element in cognition and a regulator and arbitrator of action.

ZASETSKY'S MIRROR TWINS

Some neurological patients, who have typically suffered injury to their frontal lobes, display deficiencies that are precisely the opposite of those shown by Zasetsky. Unlike him, they have no problems with conventional psychological tests and often give a first impression of cognitive normalcy. Their short-term conscious functions are normal. But they fail miserably in the conscious governance of their lives over the intermediate and long terms, and they often lose their personal identities. Whereas Zasetsky clung ferociously to his former identity and aggressively pursued an idea of what his future should be, despite his crippling abnormalities, these mirror twins lack the ability to persevere at anything or even to evaluate the consequences of their own actions. They lack Zasetsky's sense of direction, his determination, and his obsessive self-evaluation. Instead they blandly watch their lives deteriorate, and when they are confronted with the evidence of this deterioration, as when marriages break down and careers are shattered, it doesn't seem to bother them. Unlike Zasetsky, who remained firmly and passionately at the helm of his catastrophically damaged mental life, no one is apparently in control of the actions of these unfortunate wanderers. Their sensory stream is normal, but they lack depth and are simply unable to manage their lives.

Three groundbreaking cases, known as Kar, Kol, and Bred, illustrate this point. They were documented by Luria between 1948 and 1950, and introduced to the English-speaking world in 1966. They were historically important and should have had a huge influence on theories about human frontal lobe function. Unfortunately, they came to the attention of Western neurosurgeons too late to stop the thousands of lobectomies that were performed in the fifties, before the medical establishment realized the devastating long-term effects of these operations. When they first appeared on the scene, frontal lobectomy patients presented a deceptive picture to clinicians. They seemed to have normal cognitive function over the short run, were usually fully oriented, in the sense of knowing date, time, and place, and were able to plan over the short term. Their language and general knowledge were intact, and they were perfectly capable of dealing with straightforward challenges such as IQ, aphasia, or perception tests. Their short-term memory and attention were also normal. Compared with Zasetsky, they were virtually symptom-free.

Nevertheless, long-term follow-up showed that they were severely deficient in three of the most critical conscious functions, self-evaluation, prioritizing, and planning. While these were intact in Zasetsky's brain, keeping him going and giving him direction, the same was not true of these patients. While Zasetsky was constantly judging himself, tracking his progress, and striving ceaselessly to achieve his goals, they watched their jobs and marriages fall apart, without apparent concern. While Zasetsky fought hard and devised strategies to overcome his deficits, these people did not fight at all, failing to capitalize on their strengths.

Their chief intellectual problem was keeping the complex menu of priorities and possibilities in their lives arranged in the correct hierarchical order. Other species, including primates, lead much simpler lives, and this kind of deliberate prioritizing is not needed to the same extent. But as humans, living in complex cultures, we have to organize our actions around many concurrent possible scenarios. Consider a typical human occupation, such as running a small store. The owner must develop a routine, so that all the many suboperations embedded in running the store, such as renting space, selecting stock, setting hours, hiring employees, purchasing, pricing, keeping books, decorating windows, marketing, advertising, billing, maintaining the building, and so on, can be tracked and managed in a hierarchy of timetabled priorities. Confronted with this kind of relatively ordinary management task, a patient with serious damage to the frontal lobes would be hopeless, lose track of the hierarchy and of time, forget important items, dwell on trivial ones, and mix up the order in which things should be done. Some patients have proved so impaired in this regard that they cannot manage something as simple as planning a day at the beach. After all, even relaxation is complicated in human culture; you have to remember what to bring, whom to take, what to wear, whether the car needs servicing, and so on.

Thus a superficial impression of these patients would be completely misleading. On first examination, Zasetsky would appear to be much worse. They could talk coherently and performed most tests impeccably, while Zasetsky couldn't even string together the letters of a once-familiar word on a page. Their verbal answers to complex questions made perfect sense, while Zasetsky couldn't understand, let alone analyze, an intricate question. If our evaluation were restricted to immediate performance, we would have to conclude, as many did for more than a decade after the first lobectomies were performed, that there was nothing seriously wrong with frontal lobectomy patients, whereas Zasetsky was very seriously impaired. But using a different database, gathered from the real world rather than from tests, over the long run, we were able to see that frontal lobe patients were failing miserably and proved to be more fatally impaired. Zasetsky fought the good fight magnificently, against insuperable odds, like a hero. But there was no such heroism and no possible victory for those with frontal lobe injuries. They were unable to capitalize on the abilities they had retained. Theirs is perhaps the ultimate breakdown of conscious governance: failure to monitor oneself and indifference to the future. Nothing could more strongly confirm the importance of the intermediate-term governance we identify with the frontal lobes.

A LITERARY VIEW

While we are exploring the realm of clinical ethology, or natural observation, we should also consider its most highly cultivated (that is, deliberate and conscious) expression. Literature affords us a great luxury, one that we lack completely in the clinical study of consciousness because even the most experienced clinicians remain outsiders to their patients' minds and are constrained by the formal, conventional nature of their encounters with others. Fiction is not so fettered. It is entirely the product of the imagination, and therefore, writers are not so bound by convention. Their perspective provides a different kind of reality check, built from expert observations but from the inside. For this reason alone, literature must become part of our database. It is perhaps the most articulate source we have on the phenomenology of human experience.

The best writers have pushed the subjective exploration of the mind much further than would be permissible in clinical or experimental psychology. Whether they acknowledge it or not, writers always have an implicit psychology, which reflects the way the culture regards the human condition, including the condition of the psyche itself, and of awareness. Novelists in particular often explore our deepest assumptions about awareness. Their portrayals of it constitute a vast, unsystematic collection of phenomena observed from the inside and are possibly the most authoritative descriptions we have. What is the phenomenology of consciousness presented in literature? While there is no simple way to summarize it, on the central question of whether the conscious mind is active or passive, literature is generally much closer to the clinicians' than to the Hardliners' view.

With a few notable exceptions, literature did not plumb consciousness in depth until the last two hundred years or so. Most novels are related either in the first person, by characters whose stream of awareness is being exposed for all to view, or in the third person, by a narrator within whose consciousness the whole story unfolds. Novelists who explore the interior life of the mind in detail reveal their own faith in the great power of conscious governance. A passage from Henry James's *The Bostonians* provides a good impression of how the subtleties of individual awareness were portrayed in his society. In this paragraph he described a dilemma in the mind of his heroine:

> This was painfully obvious when the visit to his rooms took place; he was so good humored, so amusing, so friendly and considerate, so attentive to Miss Chancellor, he did the honors of his bachelor nest with so easy a grace, that Olive, part of the time, sat dumbly shaking her conscience, like a watch that wouldn't go, to make it tell her some better reason why she shouldn't like him. She saw that there would be no difficulty disliking his mother; but that, unfortunately, would not serve her purpose nearly so well. Mrs. Burrage had come to spend a few days near her son; she was staying at a hotel in Boston. It presented itself to Olive that after this entertainment it would be an act of courtesy to call upon her; but here, at least, was the comfort that she could cover herself with the general absolution extended to the Boston temperament and leave her alone. It was slightly provoking, indeed, that Mrs. Burrage should have so much the air of a New Yorker who didn't particularly notice whether a Bostonian called or not; but there is ever an imperfection, I suppose, in even the sweetest revenge.

There is a great deal going on in this paragraph, and this scene is only a tiny fraction of a much larger episode. Olive Chancellor is a brilliant but somewhat marginalized character whose conscious mind is constantly scheming and plotting in the social sphere. James is showing us how she can effortlessly evaluate a social situation of great subtlety, at lightning speed. Olive knows her own motives well, is quickly able to read those of the Burrages, and realizes that they, particularly Mrs. Burrage, will be an obstacle to her intentions. She reflects on her own emotions, knowing that she is not a good actor, and sees that she needs to generate genuine dislike for Mrs. Burrage in herself to achieve her ends (much like Zasetsky, she makes a calculated end run around her own perceived weaknesses). She decides to keep the Burrages off their guard in the short run. All this happens in an instant, while she continues to act in a socially acceptable manner, more as an elaborate deception than in earnest.

This picture of the conscious mind is incompatible with the Hardliners' claim that consciousness is generally passive and ineffectual. Olive has a very busy and unified conscious center. Her mind is evaluating each of the three characters in the scene in depth, simultaneously, and very deliberately. This achievement alone implies a considerable skill at assessing the mind states of other people very quickly and from a few scarce cues. Her conscious reflections on what the Burrages expect lead her to act in a very calculated way, designed to create the desired impression in her audience. She is also conscious of the fact that she is suppressing an urge to act in a very different way, as well as concealing and exerting rigid control over her emotions (guilt, excitement, resentment, frustration, and anticipation). At the same time, she is planning for the future and registering all sorts of details in the behavior of Mrs. Burrage (who is the real threat, since Olive finds that most men are simple creatures and no match for her social skills). Olive achieves all this without saying a word about it, while continuing to take part in the conversation, itself a rather complex social game going on in the room, seemingly under its own steam, since Mrs. Burrage is just as crafty as Olive herself and doing her own scheming beneath the conversational surface.

James's masterful portrayal introduces a new factor in our discussion of the conscious management of human affairs, a dimension that experimental psychologists have studied extensively. This factor is what we call multitasking, the ability to do or think about several things at the same time. As we saw in our discussion of games and conversations, human beings are uniquely good at doing this. Laboratory studies have shown that most people can perform two and sometimes three tasks at once; for instance, they can memorize a series of words while playing a video game. Equivalent examples from real life might include talking to a passenger while driving a car or mixing a salad while listening to the news. These are valid examples, but they do not test the limits of human multitasking capacity. Henry James does.

In the world of Henry James, we are given a more realistic portrayal of our enormous capacity for multitasking in the social realm because James goes beneath the surface. The objective behaviors that an outsider might observe during the exchange between Olive and the Burrages would give only the most superficial impression of what is going on. Their behavior floats on the surface of a much deeper cognition, reflecting only what is actually said and done. However, from James's narrative we know that the minds that generate and regulate this surface veneer of action are much more concerned with long-term strategic issues. His characters behave one way or another, usually only to produce an impression, while they seek weak points in the enemy camp, indulge in some fairly ruthless self-criticality, and prognosticate future outcomes.

Would anyone deny that Olive Chancellor, who is after all a fictitious character, could have had so many things in her awareness at once? It is obvious that she can keep them separate and evaluate them separately. She does all this without losing track of who she is, in the sense of maintaining her sense of self-dignity, or precisely why she is there, or how the present situation might fit into her larger schemes. We can verify the novelist's observations against our own experience and produce a consensus that is formally no different from the agreement on, say, the existence of Mach bands. We all report precisely the same sort of experience of multilayered self-management. There is little doubt that human beings routinely experience this sort of complexity in consciousness. No sophisticated critic, to my knowledge, has ever questioned the psychology implicit in Henry James's descriptions. Even his brother, William, one of the great psychological theorists of his generation, seemed comfortable with this.

Another element of conscious experience can be illuminated by another brief reflection on Henry James. I mentioned earlier the importance of working memory, which serves as the temporary holding tank for memories that are immediately relevant. Working memory is normally tested in the laboratory by means of simple tests, such as asking the subjects to recall numbers or remember the location of objects hidden in a picture. Tested in this way, Olive Chancellor's working memory would have been very similar to our own. For what it is worth, she could probably have recalled no more than six or seven digits and remembered only six or seven items from a display. But James has made the implicit claim that in the domain of social cognition her working memory is, by any standard, gargantuan. James would obviously overthrow Miller's famous limit of seven, plus or minus two, because he abandoned the sensorimotor interface and explored inner space. Within the context of conscious self-regulation, Olive's memory obviously stretches much farther back in time than the fifteen seconds or so that are normally attributed to it. Olive can instantly recall specific words, gestures, and signals minutes or hours after they were produced, even if they were not deemed significant at the time. She can imagine entire hypothetical conversations and even written exchanges and can include in her musing the detailed expressions, silent looks, subtle shifts in attitude, and hosts of other unspoken cues that give away the real intentions of other people or groups. By the consensus of a highly literate, critical, and introspective public, so can we.

James has shown us that the range of Olive's awareness is vast, cutting across time and space, as well as the more abstract realms of social geography. She is like a master chess player confronted with a huge chessboard. She assesses her available moves, while contemplating (and grudgingly admiring) the clever moves made by her antagonists. She is constantly vigilant and engaged. Moreover, as she maintains this larger overview of her life, she continues to carry out all her lesser cognitive functions perfectly and effortlessly, farming out responsibility to her automated minions, while keeping them on a short leash. These functions, which adults find so simple that they seldom need to tap their conscious capacity to perform them, are, unfortunately for experimentalists, precisely the ones that are studied to death in the laboratory because they are so easy to measure. Olive has no difficulty doing these mundane things: perceiving objects, finding words, formulating grammatical sentences, recalling facts, focusing and shifting attention, moving about in three-dimensional space, coordinating her hands and eyes, reading aloud, and so on. She often carries out several such operations simultaneously; she can balance her cup and stir her tea without spilling it, while continuing to speak, walking across the room to look out the window, and ruminating about her next move. She can do all these things without much conscious supervision, while her conscious mind is focused on the elaborate social game that is unfolding. Olive's brain is undistracted and altogether undaunted by this complexity. Her consciousness is engrossed in doing what she really came for, planning her war against the Burrages.

This picture, a novelist's crystal-clear image of a mind at work, dovetails well with many clinical neuropsychological cases in the literature. It conveys an impression of the complexity that a human brain routinely manages and gives us a more realistic idea of the sheer magnitude of the abstractions that we routinely make in consciousness. It also reminds us of the scale of the problem. Viewed from this perspective, the Hardliners' obsession with the short term, and with the automatic nature of many sensory and motor tasks, appears to be completely misconceived. Some of what they say is true, but it is horribly incomplete. It leads us away from where the action is. Olive Chancellor's awareness is not a passive funnel of sensation. It cannot be measured accurately simply by registering what she sees or tracking her performance on some trivial task.

Rather it is an active awareness, perhaps overactive, to the point of aggressivity, in the intermediate time zone in which social events unfold. In this, her awareness is very much like Zasetsky's, constantly planning and relentlessly self-correcting. But unlike poor Zasetsky's, whose perceptual apparatus had been shattered, Olive's awareness is effortlessly and seamlessly unified across modalities of sensation and across the many levels of abstraction in her many-layered awareness. That unity extends across vast stretches of time, bringing together many extended episodes to weave the mnemonic tapestry of a life. A life that never experiences a profound sense of discontinuity never loses its sense of selfhood, place, time, or purpose.

Jaded though we may have become in this century, we should not hold back our wonder at what this implies for the conscious mind. Olive, and by implication all of us, are oblivious of any supposedly crippling capacity limits. Her consciousness is more than adequate to the task at hand. She is full of assurance and races through her life full steam ahead, as indeed, most of us do. After each episode is finished, she moves on to another, and another, occasionally retreating to her suite of rooms to contemplate her victories, defeats, and future tactics. Here is a very different conception of consciousness, seamless, integral, striving for ever more integration, confident, and fully in command.

VERTICAL DEPTH AND UNITY: THE TRUE REACH OF METACOGNITION

There is no point in flogging these points further. Readers can find their own examples. But there are other nooks and crannies of conscious mental life that novelists have illuminated. Although his psychology was similar, Henry James's French predecessor Henri-Marie Beyle, known by his literary name, Stendhal, had his own way of hinting at the multilayered structure of inner awareness, the subtle shades of perception and thought that operate deep within his characters, within the scope of awareness, if not always in the spotlight. The following passage describes what is going on in the mind of Mlle. de la Mole, a central character in his novel *The Red and the Black*:

> With Julien and me there's no question of signing a contract, no lawyer at a bourgeois ceremony; with us, everything is heroic; everything will be the offspring of chance. Is it my fault if the young men at court are such conformists and turn pale at the idea of any venture that is the least unusual? For them, a little voyage to Greece or Africa is the height of daring, but even then, they have to go in troops. As soon as they find themselves alone, they are afraid . . . not of the Bedouin's spear, but of ridicule; and that fear drives them insane.
>
> My little Julien, on the contrary, prefers to do things by himself. It never enters that privileged being's mind to look at others for moral support or help! He despises other men, and for that reason, I do not despise him.
>
> If with his poverty, Julien were noble, my love would be nothing more than a shabby mistake, a vulgar misalliance. I should want none of it; it would lack that which typifies the great passion; the immensity of the obstacle to be surmounted and the dark uncertainty of the outcome.

There is enormous subtlety in this passage, but this time it is not evident in an extroverted game of complex social multitasking, such as we saw in the passage about Olive Chancellor. Rather it is in the introverted plumbing of one's own vertical depths. Mlle. de la Mole's reflection on her relationships shows that she is aware of more than she would openly admit regarding her motives for loving Julien. There is an intricate moral hierarchy at work on the fringes of her awareness, an inner self that casts a critical eye on her motives and on the abstract standards she uses to judge those motives. Her ceaseless self-reflection generates subtle metacognitive habits, all consciously constructed and consciously scrutinized. How many eyes her consciousness has, Stendhal tells us, and how many sly tricks it can play on itself! Like Zasetsky's fanatical self-discipline, Mlle. de la Mole's conscious mind confidently imposes impossible standards on her own lower cognitive minions, the platoons of modules, processors, and memories that we know (although Stendhal did not) she has in her head and that allow her to pursue her course toward an ideal, passionate, self-sacrificing, uncompromising love. Such ideas can lead people to take great risks, to insist on playing out what they have constructed as the central dramas of their lives, even to the point of destroying themselves, and all with the deliberate, considered poise of conscious command, even intentional self-deception. The author is telling us implicitly that such intricacies and such confident self-governance were common social currency, a kind of personal ideal, in his vision of society.

Stendhal is describing that aspect of human consciousness that we know as metacognition, a kind of elevated self-awareness that involves an ability to monitor our own minds and to monitor our own monitoring. But this is metacognition as it is practiced in the real world, not simply a matter of following one's success or failure in a game, for example. In this case, metacognition is elevated far beyond its laboratory or even its clinical manifestations. Metacognition is certainly one of the most powerful functions of human self-consciousness and may well prove to be one of the most difficult to explain. The intricacies of our conscious self-observations are usually unspoken, yet *potentially* spoken. They are seldom articulated in public, except by the likes of professional novelists, expert observers who make it a point to delve into such things.

Nineteenth-century novelists were very much aware of their role in this respect. They regarded themselves as psychologists. Their self-assigned mission (itself a highly contrived, highly conscious innovation) was to reveal as precisely as possible the intricate workings of the individual, which had never before been explored in such detail. They were sociologists long before there were such people in the professional sense. Honoré de Balzac, Stendhal's contemporary, subscribed strongly to this belief in his own role in society. He constructed a vast sociogram of every stratum of French society, which he tried to explore from the inside. Given the Napoleonic scale of his ambition, Balzac did very well. He managed, during his relatively short literary life, to create more than four hundred characters, drawn from every major subcategory of his world, whose lives were intricately interconnected in various ways.

The depth and confidence of Balzac's characters served as a model for both Stendhal and James. Balzac's France was full of immensely intricate, cunning, striving minds, interwoven into the fabric of their society. But few of them would have been a match for Balzac himself, whose consciousness encompassed not only the incredible drama of his own life but that of his whole culture. In my view, the indisputable existence of a mind like Balzac's challenges experimental psychology to match or at least to honor his greatness.

To put these novelistic impressions in perspective, contrast them with hypotheses about capacity limitations based on experimental findings, such as those described in Chapter 2. We argued first, from laboratory evidence, that most judgments must be made unconsciously and second, that we cannot consciously grasp more than a fraction of the information available in the briefest of moments. If this were a representative picture of human consciousness, the result would necessarily be a very tentative, fragmentary awareness, more like Zasetsky's than Olive Chancellor's. People would routinely experience a desperately disconnected existence, and given our remarkable capacity for metacognition, we would *feel* our essential disunity, as indeed, certain kinds of patients do.

Put that notion up against Balzac's testimony. He perceives no universal experience of disunity, not even a hint of such a thing. That fact has to be explained. It will not do simply to dismiss subjective accounts on grounds of method when in reality they are being rejected because they show with embarrassing clarity that there are fatal problems inherent in the Hardliners' approach. We cannot simply dismiss what Stendhal and James described, or what intelligent, critical people report as their consensus on the nature of conscious experience, when such testimony constitutes our primary ethological database. This is a basis for constructing a consensus, provided that we have clear questions. For our theoretical purposes, we must face up to the fact that consciousness is normally perceived as a connected, continuous, enduring state of mind. This is undoubtedly our canonical experience of our own awareness. This phenomenon, seamless continuity, has to be explained.

More important, perhaps, writers clearly agree on the predominance of social consciousness over sensory awareness. Our concrete perceptions of objects, time and space, as well as the world of concrete action, normally shrink into the background when we are negotiating the shoals of the social sphere. Such things become incidental, unless something goes terribly wrong. Our normal focus is social, and social awareness is highly conscious; that is, it heavily engages our conscious capacity. Stendhal and James might have been willing to concede that unconscious motives might have contributed to the reactions and desires of their characters. This insight was well developed by the last decades of the nineteenth century and was later pushed to its limits (and far beyond) by psychoanalysis. But there was no doubt that the so-called Ego, the conscious self, was firmly in control.

This commonsense view also makes eminent scientific sense. Unconscious cognitive processes could not be used to deal with most social situations because to be consistent with modern cognitive theory, they lack an updating function, one of the core underlying functions of consciousness. Consciousness is constantly reviewing and verifying our stored knowledge of the world as it is accessed and bringing it up to date. Thus, if we know Sam very well and we have been keeping track of his life for twenty or thirty years, we never make the mistake of dropping a decade or two and falling back on an earlier version of Sam. We may remember him clearly as he was then, but we have been quietly updating our stored knowledge of Sam, not perhaps with full explicit consciousness as a novelist might but by remembering what we consciously register about his life.

Nicholas Humphrey, one of the Hardliners cited earlier, acknowledged this vital negotiating role of consciousness in his seemingly self-contradictory stand on its social functions. His is a unique position in that he has steadfastly maintained both the sensory nature of awareness and the social function of consciousness. In fact, Humphrey was not really contradicting himself but simply acknowledging that consciousness has several valid meanings, which must be ultimately accounted for.

Conscious updating is vital to social life. People interact too fast, they change their evaluations of one another too rapidly, and they perceive the incredible subtleties of social life too quickly for anything but a fully attuned conscious mind to track. There are simply too many things to monitor at once, too many impressions that need correcting every five minutes. Say three businessmen, strangers to one another at the start, meet for lunch. To break the ice, they talk about baseball and politics, circling one another like boxers, getting a feel for what kinds of persons they are dealing with. To use an old commonsense expression, in such situations it is necessary to "keep one's wits about one." That means conscious wits, not automatic processes buried deep under the surface in inaccessible mental modules. To be conscious in a human social context means all that this expression implies: to be razor sharp, up-to-date, and, above all, on-line.

One might even make the case that consciousness—especially our lightning-fast, up-to-date, socially attuned, uniquely human consciousness—is the evolutionary requirement for both constructing and navigating human culture. It remains the basis, the cognitive sine qua non, for all complex human interactions.

INTERNALIZATION AND SOLIPSISTIC AWARENESS

The novelists will force us to make one more concession. Stendhal, Balzac, and James observed their subjects in the intensely focused social greenhouse of society in the nineteenth century. But consciousness can also drift in an equally intricate and solipsistic mode, which turns the conscious mind in on itself, producing a mode of awareness that is perhaps more characteristic of our time. In the twentieth century, novelists became very adept at exploring this kind of self-centered and less socially connected awareness. One of my favorite descriptions of a meditative drift might serve a useful purpose here. This is a mute soliloquy of sorts, inspired by personal memories, yet strongly attuned to the immediate surroundings. It comes from the mind of an old man in the opening chapter of Mark Helprin's remarkable novel *A Soldier of the Great War*:

> As he hurried along the Villa Borghese he felt his blood rushing and his eyes sharpening with sweat. In advance of his approach through long tunnels of dark greenery the birds caught fire in song but were perfectly quiet as he passed directly underneath, so that he propelled and drew their hypnotic chatter before and after him like an ocean wave pushing through an estuary.
>
> Had he looked up he might have seen angels of light dancing above the throbbing bright squares—in whirlwinds, will-o'-the-wisps, and golden eddies—but he didn't look up, for he was intent on getting to the end of the long road, to a place where he had to catch a streetcar that, by evening, would take him far into the countryside. He would have said, anyway, that it was better to get to the end of the road than to see angels, for he had seen angels many times before. Their faces shone from paintings; their voices rode the long and lovely notes of arias; they descended to capture the bodies and sound of young children; they sang and perched in the trees; they were in the surf and the streams; they inspired dancing; and they were the right and holy combination of words in poetry.

Despite the passage of much time and many cultural tides since Stendhal and James, Helprin still insists on the complexity of his character's consciousness. He exposes many things about his aging soldier here, not the least of which are his essentially poetic response to the world and his deep reverence for life and for people. Here is an old man in a state of grace, a mind that derives extraordinary joy and exuberance simply from being alive, above all, a mind that is physically connected to its world. In this scene Helprin achieves a beautiful linkage that closes the circuit we have traveled in this chapter, tracing a path all the way from the embodied self to the domain of meaning and back. The mind of the soldier, that fantastic semantic engine of Helprin's creation, is firmly planted in the physical world, as much a part of the old man's body as his heart. It has lived, it has digested all the knowledge that chance and opportunity have thrown its way for many decades, and it has incorporated all this experience into an extended selfhood that reaches out to embrace the familiar streets and buildings, even the trees and streetcars, of his city, Rome. But it is not anyone else's Rome. It is his Rome, planted solidly in his own representation of time, suffused with his images, in fact owned by him. A city whose very existence is a creative act on his part, an imagined theater within which he witnesses the unfolding final act of the play that is his life, with its many layers of beauty, irony, and passion, and in which he is now just a spectator, but a spectator with godlike powers of perception and interpretation.

Is this a mind disconnected from its past? Is it a fiction of its own imagination? Perhaps, to a degree. Certainly, the projected personal world of this old soldier is an imaginary one, but it reflects a striving, powerful, adequate imagination. Is it disconnected, passive, epiphenomenal, or fragmentary? Not at all. Here is the proud, assertive view of humanity that we inherited from the nineteenth century and from many centuries before that, still intact. There is a deep cultural universal here, in this experience of grace and unity, of active engagement, of representational control, an ideal of existence that most people aspire to, even if they do not achieve it. This sense of conscious power to control one's imaginary life, this striving for personal integrity in awareness, is the accepted standard by which all else in the human cognitive universe is normally judged.

CHANGING OUR MODEL

Admittedly I have selected my examples carefully. However, not many writers will contradict this view. The novels of James Joyce come to mind as a possible exception, since they obviously reject such an apparently linear approach to awareness. But one could point out that Joyce's central vision, portraying a fragmentary, evanescent awareness that seems more in line with the image of consciousness that we associate with Hardliners, was historically idiosyncratic. Moreover, we know that he was at least partially influenced by ideologies that came from outside literature, from psychology and philosophy. In the final analysis, it is not clear that he would really take issue with anything I have said. Despite his ruthless paring down of the surface of his characters' awareness, Joyce's novels were, if anything, even more inclined to place the action in their conscious minds, and nowhere else.

In literature the consciousness of the protagonist is portrayed as an arena in which the psychological action takes place, the past is evaluated, and plans are made. Consciousness is presented as an active force in the world, a force with a very wide reach, that routinely governs its own cognitive domains, within the intermediate time frame I have described. If these descriptions of consciousness are accurate on this point alone, they confirm my clinically rooted suspicion that experimental research (including much of my own early work) has seriously distorted the phenomenal reality of consciousness by placing too much emphasis on sensation, perception, and performance and not enough on active imagination and self-governance, which exist and persist in the intermediate time range.

The core functions of human consciousness cannot be properly isolated and described in the short term. Human consciousness is virtually oblivious of milliseconds and cares little for events that last for mere tenths or hundredths of a second. It is often engaged in the conventional short-term range of one to fifteen seconds, but most of the time its major focus is elsewhere, in the intermediate time frame, extending its influence over periods that endure for minutes and hours. This kind of conscious deliberation is associated with a number of mental operations that take considerable time, especially self-reflective and self-evaluative operations, such as those involved in the extensive process of skilled rehearsal, for instance, or imaginative play. The same extended consciousness is in control of the slow-moving representational processes that underlie our most elaborate mental constructs, such as telling stories, performing symbolic interactions with others, and planning lives. These activities do not take place without conscious supervision, and they cannot occur within the limited temporal zone that we traditionally allocate for short-term memory and attention.

There is certainly no evidence to suggest that we achieve our intermediate-term synthesis of experience by building unwieldy pyramids of mental images and ideas within our fifteen-second short-term span. This suggests an important new field for research: establishing a convincing phenomenology of intermediate-term awareness. We can find the beginnings of such a phenomenology in neuropsychology and literature. The next step is to push that exploration forward in a more focused manner. We need a new science of consciousness that maps out the phenomenology of inner cognitive spaces in much more detail. Until we do this, we will not have an accurate model that captures the true extent and complexity of human conscious mental activity. We shouldn't be any more surprised (or embarrassed) by discovering this than Behaviorists were to find that there were much more data on animal capacities to be found in ethology than could ever be found in Skinner's boxes.

This radical change of time frame is not the only major readjustment that must be made to current working ideas about awareness. We must rescue the process of automatization from the obscurity of the unconscious and place it front and center in our theories of consciousness. Automatization is the corollary of enhanced consciousness, a necessary complement of advanced self-governance, which is associated with an increased capacity for learning. The superplastic brains of rapid learners have to store their immense amount of knowledge, and as they store more and more learned algorithms, they must relieve these of any need for detailed conscious regulation. If we had to call on conscious capacity to manage all our acquired knowledge and procedures, the result would be inevitable overload and a severe limitation on our ultimate capacity. An efficient automatization process prevents overload and allows us to benefit from having a highly plastic brain. Although it is complementary to awareness, it is not truly independent of it. It is, rather, the child of awareness, the end product of conscious rehearsal and review.

Finally, we must be careful that we don't go overboard in this inclusive definition of conscious function and lose track of the central unity of conscious experience at all levels. That unity must be explained. It is no accident that we continue to use the word "consciousness" to group together so many disparate phenomena. These phenomena share important properties, but above all, they all tap that mysterious entity called conscious capacity. Ultimately capacity may prove to be the critical unifying piece of the puzzle.

Until now we have collected only a few pieces of the puzzle, tantalizing little clusters of growing clarity here and there. We have not entertained anything like a Grand Unifying Theory of consciousness. This is hardly surprising. Unlike stars, atoms, cells, or ecosystems, our quarry is intelligent and elusive. Consciousness has insinuated itself into the very process we are investigating; that is, if science is a conscious undertaking, as seems a fairly reasonable assumption. It is capable of spying on us and constructing clever conceptual obstacles to its own discovery. Above all, Minimalism is a conscious construction, a philosophical apple, tempting us to accept it. It is thus that consciousness urges us to give up, consciously, our quest to understand it.

So we box, and consciousness bobs and weaves. This is not some mindless chunk of ice hurtling through space, waiting for a satellite to give it fifteen minutes of fame, or some empty clockwork of chained enzymes, twisting slowly in the wind, waiting for a microscope. This is an intelligent adversary, as clever and elusive as Lewis Carroll's Cheshire Cat. Like that ridiculous cat, it grins back at us, holding our attention, despite our best efforts to be indifferent. And that ironic, irritating grin stays with us long after everything else about the debate has faded from memory.

4

The Consciousness Club

Much of twentieth-century science has gradually slipped into an attitude that belittles nonhuman animals. The acceptance of biological evolution and the genetic relationship of our species to others was a shattering blow to the human ego, from which we may not have fully recovered. This may help explain why so many appear so certain that consciousness and language are uniquely human capabilities.

—DONALD GRIFFIN

On Earth there is nothing great but man; in man there is nothing great but mind.

—SIR WILLIAM HAMILTON

In Chapter 3, I seized upon Henry James's fictional character Olive Chancellor as an icon of fully conscious humanity, not because she is particularly likable (personally, I find her altogether too calculating for my comfort) but because she rings true in a qualitative sense. Like most humans, she is clever, passionate, complex, and resourceful. Her conscious mind embraces layer upon layer of meaning and intention. She is a great cauldron of conflicting ideas and motives. Also, like most of us, she takes her own complexity for granted, experiments with her psyche as an artist might experiment with paint on canvas, plays with her involvements with the rest of society, and even tests her relationship with her own changing persona.

Olive stands at the near, familiar, human end of the spectrum, very far from the mindless mechanical spheres of protozoa, insects, and eels and distant from the social mammals, even from our close cousins the Great Apes. We may feel great empathy with apes, but we have no illusions that they could navigate a social world that included the Burrages and Boston society, let alone, God forbid, New England Transcendentalism.

We cannot help suspecting that the perceived distance between Olive and the rest of the natural world has something to do with what we call consciousness. Somehow her conscious mind seems simply more *aware* and more *in control* than the rest of the universe. Her unconscious mind, powerful though it may be, does not garner the same respect. Olive's unconscious is undoubtedly rich, and Freud would probably have dined out on it for decades, had he discovered American fiction. But Olive's unconscious is driven by blind chance and association, just like Darwin's world. Her unconscious mind is simply not *in charge.* James realized that fact in the way he portrayed her struggles. He wrote them as dramas, played out in the arena of her own awareness and those of others. Despite the pressures emanating from her unacknowledged and repressed conflicts, which he knew about, he perceived that her conscious self was the source of control. It alone gave her special status, in a universe of chance. Hence his emphasis on what his characters feel and think and how they interpret events. This would have been applauded by his brother, William, who had his own, differently expressed obsession with consciousness. This Jamesian focus still rings true.

This is the point at which the scientific problem of consciousness becomes a great deal more complicated. Faced with this prospect, the reader might be tempted to grab the channel changer and switch to something less convoluted or to become a Mysterianist, like the philosopher Colin McGinn, who has proclaimed at great length that human consciousness is an unsolvable mystery. I am not so tempted. I think that Mysterianists should emulate their predecessors, the ancient Greek and Roman Mystery cults, by pooling their considerable financial resources, building a temple in a beautiful place, and holding secret rituals on the summer solstice. It seems eminently clear that declaring that something is a mystery doesn't make it go away. The structure of matter was once a mystery. So were energy, and light, and life. These things are no longer mysteries, yet in another sense, they remain more mysterious than ever. None of this has any bearing on science.

This is easily said, but our inability to solve consciousness haunts our current intellectual banquet like Banquo's ghost. David Chalmers has suggested that some problems related to the theory of consciousness might prove easier to solve than others. He has declared that the neuroscience of consciousness should be one of the Easy Problems to solve (needless to say, he isn't a neuroscientist). In Chalmers's view, the Hard Problem is the philosopher's endlessly convoluted word quest for an understanding of phenomenal awareness. But he avoids the key issue, which is neither purely empirical nor purely philosophical. The debate over the so-called Hard

Problem is nothing more than a local cultural squabble between members of a species who are already able to represent what they know or don't know in words. The thing that really needs explaining is how a particular species (humans) came to be able to have such squabbles in the first place. The central question is the nature and origin of our peculiar kind of consciousness, which seems so different from that of other species.

This widens the debate and burdens us with additional challenges. If human consciousness evolved from a different kind of consciousness and acts both in the short-term and intermediate time zones, as well as presiding over every aspect of higher mental life, including symbolic thought, it is surely unique in the biological world. In its Minimalist definition, human consciousness does not seem unique, and so discontinuous with animal awareness, because many species share its elementary features to some degree. All mammals have perceptions and some ability to attend to the world in a selective manner. This surely qualifies them as apprentices in the mental games that people play. But in its more inclusive definition, humans play both sides against the middle. We play the same games as most mammals, but we also engage in some that are uniquely our own. This calls for a different kind of theory, with a focus on evolution.

Our awareness presides over an inner landscape so rich that it seems to defy reduction. Many theoreticians are convinced that the qualitative aspects, or qualia, of our subjective experience can never be subject to a materialistic theory of origin. It is logically impossible to refute that assertion. We can never prove that another person experiences the color red as we do, but to paraphrase the Positivists, with whom I agree on very little else, so what? Down on the ground, this has no consequences. We must employ our mindreading skills as a matter of survival. Of course, we infer awareness in others, and of course, we share a consensus on what our experiences are like. That is the foundation of human culture.

Moreover, our uncertainty about animal awareness is no greater than our uncertainty about the subjective experiences of other people. If we assume the practical necessity of accepting the consciousness of other humans, how can we reject doing the same for the rest of the biological world? It is intuitively obvious that some animals experience qualia similar to our own, and this fits well with our empirical science. Many animals have virtually identical sensory physiologies, with sensitivities that are similar to or better than our own and a range of perceptual and cognitive abilities that indicates considerable awareness of what is around them. They also have detailed memories for past events, whether they consist of smells, sounds, or sights, and their experiences are stored in the form of episodes. Their awareness breaks down after neurological injury in ways that are similar to ours. These facts cannot be disregarded, and neither animals nor people can be denied awareness simply because we cannot climb inside their heads and share their experiences. At the same time, we would be missing a lot if we did not acknowledge the specialness of human consciousness.

A more inclusive definition of awareness is crucial to sustaining our belief in human purpose and intentionality. It opens the door to consciousness in other species, so that we will no longer be able to see awareness as an all-or-none representational issue or a simple matter of achieving a certain level of civilization. Once we broaden its domain and allow many grades and shades of awareness, in both quality and breadth, we must inexorably move toward an evolutionary theory of consciousness. Our capacity for consciousness must have evolved, in itself, not just as an irrelevant add-on, but perhaps as an exaptation (Stephen Gould's word for an adaptation with accidental spin-off gains), or as a series of adaptations. This is presumably the way vision, the prehensile grip, and flight evolved, and this resulted in many different evolutionary variations on seeing, grabbing, and flying. If we accept the possibility that conscious capacity evolved, we are left with the curious conclusion that even our capacity for purpose may not be as unique as we think and that it must have deep roots in our evolutionary past. The germ of human conscious capacity may exist in many other species. There may be forms of awareness out there, on some remote evolutionary branch, that we would find very difficult to understand.

Most traditional theorists still cling to the implicit belief that purpose emerged only during human evolution and that it appeared suddenly in a universe otherwise completely lacking in it. In this strange genealogy, purpose must have been born of chaos, out of chance. Not only that, but it must have been conceived in our species cold turkey, with no precedent. This is a bit of a stretch, but Julian Huxley, one of the twentieth century's most visible neo-Darwinians, held firmly to this belief. Perhaps he had no alternative. He accepted evolutionary determinism and at the same time asserted our capacity for purpose, without so much as blinking an eye. Apparently, neither he nor Charles Darwin, for that matter, ever seriously doubted our capacity for formulating purpose, freely and consciously, while adhering to their materialist theories of origins. But can purpose and materialism live in the same house? That nagging Hardliner voice, irritating and persistent, suggests that they cannot.

It doesn't help that Darwin and Huxley may have resolved their conflicting views not so much by dint of any profound reconciliatory logic as by an old-fashioned Victorian double standard, insisting on strict determinism in every other realm of inquiry while granting an exception for the human mind. But they could get away with it. Aside from the totally unjustifiable deference shown to professors in those days, which gave them a place to hide, it was far easier for them to discard the feeble materialist theories of their times. Materialism has become more sophisticated than it once was, and cognitive science has far better tools at its disposal. In light of this, is the Hardliner account still the only way to reconcile consciousness with evolutionary materialism? Will the pursuit of truth in this case cost us our belief in the freedom and purposefulness of our existence? Because, despite furious denials by those who reject the efficacy of consciousness, that is precisely where these ideas are leading us. If unconscious demons do all our mental work, what is left for us? What *are* we, in that case? Before we accept such a doctrine, we should consider the alternatives.

THE MATERIALITY OF MIND

There is no doubt that the context in which consciousness came into existence is an overwhelmingly material one. If we know little else about its origins with certainty, we must concede that consciousness is an aspect of complex life. This concession has far-reaching consequences. Since we already know that complex life evolved from inert matter, it follows that consciousness also evolved from inert matter. This is Emergentism, an approach that is inherently credible because consciousness is, in every known instance, both alive and embodied and therefore an aspect of the natural world. As such it must have evolved, like every other property of life.

The all-encompassing context for the evolution of consciousness is the breeding ground of life itself, the biosphere. Averaging only half a mile in depth, the biosphere exists near the surface of the Earth's crust. It is the realm of all terrestrial life. It harbors hundreds of ecologies and up to twenty million different living species. Above and below it there is no life. Above the Earth's crust stretches an apparently endless vacuum whose temperature approaches absolute zero, and below it, there is a molten planetary core whose temperature exceeds seven thousand degrees. Sandwiched between this metallic furnace and the frozen void, the biosphere is a gossamer thing, hovering on the edge, indeed on two edges, of extinction. Life on Earth appears vulnerable, almost invisible. Nevertheless, the biosphere has endured, providing a nurturing environment that allowed terrestrial matter to self-organize, over aeons, into living organisms. In this planetary petri dish we were grown, along with many other species, like so many bacteria in culture. So was consciousness. There is no credible alternative.

The actual physical presence of mind is very small. The phenomena that we call mental are caused by a tiny fraction of supercomplex living matter called nervous systems. Evolutionary theorist E. O. Wilson has written that the total biomass of humans, even after our recent population explosion, is only a fraction of a fraction of a percent of one ten-billionth of the mass of the planet. Moreover, our brains are only one-fiftieth of that, a minuscule presence by any standard. In most species the nervous system is an even smaller fraction, usually no more than a hundredth or a thousandth of each species' total biomass. This delicate reticule of nervous systems stretches across the biosphere like a spiderweb, apparently too fragile to support the weight of a universe. But it does, nevertheless. All things cognitive emanate from that infinitesimal fraction.

Regardless of how small a thing it is, the physical presence of mind is crucial to our story because it allows us to track its history on Earth from tracings of the shapes and sizes of ancient nervous systems contained in the fossil record. Judged from this record, primitive vertebrate nervous systems emerged about five hundred million years ago, and the forerunners of advanced nervous systems, in archaic birds and mammals, about two hundred million years ago. The basic design elements of the nervous systems of higher mammals, including our own line, the primates, emerged approximately sixty-five million years ago. Human beings are relative newcomers, having shared a common ancestor with chimpanzees as recently as five million years ago. Our particular species only emerged very recently, less than two hundred thousand years ago. We are newcomers, but our species' history is very old, and like other species, we have a long-standing relationship with the ecology in which we evolved. We may imagine ourselves to be the exclusive caretakers of mind on this planet, but we are also part of the larger terrestrial biosphere. Our remarkable capacity for conscious awareness emerged within the constraints imposed by this all-encompassing physical frame.

There are circles within circles of implication and meaning in those loaded words. They summarize our collective discovery of human origins during the past two centuries. The neo-Darwinian reconstruction of the deep past is now so detailed and convincing that the brilliant generation of thinkers that immediately preceded the explosion of modern science, which included minds as diverse as those of Blake, Newton, and Voltaire, would have found our ideas about mechanism quite startling, and so should any reflective person, even now. How could a mechanistic universe, evolving blindly, become any more conscious or responsible than it was before our species emerged? It is entirely reasonable to suspect that it couldn't and that our impression of having conscious choice is a cruel illusion. This would bring us in line with Hardliner beliefs.

Some schools of scientific materialism have compared the problem of consciousness with that old philosophical workhorse Bergson's *élan vital,* the life essence, an imaginary force that supposedly infuses life into an organism. But the concept of a life essence was doomed from its conception because the term was tautological, circular, and scientifically unnecessary. If something was alive, it had within it, ipso facto, a life essence, and vice versa. This led nowhere. Is consciousness a similarly circular idea? No, it is not. Unlike Bergson's idea of a life essence, the concept of consciousness is neither circular nor tautological. Rather it is an empirically sound category that unifies a cluster of phenomena. It is also a scientifically necessary concept, without which we would have to invent another term with a similar function because it accounts for things. It serves a purpose more like that of the concept of energy in physics than like Bergson's misconceived tautology. Like the physical notion of energy, consciousness has several distinct meanings and requires several different operational definitions (energy has at least four, and consciousness, as we shall see, at least three). Nevertheless, it retains a conceptual unity across its various definitions.

There is another critical difference between our nineteenth-century ancestors' musings about a putative life essence and our ideas about consciousness. The life essence was by definition a property of every living thing. But consciousness is not an inherent property of every mind, nor is it inherent in every nervous system or in every component part of those nervous systems that are capable of consciousness. This is abundantly clear, even if we restrict our database to the evidence reviewed in Chapter 3. From the study of brain injury, we know that certain neural networks in the human brain are essential to conscious function, while others aren't. It follows that consciousness is not an inevitable product of nervous activity or even of complex nervous activity. The activity of a brain in deep sleep is very complex, but it is not conscious. The computations of the visual system are complex yet unconscious. There are many nervous systems of great complexity (for instance, honeybees) that, by any criterion I know, do not possess consciousness. Thus one cannot extrapolate directly from neural complexity, or neural activity, to awareness. The physical foundations of awareness cannot lie upon some vague general principle that applies to all nervous systems but rather upon some very specific design features of particular kinds of nervous systems.

In sum, the phenomena by which we define consciousness are correlated with certain configurations of activity in certain nervous systems and not with others. Most neural activity doesn't generate consciousness, even in the supremely conscious human brain. Moreover, the activities that do generate consciousness do not produce it by accident or in a happenstance manner. It isn't just any "draft" of sensory data that happens to float to the surface of some mysterious cognitive soup that becomes a conscious experience. On the contrary, consciously processed events in the nervous system have a very clear physical signature, in the form of characteristic brain activity. There is good empirical evidence that consciously registered events leave distinct traces in the brain and are processed in special ways within the brain's networks. This will be reviewed in the next chapter.

ELIMINATING THE SCALE PROBLEM

Before carrying this logic any further, we need to address a very strong conceptual obstacle to even attempting a materialistic explanation of consciousness, however humble its initial objectives. This prejudice is based on a widespread misperception of the scale of the human brain. Most people cannot grasp the possibility of placing the entire mental universe of a human being into three miserable pounds of protoplasm. This is a difficult thing to accept even for a professional neuroscientist. I suspect that many dismiss physicalist accounts of mind out of hand simply because the brain appears to be such an unimpressive organ. How could all the experiences of a lifetime be contained within a soft grayish mass no bigger than the liver? The suggestion seems ridiculous in the extreme. Even an introductory course in human neuroanatomy doesn't usually succeed in completely eliminating this feeling. Under a microscope the brain looks too simple in design to contain exotic things like experiences, personalities, and ideas. All of Shakespeare's, all of Newton's, all of anyone's lifetime, even the pleasures of a single hot summer afternoon, packed into a pathetic, squishy, little parenchymatic blob? Absurd.

The main obstacle to a suspension of disbelief in this case is undoubtedly our own imagination. Our new scanning and imaging technologies might someday solve that problem by helping us visualize the microscopic details of the brain's operations, but we still have to care enough to pay very close attention to what such illustrations really show. On a technical level, we have already solved the problem of measuring the true scale of the nervous system. This is a great achievement, but like our discovery of the scale of outer space, it is a discovery whose implications can easily be missed. Just as we struggle with the sheer magnitude of interstellar light-years, so we have difficulty with the immense miniaturization of function that the nervous system has achieved. As we descend into the microscopic domains of the brain, we find an infinity of smaller microworlds, and with each additional power of magnification, other worlds reveal themselves. This tangled domain is the realm we inhabit, the inner space that defines our universe. This is our true home, not that other one, of outer space and stars and supernovas.

The brain's three-dimensional complexity makes its examination and its visualization very difficult. In comparison, the ancient astronomers had it easy. They recorded and rerecorded the same familiar pattern of points of light in the night sky for thousands of years. Their database was stable, reappearing in the same cyclic configuration over and over, night after night, year after year. They had the leisure to try one geometric solution, then another, and then another for millennia. All the while, well into the nineteenth century, their theories evolved, but their basic database did not change very much. One is tempted to wonder, What took them so long? Compare the simplicity of their task with the dilemma of a modern neuroscientist trying to construct a model of a living brain. Unlike the relatively stationary stars, patiently emitting light for the leisurely delight of endless generations of peripatetic astronomers, the electrical patterns of the brain are constantly moving, changing, flashing codes and rhythms that are harnessed to the actions of living organisms. It is as if the stars of the astronomers had come to life, moving around, bursting and recombining into various functional clusters, their actions embedded in, and determined by, countless undeciphered codes. Unlike the near vacuum of interstellar space, the neuronal heavens are alive and intelligent. They have their own mind, so to speak.

This makes the brain hard to study. It is a world of incessant activity and filled with structural detail, including cells, microtubules, chemical and electrical pathways, an infinity of structures, depending on what scale of anatomical analysis we might want to single out. When we view a magnified image of a clump of neurons, depending on the scale we choose, those innocent-looking black blobs and stripes in the picture might reflect the presence not only of single neurons but of entire globular clusters of neurons, known as nuclei and ganglia, or of the tangles of interconnections, neural plexuses, that link them together. Each cluster is a world unto itself, with its own unique connections. And on it goes. Nothing in the universe, natural or man-made, is more beautiful or mysterious. A lifetime is not nearly enough to penetrate this inner realm, and if I had the time to start again, I might become a neuroanatomist. It is simply the most elegant of all disciplines.

What cannot be visualized, because we still have no way of imaging them directly, are the millions of widely distributed networks of neurons that extend their tentacles throughout the brain, sometimes mediated by the tiniest of connection pathways. The most difficult to envisage are the very fine, long networks that reach across both sides of the cortical hemispheres, some going all the way from the back to the front of the brain. The white-matter regions that pass between nuclei and ganglia contain hundreds of millions of insulated long axons that form the widest networks in the brain. We cannot yet create accurate images of such complex assemblies. The technology to do this will come, but meanwhile we can only observe the effects of these circuits indirectly, as physicists must observe the effects of subatomic particles without being able to look at them directly.

As we decrease our scale, the functional units of the system proliferate. Say we locate a brain nucleus and can determine that it sends its communicating cables, or axons, to some area of the cortex. On closer examination the nucleus itself might be reduced to another, even tinier level of highly structured architecture, perhaps several smaller nuclei wired into a highly ordered complex. A good example of this is the hippocampus, an ancient precursor to the cerebral cortex that still serves an important function, converting experiences into accessible memories. The thalamus, a ganglionic cluster in the center of the brain, has even more internal complexity, with many subnuclei, each with its own internal architecture. Microstructures like these often have destinations in many areas of the cortex.

Even the cortex, which appears to be a large, fairly uniform structure, can be broken down into many functional regions, each containing thousands of cortical columns. The cortical column deserves particular mention here. It has a special status in cognitive science because it is widely regarded as the elementary functional unit of cognition. This is a fairly recent and very major breakthrough in our understanding: that the unit of cognition is larger than the single neuron and smaller than an entire cortical convolution, or gyrus. The now-mythical Broca's region, once believed to be the language region of the dominant hemisphere, consists of a major chunk of an entire gyrus of the frontal lobe. But today the focus has shifted. It is less on the gyral unit than on the much smaller columnar unit. The real unit of cognition is thus in the middle scale of magnification, in the column.

Columns are typically about half a millimeter in diameter and contain about one hundred thousand neurons. They could perhaps be renamed Szentágothai machines after one of their pioneering investigators. This term captures that great Hungarian's central idea that columns are analogue computers with a standard anatomical structure. The physiologist Vernon Mountcastle has suggested that the human cortex alone contains at least six hundred thousand of these very powerful and widely interconnected columns, woven into various brain-wide networks by millions of long communication fibers. These networks can respond to complex intellectual and emotional challenges with amazing speed and unity. The sheer number and complexity of these agglomerations, all contained inside a single human brain, exceed those of the entire global electronic highway, by many orders of magnitude. Furthermore, if they exceed the highway in complexity, they are simply in another universe when it comes to speed and flexibility. This tangled thing, this frighteningly clever tissue, registers the one thing no symbol machine, no computer, can emulate about us, conscious experience. It allows us to experience the world, parse its experience neatly into episodes, and remember those episodes.

Parenthetically, I seriously wonder if the use of the word "compute" is really appropriate here. It is only a working allegory, and I think there is a much better metaphor in nature: digestion. The neurochemical and neuroelectric actions of the brain are much more like digestion than number-crunching. They are so rich in structure, with so many more potential ways to communicate and coordinate than digital symbol machines, that we still lack a formal notation, or even terminology, that can capture a process of such density and complexity. In its style of operation the brain is really not at all like a digital computer. It may be more like a very large network of extremely fuzzy analogue computers. The difficulty with this is that analogue computation is not well understood. Analogue devices do not employ symbols in the classic definition of that term. They acquire impressions, make transformations and comparisons, and integrate actions, without explicit labels.

Having acknowledged the limitations of this metaphor, we can still say that an apparently tiny column of the cortex has tremendous computational power and that it can extract abstract patterns from concrete sensory stimuli. But this can only vaguely hint at the power of the human cortex as a whole. If the amount of intricate wiring compressed into a single Szentágothai machine is quite staggering, the entire cortical mantle begs description. According to neuroanatomist Walle Nauta, a block of human cortex one millimeter in diameter contains about 300,000 neurons. If we assume roughly four columns per millimeter and an average human cortical surface area of about 150,000 square millimeters (or 1,500 square centimeters), there are about four times 150,000, or 600,000, columns, as Mountcastle suggested. We could conceive of this fantastic arrangement as an enormous local area network of more than half a million analogue minicomputers, wired into an intricate complex of dedicated modules that reside elsewhere in the brain, any of which may exceed the size of most entire nervous systems. But this surely underestimates the real power of such a network. Neurons are far more flexible and have many more ways of communicating, both electrical and chemical, than the components of any known computer. Each column has much more power than its number of components might suggest, with an unknown multiplier effect on the entire network.

On a local scale, each column is a self-contained network in itself, with a fixed internal wiring diagram, or architecture. But it has no fixed pattern of connectivity to start. That distinction is important because while the basic columnar architecture is innate, its connectivity pattern is set by experience. Each of these columnar devices has countless interconnection points, or synapses, which connect neurons to one another in various patterns. This allows them to be wired up in various ways as the brain grows and develops and introduces a degree of flexibility or plasticity in their development. An average cortical column receives input from about two hundred thousand synapses on its input fibers and sends out about fifty thousand output fibers. Its neurons also influence one another in many other ways that bypass conventional synapses. Some columns seem to be subdivided into smaller working clusters that are about one-quarter the diameter of a full column, and these in turn are sometimes reducible to very small working minicolumns that contain as few as two hundred neurons. The existence of these little working networks within columns shows that with sufficient experience, columnar tissue can self-differentiate in a variety of ways and create order from the initial randomness of its interconnections. As the brain develops, its main achievement is one of creating macroorder from microchaos. As it accumulates experience, it acquires a deep functional structure that reflects its synaptic patterning.

On further reduction, down to the cellular level, more detail starts to emerge. Believe it or not, each neuron has a sort of internal frame, a so-called cytoskeleton, formed from a network of tiny microtubules, which might have an important memory storage function. There is a virtual neural ecosystem of cell types, in all shapes and sizes, wired into various recurring patterns. Still farther down the scale, on the biochemical and genetic level, there are many other, smaller, material structures inside and outside the cell membrane. Dozens of chemical paths run through the nervous system, not to mention the genes inside neurons that are switched on and off by various neural and metabolic events, influencing the growth of new connections and the production of transmitter chemicals. These genes can guide the growth of synapses, influence memory storage, and direct the wiring of the growing, ever-changing brain.

The machinery of conscious processing cuts across all this complexity. It must connect in some way, on every level of magnification of the nervous system, from the very large brain-wide neural networks to the incredibly tiny but complex membranes that define the physical boundaries of each neuron. Some estimates hold that neurons have a degree of miniaturization that exceeds present-day computers by a million or more times. Whether that is an accurate estimate is anybody's guess since we have a very long way to go before we can crack the analogue macrocodes of neuronal populations. But the scale of the cortex can be appreciated with a simple example. Take the area known as PG, in the left parietal lobes, thought to be important in language. In an adult human brain the surface area of PG is about 2,500 square millimeters (about 4 square inches). That implies the presence of about 10,000 columns, containing about 750 million neurons, with connections that number in the hundreds of billions, for just this one small focal region of the cortex, on only one side! It is hardly surprising that a bruise to this region can play havoc with the mind's intricate circuitry. What is more amazing is that such a bruise doesn't completely shut down the brain. Many people who have suffered strokes and head injuries to this area have recovered completely. Despite its delicate and condensed design, the brain is surprisingly resilient. This results from its high redundancy of structure. It can survive a degree of injury that would completely destroy a computer.

Once we have acclimatized ourselves to the extreme condensation of function that the brain has achieved, the scale problem no longer presents a serious conceptual obstacle. Instead the human brain looms as something unimaginably huge. This applies to the brains of many nonhuman species as well. We have the same transmitter chemicals, the same network structures, and, most important, roughly the same internal scale as the brains of most other mammals. The scale of nervous systems is fairly universal, and even an insect as small as a fly might have hundreds of thousands of neurons arranged in complex networks. Furthermore, even these relatively small networks can learn, to a degree. Fruit flies, with their microscopic nervous systems, can learn in at least three distinct ways, implying serious network complexity even in such a tiny organism. A larger insect, say, a honeybee, has a very high order of neural complexity, well beyond the modeling powers of existing computer simulations. This puts the problem of scale in its proper context. In comparison to insects, even the smallest vertebrate nervous systems are gigantic. For example, the smallest nuclei of fish brains are larger than most entire insect nervous systems, and we usually classify fish brains as primitive, when compared with those of birds and mammals.

In this company, some birds, and all higher mammals, stand out as overcerebralized geniuses. A typical dog has a brain large enough to accommodate several fish brains. But the dog's brain is not particularly large compared with the ape's or dolphin's brain. Scale may not be everything, but it is foolish to deny that it matters. It matters because it tends usually, if not always or necessarily, to run with complexity. The chimpanzee has a superlarge brain, with the same complexity of design as our own. We could tuck most of a dog's brain into one of the larger subcortical regions of a chimp's brain, the cerebellum. Such superlarge neural structures exist in only a few places in the biosphere. Elephants have them, as do whales, all the Great Apes, and humans. Elephants and whales have the largest brains by far, but they are gigantic creatures whose bodies are equally outsize, and when their body size is taken into account, their brain-body ratios are not much above the average for mammals. It seems that their enormous brains, which can measure up to five times the size of our own, are heavily occupied with servicing a body that weighs tons, rather than pounds, and has tens of square yards of skin, as opposed to a square yard or so. They also have to move and coordinate thousands of pounds of flesh and bone. The size of the body directly affects the size of the brain because the amount of brain representing a given part of the body determines its sensitivity. Highly sensitive areas, like the tip of the tongue, demand more cortex than low-sensitivity areas, such as the back. To maintain sensitivity in their skin and deep body sensations comparable to those of other mammals, very large creatures need considerably more cortex than smaller ones, just to map out their vast sheets of skin, great slabs of muscles, and miles of tendons, all of which are equipped with sensors and feedback circuits.

In strictly cognitive respects, however, the size of the body shouldn't matter. There is no particular reason why a whale's visual cortex should be larger than a monkey's just because of its body size. Nor should the areas of its brain concerned with thinking have to be larger than ours (which they are probably not, although our evidence on this point is understandably sketchy, given the difficulty of convincing whales to sit still for a functional brain scan). Certainly, their large brains cannot be explained entirely in terms of body size. They also reflect the presence of a formidable intelligence. But we cannot say with any precision how higher intellectual functions are organized in their brains.

Even while we concede the remarkable intellectual capacities of whales and elephants, in the brain-body metric, humans stand out. We have about seven times as much brain as we should if we were average mammals. That means that our brains are seven times larger than they would be if we were mountain lions or German shepherds of similar body weight. That is based on a crude estimate of volume, of course, which underestimates the real gain in brain capacity that this implies. Given that the microscales of all nervous systems are approximately the same, the size of the human nervous system seems daunting. Why should humans need six or seven times as much brain as a dog? Or three times as much as a chimpanzee or gorilla? Keep in mind that on average, our bodies are not larger than theirs and that our gain in brain size may be tied entirely to increased cognitive powers. Couched in these terms, the problem of scale does not seem so daunting.

Of course, brain size is not the whole story, but combined with what we know about miniaturization, it is important. Each human brain is effectively a universe. This may look like a rather transparent attempt (which it is) to root out a deep prejudice. Yet it is important to do so because Dualists of every stripe are prone to dismiss physicalist accounts of mind on whatever basis they can find, and the scale problem has been used as a kind of academic coup de grace. Before abandoning Dualism, we must acknowledge that the brain may be at least potentially equal to the task. My point is simply that it is. The scale problem should be set aside, along with its accompanying prejudices. On close examination, the scale of the human brain is, as the French would say, eminently *digne de soi*, worthy of its role, even though we do not yet understand in detail how it achieves all the marvelous things it does.

VESTIGIAL BRAINS, NOT SO VESTIGIAL MINDS

There is another prejudice that stops many people from accepting any strong form of mental materialism, and this one is more difficult to eradicate. It is a visceral resistance to one of the key tenets of Darwinian theory, the notion of vestiges. Our brains have what may be called a vestigial structure imposed by hundreds of millions of years of evolution. It is possible to pursue the humanities and social sciences without confronting the idea of vestiges because these fields naturally assume that humans are completely distinct from other creatures, as if it were a viable philosophical option simply to think our biological origins out of existence. This is one very good reason for the continuing rift between the humanities and the natural sciences. But we ignore vestiges at our peril. Vestigial structure is an overwhelming reality in the study of the human nervous system, and it is therefore central to our argument. We have inherited a deep mental structure that is a direct reflection of the evolutionary history of our species. This fact must be taken seriously by any discipline that pretends to study the human mind.

Vestiges are traces of the deep past, passed down from evolutionary ancestors, but whose original function has either disappeared or been drastically changed. One common example of a vestige in humans is the tail at the end of the spinal cord, a useless appendage left to us by our monkey ancestors. One of the best-known examples of a neural vestige in humans is the propriospinal tract, one of the spinal cord pathways that connects the brain to the hand. We have inherited this primitive hand-control pathway from a very ancient ancestor. It conveys messages from the motor brain to all the fingers of the hand at once, allowing only the diffuse contraction of all these fingers together. If it is activated, the result is a crude prehensile grab. This is the only way primitive monkeys can flex their fingers. Apes inherited this tract from distant monkey ancestors, but they (and we) are capable of much finer hand control because apes evolved a second spinal cord path for finer control, via the pyramidal tract. This newer pathway projects independently to each finger, enabling apes, and humans, to move one finger at a time and generate much finer patterns of coordinated hand movement. This is where our much-vaunted "power grip" comes from, not to mention our ability to play the piano or string a guitar. In this case, function follows anatomy very closely. In humans the propriospinal tract remains useful for a brief period during infancy, supporting a limited use of the hands for grabbing hold of things and people, until the pyramidal pathway matures. The point is that more than fifteen million years after a newer, more powerful pathway evolved, the propriospinal tract is still here. This vestige not only refuses to die but won't even fade away gracefully!

The strictest biological meaning of the word "vestige" defines it as a nonfunctional trace of something that once existed or was once more important than it is now. But the term also has a wider meaning, referring broadly to any remnant of the past that endures, whether it is functional or not. The word thus conveys a sense of evolutionary continuity. This is more faithful to the etymology of the word, which comes from the Latin *vestigium*, for footprint. The brain is loaded down with footprints of the past. Many of these footprints are still functional, although their functions may have changed. Some nuclei and ganglia serve exactly the same purpose that they serve in distantly related species that evolved long before we did. This includes many basic reflexes, such as coughing, sneezing, vomiting, balancing, and blinking when the eye is touched, and many complex instinctual systems, such as those involved in hunger and thirst, eating behavior, and eye movements, which involve both newer and older brain structures. Vestigial structures are often embedded in new neural networks, such as those that determine the way we perceive space, move our limbs, and express emotion. In reality, much of the underlying architecture, or blueprint, of the human brain is in a broad sense vestigial, because evolutionary change must always be based on preexisting neural structures. Nonfunctional vestiges, such as the propriospinal pathway, are thus not the only ones that are retained. The archaic structure of the brain evolved over hundreds of millions of years. It was slowly sculpted by natural selection, and the underlying lesson is one of sameness and continuity in the form of the brain, from each parent species to its descendants. The core design of the vertebrate nervous system evolved very slowly, and the core of the human brain is thus very, very old, much older than the primate family and older than the entire mammalian phylum. It has some features that are even older than vertebrates.

However, despite this slow-moving evolutionary conservatism, the nervous system has undergone radical changes at various points during its evolution. This has sometimes involved the addition of new anatomical modules to the nervous system, which have typically taken the form of novel cellular clusters; that is, new nuclei and ganglia that evolved out of older structures and added new capacities to the nervous system. This is an effective evolutionary strategy because in this way, new nuclei can evolve without interfering with the work of the previously existing, and adequate, structures that continue to do their work perfectly well. In this way, a species can add new sensory or motor functions, such as binocular vision, without destroying the hard-won efficiency of preexisting modules that regulate older visual functions, such as pupillary constriction to bright light. The latter is a very basic reflex mediated by a tiny nucleus in the brainstem that has a very ancient ancestry. To give some idea of what this means in terms of time depth, that same nucleus exists, and serves the same function, in the brains of rodents, which split from our evolutionary phylum more than one hundred million years ago. We have many ancient structures such as these buried deep in our brains.

Once a brain design is established in evolution, it is very slow to change, but its blueprint is usually flexible enough to allow for considerable variation between species. Closely related species (for instance, different songbirds, monkeys, or cats) with slightly different skill profiles tend to have very similar brains. The behavioral and cognitive differences between kindred species are usually explainable in terms of comparatively minor differences in the relative size of different nuclei or on the basis of a slightly different chemistry or a barely altered sensory organ. Fernando Nottebohm has given us an elegant example of this: songbirds with many songs in their repertoire have much larger song nuclei in their brains than those with fewer songs. In this case their capacity reflects a slight variation in the relative size of a neural module, with no change in the basic blueprint.

In the evolutionary history of the vertebrate brain, truly radical design changes, involving many new modules, in a new brain design occurred only a few times. Always coinciding with major shifts in the nature of vertebrate behavior and cognition, these are the most extreme examples of evolutionary change in the brain's prehistory. The greatest of these changes undoubtedly happened simultaneously with the colonization of land by vertebrates. This was the great terrestrial shift, when vertebrates converted their field of action from an oceanic to a terrestrial environment. The vertebrate nervous system underwent a truly radical redesign on this occasion, acquiring many new modules as new species adapted to the demands of living and moving on land. The scale of this change is illustrated in Figure 4.1, which focuses on the major motor nuclei of the vertebrate brain (their names are given in the endnote for this page). The basic layout of the aquatic brain is shown in the small, shaded ovals. It is similar in all aquatic creatures, such as fishes and eels, that evolved before the evolution of land animals and survives intact in the brains of their modern counterparts.

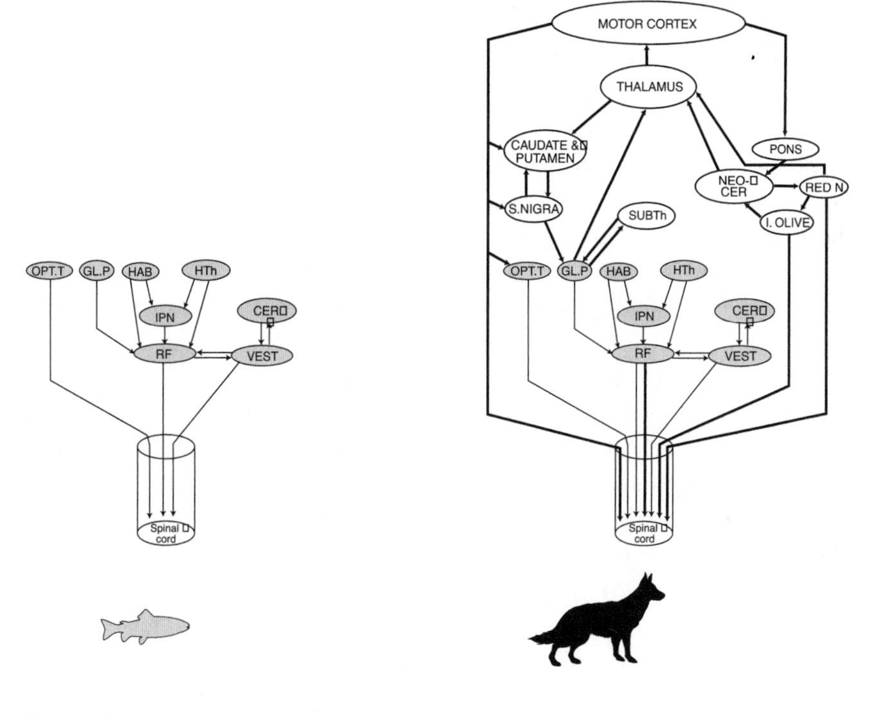

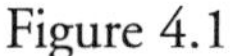
Figure 4.1

When vertebrates moved onto land, the aquatic brain design was eventually surrounded by and subordinated to a new set of brain nuclei and ganglia that are found today only in land vertebrates. These nuclei were the shock troops of a terrestrial cognitive revolution, the new kids on the block. They are shown in the larger, unshaded ovals. Note that these new modules were added to a blueprint that already existed. This is a very clear example of modular evolution. The principle of evolutionary conservatism asserted itself, loud and clear. The older aquatic brain was retained within the core brain architecture of terrestrial vertebrates because it still worked, as far as it went. There was no need to fix it. Its functions were inevitably altered by being embedded in the vastly more elaborate neural architecture of mammals, and they became in that sense vestigial. New modules evolved as outgrowths of existing structures and eventually enveloped them and changed their functions. Like Caesar's armies, they encircled and eventually conquered, although this particular conquest took somewhat longer than Caesar's, some two hundred million years.

There are many reasons why the vertebrate brain had to undergo such a drastic change in design when animals moved onto land. Land animals had to develop limbs instead of fins, including legs, arms, paws, and hands, and such things as independent fingers. This required new mechanisms of feedback control, new drives and motivational systems, and new patterns of movement, including prehension, climbing, walking, and running, all subject to different conditions of gravitation, strength, and balance. This massive change demanded a new brain organization, a modified vertebrate blueprint. New sensitivities also emerged, as many land animals colonized a variety of new environments, such as bogs, caves, jungles, deserts, ice fields, treetops, and forest floors. As the competition for new niches grew, some of these niches demanded more central control, adding such capacities as focused attention, vigilance, and working memory, which are generally absent, or very limited, in fishes and eels. The human brain inherited most of these structures and the cognitive capacities that came with them. Many older neural structures stayed with us because they work, even if they work in a different way. Like the propriospinal tract, even those structures that are not used very much, or might have been designed differently in an ideal world, with the help of 20-20 hindsight, were retained anyway.

Surely this slow-moving pattern of brain evolution could not have predicted the scale and rapidity of human cognitive and cultural change. After all, we are the only species that has evolved a symbolic mind with symbolic cultures, and from a purely cognitive standpoint, this constitutes a revolution that is easily as radical as the vertebrate conquest of land. This raises a huge question: Where are the new neural modules that supposedly allowed us to achieve this radical change? If we were to take seriously most of the "pop" evolutionary psychology that fills the airways and the Web, our brains ought to be shot through with new neural modules that proclaim our uniqueness. This is an understandable, if naive, expectation. We have broken with the key patterns of mammalian cognition. That fact in itself surely calls for new modules in the human brain. We should expect at least a module or two dedicated to speech, visual imagination, and complex toolmaking, which are so characteristic of humans. Also, what about all those vocabularies, grammars, and memory retrieval paths whose absence we can often see in the shattered minds of neurological patients? Aren't they the products of newly evolved modules?

This has been the great driving question behind the neuropsychology of language for at least a generation. Many researchers trained in the sixties, including me, sought to discover the neurological "magic module" that might explain human language and symbolic thought. This tradition extended back to the early heroes of the Great Module Hunt, with explorers such as Carl Wernicke, Pierre-Paul Broca, Lord Brain (an appropriate name if ever there was one!), and Norman Geschwind. However, the results of that search were largely negative. Every conceivable anatomical comparison has been made between chimpanzees and humans, in the hope of finding the critical structure that explains the gulf between us and our closest relatives. But this has yielded very little. Essentially every structure we can describe in the human brain has an equivalent, or homologue, in the chimpanzee. It is thus virtually certain that our common ancestor five million years ago must also have had the same brain architecture. This in turn implies that no radical modular redesign of the human nervous system has occurred during our evolution. If we are looking for a modular "table of elements" to explain our uniqueness, we had better look somewhere else. It is not there.

So the answer is no, these purported new modules simply do not exist. We have a very typical primate brain design. It is true, however, that our brains are very special. As Figure 4.2 shows, the most obvious distinctive characteristic of the human brain is its size. We have much larger brains than any other primate, relative to our body size. If you and I were apes, we would have to get by with one-third of our brain volume and a little more than half of our neurons.

Figure 4.2

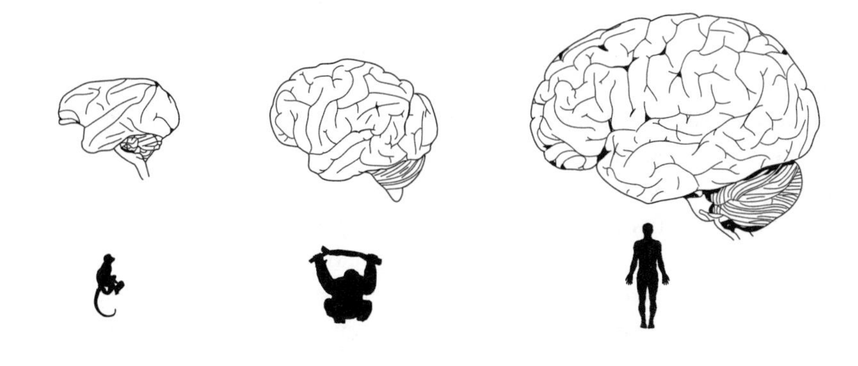

However, despite its great size, our brain has followed a time-honored primate trend. Its history of expansion can be traced back by following the fossil record, which shows that it has been changing for more than two million years. During that time its growth in size followed the same general pattern followed by several previous expansions of the primate brain, in both monkeys and apes. In addition, it was restricted to the same neuroanatomical structures that grew larger in those expansions. Apes and certain fruit-eating monkeys expanded their brains in the same way during their evolution from primitive monkeys. In apes, the greatest expansion, relative to monkeys, occurred in certain regions, such as the neocortex, hippocampus, cerebellum, and basal ganglia, with virtually no change in the sensory, motor, or brainstem regions. This was exactly the pattern that humans followed, with a few changes in the scale settings. British neurobiologist Richard Passingham plotted the increases in primate brain size across various species and went so far as to claim that the expansion of the human brain fell smack on the extrapolated curve, not deviating at all from the primate pattern. Boston neuroscientist Terry Deacon has challenged this, but only in a small way. He claims that in humans, certain parts of the frontal cortex expanded considerably more than they should have and extended their range of interconnections. But these are quibbles over small facts. The Big Fact is one that should be inscribed on every cognitive theorist's door: NO NEW MODULES.

This has major implications for any theory of human cognition. It means that we have not only kept our vestigial primate brain but also retained many of the underlying cognitive traits that go with it. Humans have primate brains, and it does not make sense to postulate a complete rewrite of primate cognition in our case. But if we dismiss radical modular redesign of the human brain as an explanation, how could our newfound cognitive powers have evolved? Something else that was also radically new must have been introduced into the cognitive equation. If it was not a redesign of the basic modules of the primate brain, what was it? How could a relatively straightforward expansion of an existing primate brain design have conferred symbolic capacities on a seemingly minor phylogenetic variation on the apes, the hominids? There is a hint about what might have happened in the specific anatomical structures that showed the greatest expansion. The regions of the primate brain that expanded most dramatically in humans were those associated with executive function. We know, from clinical research in particular, that these vastly expanded areas are important for supervisory, governing, and metacognitive functions.

Supervisory, governing, and metacognitive functions, you say? Are these not the very terms we already associated with consciousness, especially with self-regulation in the intermediate time zone?

AVOIDING THE *SCALA NATURAE*

It is common to suggest that there is a cognitive hierarchy of animals with us at the apex. Since we tend to identify intelligence with consciousness, we might be tempted to construct a pyramid of consciousness on which we stand at the top, as high priests of awareness. When we think in terms of a continuum instead of a simple dichotomy, we are inevitably left with the concept of a *scala naturae*, or scale of nature. There have been many such proposals, quasi-empirical versions of the great chain of being, and they trace back to the earliest attempts to construct a classification of life. In all of them the simplest insects are usually placed at the bottom of the scale, and primitive vertebrates, for instance, sharks, only slightly higher up the ladder. Birds and primitive mammals fall into the upper middle range, and primates are near the top, along with other social mammals, such as dogs, elephants, and dolphins. Humans are always the alpha species.

In its most simplistic form this idea has been roundly rejected by modern Darwinians (although it was not rejected by Darwin himself), who prefer a relativistic classification. In their view, a bat is just as well adapted to its environment as human beings are to theirs, and there is no justifiable basis for a scale of nature. This is the essence of relativism: Each species has its own special niche and should be judged according to that niche, not to that of another species. Certainly, from a bat's point of view, humans aren't really so smart, since we couldn't survive for very long in a bat's ecology. Most of us would fail an IQ test designed by bats. They would ask us to fly around in the dark without hitting things, hang naked from the ceiling, live in caves surrounded by dung, and perhaps catch insects in our mouths while flying at high speeds in the dead of night. Most humans would fail these tests in a most spectacular manner. We are, after all, specialized for a very different environment. In the same way, we would fail IQ tests devised by many other species, including fishes and insects.

The central idea of species relativism is simple, but like all ideas, it can be carried too far. This is what happens when extreme relativism is applied to cognition. In this application of what is otherwise a good idea, it is forbidden to declare that other species have less cognitive capacity than humans. To do this is to risk being labeled anthropocentric. Many researchers feel obligated to repeat this pious relativistic caveat every time they discuss this topic, lest they be ostracized as ideological deviates. They argue, in effect, that no species is more intelligent than any other and proclaim a phony biological egalitarianism, in which frogs are no less well adapted to their environments, and thus no less intelligent, than nuclear physicists, for example. By this criterion, corporate CEOs are no more or less intelligent in an adaptive biological sense than, say, maggots, a conclusion that may have a certain emotional resonance for many but falls a bit short on the evidence.

Now I don't want to appear, heaven forfend, even slightly anthropocentric; I don't want to be banished to the bottom level of some sociologically conceived Inferno; and I certainly do not subscribe to any old-fashioned *scala naturae*. But I would venture a guess that humans are a tad more intelligent than frogs. We can all become relativists anytime we wish, simply by redefining "intelligence" so that the term becomes meaningless. This achieves nothing and fools no one. Intelligence, or cognitive capacity, is one thing. Adaptability to an environment is something else altogether. Sometimes adaptability demands intelligence. Most of the time it doesn't, which is fortunate indeed, for most species.

All this is essentially a bit of political theater, of course, and despite the persistence of these ghastly ritual pieties in academe, we have no choice but to persevere in making careful comparisons of cognitive capacities across species. We can compare the memory capacities of different species, just as we compare foot shape, thermoregulation, or eyesight. Careful comparative research stands up to scrutiny, even if specialized skills in certain species, such as echolocation in bats, long-distance navigation in birds, and infrared vision in fishes, seem to be qualitatively different from the abilities of humans and make direct comparisons difficult. The importance of comparative research is that it makes clear that there are commonalities, as well as differences, between species and that the emergence of both can be tracked over time, giving us crucial clues to the relationship between anatomy and function.

The path to higher conscious function was long and indirect. It was based on the emergence of domain-general regulatory capacities, such as attention, which enable the brain to oversee information from various sources. These capacities are special. Intelligence is present in some form even in the simplest creatures, but it does not necessarily involve domain-general or conscious processing. Spiders, for instance, which reached their present form about five hundred million years ago, can construct instantaneous mental maps of the local environment and use them to guide their movements when stalking prey. The jumping spider is perhaps the cleverest at this. It hunts insects in flowers and bushes and will often "ambush" a potential quarry. Having spotted its prey perched on a flower stem, it will move away from it, rather than toward it, drop to the ground, and climb up the other side of the plant, out of sight of its victim. It will then approach it stealthily, always from behind, and, once close, suddenly attack.

This is not a simple form of cognition, and it cannot be entirely preprogrammed, since it depends upon remembering certain parameters of the current terrain and locating a particular prey within that complex space. What is going on inside the nervous system of this tiny creature, which doesn't even have a proper brain? There might be several possible explanations, but there is no way to avoid the fact that the spider could not carry out such a subtle ambush without having a good fix on its prey's location in space and maintaining it, independently of which way it was looking. Otherwise the spider would lose track of where it was when it dropped to the ground. The world looks very different from the other side of the plant, which is as big, for a spider, as a tree might appear to us. This example shows how much can go on in the simplest mind. But does the spider's ability to maintain a fix in space over time necessarily imply the presence of voluntary attention or working memory? This is another matter altogether. The spider has the intelligence to carry out its highly stereotyped attack strategy. However, most scientists would not grant it conscious capacity. The question is, Why? Why are we inclined to deny this clever little creature even one iota of awareness?

The answer to this question tells us as much about ourselves, I suspect, as it does about consciousness. The spider is too narrow, we might suggest, too rigid, and too fixed in its repertoire. It seems blissfully unaware of the most significant objects in the larger environment. It goes on weaving webs and ambushing anything that resembles a prey, no matter where it is. It shows no signs of adapting its behavior to the larger scenarios that might be imposed by a wider world. The spider's world is tiny, restricted to a small number of players and situations. It misses any feature that might demand significant executive capacities to perceive or remember. This suggests a simple continuum of awareness, along which the scale and complexity of the spider's mental models can be placed, relative to our own. Our implicit rule is that as the executive system of a given species increases in power, its mental models tend to be wider and more like our own. This makes it appear smarter, and more aware, to us.

Although cross-species comparisons are possible, it is much more effective to do this within a single line, or phylum, of evolution because in that case, we can quantify similar capacities across different subspecies. In this way, the special nature of human cognition can be approached by comparing ourselves with other primates. Humans are broadly equipped with more powerful mental models than most primates. We also have more communicative ability, more symbolic capacity, and more advanced tool technology than they have. In this sense, there seems to be a hierarchy of primate intelligence, and it probably parallels our global impression of the hierarchy of primate consciousness. The underlying mechanisms of this putative hierarchy should tell a story of conservative, gradualistic change, regardless of what interpretative spin we might wish to place on it.

This raises an important cautionary point and brings us back to some of the virtues of relativism. There are limits on the effective use of comparative research. We might feel that we are very similar to apes and that apes are more intelligent than any bird. But is a very bright bird, say, a parrot, really any duller (or brighter) than a chimpanzee? We cannot answer such a question. Birds do not play the same cognitive games as primates. When applied to birds, words such as "intelligence" and "consciousness" might take on subtly different meanings from their application to apes and humans. Comparisons extended across such vast evolutionary distances don't work very well. We might more easily agree that some birds are more intelligent than others—for instance, that crows and other scavengers are generally brighter than songbirds (of course, they can't sing as well). With their usual perspicacity, science fiction writers have intuited this. They seem to have agreed among themselves that if birds were ever to become superintelligent, superbird minds would be very different indeed from those of humans, as would superbear minds or superant minds. There are huge differences between species in their emotions and social strategies, and these have far-reaching effects on their cognition.

However, certain elements cut across those differences. The same science fiction writers give away the game beautifully because whenever they create supersmart creatures, they assume that no matter what kinds of specialized intellects they have, to become supersmart they would need to be able to think and communicate symbolically. Supersmart birds might remain birds, but they would also be able to think, plan, and scheme. Moreover, they would have to be able to transmit their thoughts, plans, and schemes to other superbirds. In other words, they would need something like language. They may achieve this differently from humans. But no matter what their basic cognitive profile might be, they would need a capacity to trade representations with one another and to create a public representational space that is in principle similar to human culture. It needn't resemble human culture on the level of operational detail. But it would resemble it in the sense that it must be based on some kind of intentionality. Otherwise the species couldn't communicate, think symbolically, or operate a complex culture. It simply wouldn't fit our definition of supersmart.

Science fiction writers have hit the nail right on the head (so, by the way, have many psychologists and philosophers who have resisted the more simplistic forms of relativism). Any species aspiring to a high degree of intelligence must acquire symbolic skill. There is something different and powerful about symbolic capacity, and intentional representations in general. According to our present indications, only humans have this ability. Thus, in at least this one sense, for the sake of argument, we are justified in thinking that humans are brighter and probably more aware in the representational sense than any other species we know. But this does not imply that we are smarter across the board. The ability to construct and communicate our mental representations did not necessarily alter the basic set of mental capacities we inherited from apes. Down deep, we are still primates and retain the primate kind of mind, which lacks some of the specialized capacities of bats, fishes, birds, cats, and so on and has its own peculiar profile. Moreover, we are a special kind of primate, with a very particular pattern of emotional and social behavior. This is a deep design feature of being human, and our representational capacity cannot alter that fundamental constraint.

Regardless of the cognitive design with which a species may start, intentional representations are the only known way to jump-start its cognitive process to the cultural level. These intentional representations must evolve within the envelope of a given species' basic design. But no matter what specific form they take, their effect would be the same in every case, once they were adjusted for the idiosyncrasies of each species. Symbols can radically amplify the power of any mind, whatever its design. They can alter awareness as well.

Or is it the other way around? Perhaps symbolic skill is the result, not the cause, of increased conscious capacity. This is the heart of the matter.

DEFINING THE DOMAIN

Before trying to describe its evolution, we must mind our definition of consciousness. It is not really a unitary phenomenon and allows more than one definition. In fact, it encompasses at least three classes of definition. The first is the definition of consciousness as a *state*. We use it when we speak of "altered states," such as trances, comas, and dream states, or specify a state of readiness or concentration. Most mammals, and many nonmammalian species as well, have this aspect of conscious capacity. All mammals alternate between states that can be labeled as asleep, awake, and alert. Several species of mammals even seem to dream as we do and show rapid eye movements and brain waves that resemble our own during certain phases of sleep. Most mammals also show daytime variation in arousal; that is, they alternate between what we call active, vigilant, and wide-awake states and those that are passive, unfocused, and marked by a reduced level of activity.

Moreover, most mammals are affected by anesthesia just as we are, recover their conscious functions gradually as the anesthesia wears off, and pass through various stages of wakefulness that are very similar to ours as we recover from the effects of the same anesthetic substances. Many mammals respond to drugs with state changes that appear qualitatively similar to those of human subjects who take the same drugs. Many animals also show many of the same reactions as humans, both physiological and behavioral, to hallucinogenic drugs, such as LSD and psilocybin. These similarities of state manifest themselves in the way drugs modify the startle reflex and alter sensory and motor responsivity, emotions, drives, and general activity levels. We might be forgiven for calling the normal waking states of other species conscious, as opposed to unconscious, when they correspond very closely to those that we invariably experience and label as conscious in ourselves.

A second class of functional definition takes an *architectural* approach, whereby consciousness is defined as a place in the mind. This is where the cognitive buck supposedly stops. As I mentioned in Chapter 2, like the office of the CEO in a corporation, this putative central processor is located high in the hierarchy, usually at the top, and is not directly concerned with the fine details of gathering sensation or controlling muscles. Rather, it sits in a position of command and control, concerned with the overall orchestration of thought and understanding. This does not imply that there is a single neural locus, a "consciousness module" somewhere in the brain. We have already dismissed that possibility. The concept of a central processor is a strictly cognitive one, and the evidence for it is cognitive, not anatomical. Its physical counterpart might conceivably be distributed across several subsystems of anatomical structures, rather than confined to any single part of the brain (just as, in the electronic universe, a business might be anywhere). The physical location of the central processor is not important, as long as, from a cognitive standpoint, it is at the top of the control hierarchy.

This definition acknowledges the very wide reach of awareness and the fact that it can bring cognition, emotion, and action under a unified command. The unity of experience and control is a direct consequence of this kind of central control, which is absent from the unconscious. It gives an organism more focused intellectual power. It can also enable it to self-regulate to some degree. This definition is a common element of many classic models of mind that have come out of early cognitive science. They all postulate a nonmodular zone of mind that controls attention, gains access to all the specialized domains of cognition, and unifies experience. It also borrows the very old notion of controlled processing, mentioned in Chapter 3. Controlled processing is concerned with deliberate self-regulation. By this definition, all the higher mammals, such as dogs, cats, dolphins, and apes, as well as many species of birds, would have to be classified as conscious. Many of these species can attend selectively, pursue a goal over the short run, and behave as though they possessed an essential unity of experience.

The third definition of consciousness takes a frankly human-centered view of cognition and has more to do with enlightenment, or illumination, than with mere attention. This is the *representational* approach, in which consciousness is made dependent upon our human capacity for symbolization, and becoming aware of something is synonymous with capturing it in symbolic form. Finding the right word, an apt metaphor, or the appropriate phrase reflects awareness on the part of the speaker. This rigorous standard of awareness invariably excludes animals from true consciousness, primarily because we have language and they don't. This is a bit circular, however. If awareness is defined in advance as a direct product of language, it is hardly surprising that it should be special to humans. Moreover, it is not obvious that language is a good criterion for awareness. The use of language is driven by many agendas, very few of which originate in language. Language is an add-on, a Johnny-come-lately in the evolutionary sequence, and it gets most of its material and its content from much older parts of our mental universe. Despite this, many scholars adhere fiercely to this definition, and some have claimed that searching for evidence of animal consciousness is simply a waste of time. Consciousness is exclusive to human mentalities, they think. There, with a wave of the wand, you have it. Animals are, well, automatons. And, of course, unconscious. So much for that.

The Postmodernists adhere to this idea in its most exclusive and censorious form. They define awareness entirely in terms of an imaginary ongoing text that characterizes a particular culture. This idea is a return to the old canonical definition of consciousness, in which it is once again conscripted into the service of rabid a priori tribalism, in this case anthropocentrism. Taken to its logical conclusion, this text-bound definition denies consciousness not only to animals but to many humans as well. Pity the poor preliterate Mesolithic humans; they were unconscious no less! They struggled with survival in conditions that the typical modern person couldn't negotiate for an hour, and they did this unconsciously! Remarkable. These are the opinions of ideological high priests in secular garb. So, who was the first conscious human? Foucault? Derrida? The mind boggles.

But there is a core insight that we must salvage out of this human-centered definition. It is certainly true that in general, animals lack detailed declarative knowledge about the world. Whatever knowledge they do have remains implicit because they cannot express, or declare, their knowledge in any form of symbolic representation. They cannot bracket their own experiences or re-present them in memory. This leaves animals immersed in a stream of raw episodic experience, from which they cannot gain any distance. Their mode of awareness is enormously restricted. Somehow, symbolic capture in itself makes us more aware of whatever is being represented. This may be an important clue to the true functions of languages and symbols if we accept the possibility that the "whatever" in question might have the same episodic basis in both humans and nonhumans.

However, things are never so simple. Symbolic skill in itself neither changes the nature of animal awareness nor transcends its limitations. Some chimpanzees and bonobos (a distinct species of pygmy chimpanzee) have learned to use symbols intelligently, and some can even understand spoken English. The best-known example of this is Kanzi, a bonobo raised by Duane Rumbaugh and Sue Savage-Rumbaugh on a quasi-naturalistic reserve just outside Atlanta, Georgia. They have raised both bonobos as well as common chimpanzees in captivity, exposing these animals from infancy to the regular use of symbols. Reared in a hybrid bonobo-human culture, Kanzi is their star. He has mastered a vocabulary of several hundred symbols and can understand a significant amount of spoken English. He has been tested with rigorous control procedures to guarantee that he does not receive any inadvertent cues from the experimenter. These controls involve some fairly exotic measures, such as having the speaker wear a welder's mask, to conceal any facial expressions or eye movements that might hint at what the correct response might be, or having Kanzi wear earphones, so that he can listen to someone reading word lists in another room. Even under these conditions, Kanzi has been able to respond appropriately. He can identify photographs, point at objects, and understand completely novel sentences, spoken at a normal rate, such as: "Take the vacuum cleaner outdoors," or "Give Pinky some water."

Kanzi understands the meaning of many nouns, verbs, and prepositions, as well as word order, to a degree. He has difficulty with elaborate constructions, such as sentences with embedded clauses that separate the verb from its object. But he performs correctly with reversible sentences, in which meaning depends directly on word order. This shows not only that he can segment the speech stream (thought to be impossible for an ape, only ten years ago) but that he can also comprehend the grammatical relationships between some of the words in the sentence (still thought to be impossible now, by a few diehards).

However, despite his linguistic achievements, it is questionable whether Kanzi's awareness has been fundamentally altered by his remarkable skill with symbols. He is a member of a species that has not produced any symbolic invention in the wild. He has learned to use human symbols in very clever ways, but on current evidence, even after intensive enculturation into human society, he still cannot invent symbols, and this is crucial. Presumably this reflects not language but the deeper motive force that generates language. Kanzi's mind lacks the fundamental defining capacities that make human language happen. One of those capacities is a certain kind of self-awareness that provides the motive to describe one's mental states. Kanzi has never tried to describe his own experiences or feelings, using symbols. He cannot construct the kinds of memories that we call autobiographical, even though he has been given a set of symbols that could in theory enable him to do this. Human children use language in this way very early in their development, even before they can form grammatical sentences, but Kanzi has never tried this kind of self-description, despite his considerable symbolic and grammatical skills. He does not say "I think" or "I feel" or "I want." This suggests that his self-awareness is not like that of most humans. He seems to have no natural state that would motivate him to construct such self-referential expressions.

Language per se has no magical power to transform the awareness of an individual beyond its innate capacity. In itself language cannot bestow self-awareness, although it undoubtedly has a major impact on self-awareness, given a basic capacity for it. The ability to use a set of symbols does not amount to a capacity to exploit them fully. To borrow a term from psychologist Katherine Nelson, symbols can "mediate" certain forms of thought and make it possible to formulate new kinds of concepts, provided that the underlying capacity for understanding is already there. The mere possession of symbols will not alter basic ability. The capacity to take a perspective on one's own mental states cannot be changed simply by one's possessing a lexicon or vocabulary. This is true of Kanzi as well as of humans. The same applies to other cognitive domains, where the dissociation of symbolic skill from understanding is perhaps more obvious. Simply giving Kanzi equations will not make him (or, for that matter, most people) a mathematician. Many humans can never grasp such abstract ideas as infinity, the gold standard, and electrophoresis, no matter how hard their teachers try. Symbols can point, nudge, guide, and focus, but they are no substitute for the underlying capacity to understand what the symbols represent.

This logic applies in reverse. Just as the presence of symbols alone cannot generate an awareness of meaning, so the absence of symbols does not necessarily indicate lack of awareness (although in some kinds of meanings, such as certain mathematical ideas, it probably does). When people cannot communicate or think symbolically, for whatever reason, we should not assume that they lack conscious mental lives. We should demand substantial proof before suggesting that someone without self-referential language actually lacks self-awareness of any kind, and if we are to remain consistent, this standard should apply to animals as well as humans. We know that Kanzi can recognize himself in a mirror or a photograph. In that sense he has some self-awareness or self-recognition of his body. But self-referential language requires an awareness of inner states, and this apparently is absent or insufficient to enable Kanzi to use language in this way.

THE CONSCIOUSNESS CLUB

Given this range of definitions, which species shall we admit to the Consciousness Club? Where, in the long history of life on this planet, did consciousness begin? The bar can be set high or low, depending on what definition we favor. My own inclination is not to draw any lines in the sand. I prefer to take an inclusive, pragmatic stance and list as much evidence as we can find on the presence of consciousness in as many species as we can observe, using all three of the definitions—state, architectural, and representational—discussed in the previous section. I prefer to toss in everything that might prove relevant and look for patterns.

My primary reason for doing this is that I lack confidence in any of our existing, one-dimensional definitions of awareness. Human consciousness is the product of many functional systems working in cooperation, and we are in no position to throw away any evidence on a priori grounds. Awareness comes in many different forms, and there is no basis on which to give precedence to any of them. Is a scientist more conscious than an artist or a king more aware than his court jester? Is body awareness inferior to logical awareness or superior to emotional awareness? Are we obliged to grant precedence to linguistic expression, over perceptual acuteness or social subtlety, in defining awareness? I can see no rationale for excluding any of these things. Yet every school of thought that is locked into a fixed definition of awareness has excluded something important. The philosopher lobbies for a philosophical definition, the linguist for a linguistic one, the cognitive scientist for a computational one, the religious thinker for a mystical one, and so on, like so many paid lobbyists begging Congress to heap favors on their constituencies.

What we need instead is what scientists call stamp collecting, a compilation of all data that might be relevant, as free of bias as possible. This might be a useful first step in building a massive database of comparative evidence. There may be many species that can meet some of the criteria for conscious thought and behavior. This is very likely to be true because it is true of every other aspect of cognition. Memory comes in many forms, in many different brains. Vision has been achieved in a variety of organisms with very different eyes and nervous systems. The same applies to hearing, taste, and movement. In the same way, conscious capacity is likely to have many distinct instantiations. If we accept all three definitions of consciousness, we can evaluate the existing comparative evidence in a new light. Psychologists John Newman and Bernard Baars took a step in this direction, with their comprehensive theory of consciousness. They included all three definitions, and tried to deal with the state, architectural, and representational aspects of awareness, in their proposed model of the conscious process. They set an important precedent by taking such a comprehensive approach.

In evaluating other species, we have no choice but to use ourselves as the normative species. This may introduce bias at the entry level, but it is our experience and behavior that have germinated the debate over consciousness, and we can only evaluate other species with reference to that norm. All our definitions are tainted by this fact, but there is no choice in the matter, and this need not throw us off course, as long as we do this with our eyes open. Combined with what we know about brain evolution, a new generation of comparative studies might eventually give us a taxonomy, and a cladistic tree, of conscious capacity. This will leave us with something upon which we can build.

The simplest commonsense standard would be to grant a modicum of consciousness to any creature that can achieve an independent mental model of the world, one that transcends its immediate environment. A creature that responds to the environment in a completely predictable manner would be excluded as an automaton or a reflex machine. A species with awareness would be, by this standard, something more, capable of independently assessing what it sees, tastes, and feels. Some surprisingly small creatures meet this criterion. Ants and bees have discernible agendas of their own, to judge by the complexity of their mating, food finding, navigational, and social behaviors. They routinely take shortcuts and navigate around obstacles in highly efficient ways. This requires a kind of judgment and a goal. Nothing in the environment tells them to take a shortcut or how to take it. Some readers may balk at the idea that such creatures are aware. Conscious bees, beetles, spiders, and dragonflies? But there is no doubt that they can meet this standard. Some of them can perform amazingly complex feats of navigation, micromanage long migrations, or adjust to local changes in circumstance, which demand many interruptions and may take days or weeks to achieve. Not famous for their brain size, new generations of monarch butterflies can nevertheless find their way from the shores of Lake Ontario, where I am writing this, to central Mexico, year after year, with great precision. If consciousness can be graded along many parallel continua, perhaps this kind of navigational skill and situational adaptability might be worth a few points, and these creatures should be conceded some small quantum of awareness.

Outrageous? Only if we want to raise the bar. A more demanding standard may be the ability to perceive complex objects and events. Some creatures lack this ability. For instance, worms seem sensitive only to very concrete dimensions of the immediate sensory environment. They may respond to a particular taste, dampness, or darkness but cannot synthesize these sensations into a complex perception, such as a person or object. At first glance, many insects, and even fishes and reptiles, might be excluded from the Consciousness Club if we use this norm. They have finely tuned sensory systems but cannot perceive any feature of the world that they were not specifically designed to detect. Frogs are designed to detect bugs, but not cows, oak trees, or Chevy Malibus. But even such humble creatures are able to perceive objects and events, within a limited framework. This is evident in their very complex mating rituals, in which they must recognize not only their own kind but also the correct forms of the ritual itself. This is a matter of life and death. They have to get it right, and they do. It is true that the rituals are rigid and stereotyped. Nevertheless, the events that trigger them are often very abstract and involve a sequence of actions and replies that can vary tremendously in their precise form. A given sequence must be perceptually "solved" by the nervous systems of the creatures involved. Brains that can do this may not be intellectual giants, but they are not dummies by any means, on a perceptual level. They can deal with serious complexity in finding food, building nests, defending territory, hiding from predators, finding escape routes, and so on. Cognitive science has taught us how complicated such perceptions are, and it should also teach us respect for the resolving power of these tiny nervous systems.

It is a hard sell to convince people to grant conscious capacity to insects, fishes, and reptiles, even though they can obviously satisfy one or two of our criteria for consciousness. Perhaps the main obstacle in this case is the lack of flexibility in such organisms. They are designed to carry out certain specialized operations with great efficiency. But they cannot move beyond this and adapt to novel situations. A frog may perceive its mate, but it cannot create categories for the other players in its environment. It is inflexible. With birds and mammals, the picture changes. They are flexible, knowing creatures with a wider view. This introduces new criteria for awareness: flexibility and adaptability. Birds follow what is happening in their world intensely and register the events around them "like a hawk." They can also follow and remember many dimensions of that world. Similarly, a rat can navigate a very large and complex territory and react with considerable flexibility to many kinds of challenges. Such creatures are much faster learners, have more powerful memories, can better map out and retain the significant structure of a large environment, and make finer distinctions than any insect, reptile, or fish. They can also carry out a protracted chain of actions to achieve a goal, focus on the complex objects of their perceptions, and solve tricky problems.

In fact, we might say that a love of solving tricky problems is a special mammalian trait. Small mammals can be irritatingly persistent in trying to get into a source of food and use many different solutions. Squirrels might spend a week patiently solving a supposedly squirrel-proof bird feeder. Raccoons are fiendishly clever at surviving in an urban environment and obviously carry around elaborate cognitive maps of entire neighborhoods. The locations of sources of food and places of danger, and safety, are noted and stored for future use. Significant objects are also understood in terms of distinctive characteristics. These are complex, formidable minds that can "parse" their environments into landscapes, agents, actions, and objects, each of which can be recognized and tracked without lumping or smearing the picture into a vague assemblage of specific bits and pieces, as simpler minds do. Reptiles and fishes are apparently not able to achieve such a subtle parsing of events, and because of this, they are often judged to be less aware of what is out there. By this somewhat more demanding criterion, breadth of knowledge and flexibility, rudimentary consciousness would be largely restricted to land animals and birds and would be very faint, if it existed at all, in any species that evolved before the emergence of terrestrial vertebrates.

The bar can be set even higher. Neuropsychologist Donald Hebb argued that mental autonomy from the environment is the criterion of choice for judging the existence of any advanced mentality. In his opinion, the best standard for mental autonomy is a capacity for delayed response. This is a fairly pure criterion in the sense that it singles out one distinctive, measurable feature of cognition. A delayed response is incontrovertible evidence that an animal can carry around an image or idea in its head independently of the environment. This is a more difficult standard for independence of mind than taking a shortcut, because the image, or idea, is more than an immediate response to a challenge. By this standard, insects are out of the race. Even in their most sophisticated action patterns, there is no delay. A forced delay instantly "resets" an insect's mind. The famous tail-wagging dances of bees, used to signal the direction of a food source to other bees, are entirely predictable, to the point of stereotypy, and always occur in response to an immediate stimulus. The cognitive system of the bee is fixed to respond automatically, with no possibility of withholding a response and keeping it in mind, for even a few seconds. The same applies to many other species, including fishes, reptiles, and most birds. It may also apply to the most primitive mammals, but the only way to prove this would be to test these species on delayed response tasks, under controlled conditions, which has not been done. We know that rats are capable of delayed responses, and thus, on Hebb's criterion, we should admit them to the Consciousness Club.

Higher mammals seem able to delay their responses in the natural environment for long periods of time. Say a wolf notices a nest of mice during his travels but is too preoccupied with other business to do anything about it immediately (perhaps it is running from hunters). Nevertheless, he registers the event in memory. A day later he returns, looks for the nest, and has himself a tasty meal. This is a delay, but it doesn't quite meet Hebb's criterion since its recall of the event might have been cued by something outside the wolf, such as a sight or smell. Laboratory tests are more rigorous. The best laboratory demonstrations of delayed responses in animals have been done using primates. A typical test is to show them where food is located and then require them to wait for a period of time before opening a door, to let them choose the correct location. Primates can even perform a more difficult task known as delayed alternation, in which they have to base their choice on their previous response (for example, the rule might be, if they responded on the left on the last correct trial, the next response should be on the right). They do this very well, even after a delay of ten minutes. Rats are also able to do this.

A capacity for delayed response is taken as evidence that a species has a short-term memory system that is able to keep experience autonomously active. Hebb identified this particular function, short-term memory storage, with awareness, and consciousness was identified with a particular place in the mind, the working memory system. Aware minds are, by this definition, those that can carry around an idea or image for a period of time, in temporary storage, regardless of what the surrounding environment looks or sounds like. Such a brain carries around a version of reality that may or may not match the immediate environment. This tough criterion admits mammals to the Consciousness Club, and only very few, if any, other species.

Harvard biologist Donald Griffin accepted the Hebbian notion of mental autonomy in principle but carried it considerably further. He summarized the evidence for consciousness in various species, using a variant of the same criterion. His standard for birds was their clever adaptability to environmental novelty. As in the case of delayed responses, complexity and flexibility imply an internally controlled, autonomous response to the environment. Some might claim that the same data could be used to argue merely for intelligence, not necessarily conscious intelligence, but Griffin does not accept this distinction. For him, intelligent behavior, which is, by his definition, conscious, is always selective, voluntary, guided, informed, and exquisitely sensitive to feedback. This is a reasonable position for an ethologist because in the natural world that Griffin studied in such detail, automatisms and other kinds of unconscious behaviors are always defined as involuntary, unselective, and insensitive to feedback.

Adaptive flexibility is the real giveaway for Griffin. It always implies conscious management. Automatons are, by definition, stereotyped and unimaginative in their response to novelty. In this sense, insects and fishes behave like automatons. Birds, however, do not. At least not all birds. The most striking example is probably Irene Pepperberg's famous talking parrot Alex, which has not only learned to pronounce many English words but uses them in a flexible and appropriate manner. His abilities are somewhat startling, especially to rigid anthropocentrists. Alex can answer conceptually difficult questions. In one experiment he was shown an array of differently colored objects and asked, "Alex, which one is yellow?" His answer was to speak the name of the object that corresponded to the named color, in this case "banana." To achieve this, Alex had to make a disjunction. He had to isolate conceptually the form of the object from its color. Significantly, Alex learned to do this completely outside his natural environment, in a world that is dramatically different from the rain forest treetops where his species lives. According to Griffin's norm, this kind of flexibility is surely a sign of conscious capacity. Many animals have this versatility and respond to novel situations with imagination and panache.

For Griffin, another key criterion for mental autonomy is selectivity of attention. In order to shift attention between several equally salient attractors in the environment, the mind must generate a preference, or an attentional bias, on its own, independently of the environment. Some researchers think of attention as a searchlight, and others as an internally directed filter. In either case, the animal must focus its mind independently of the physical attractiveness of competing stimuli, so that it becomes more sensitive to one aspect of the world over another. Many mammals do this very well. This is especially evident in the single-minded vigils of predators looking for prey. There might be no immediate evidence of prey around them when they start, but they initiate their vigil autonomously of their environment because they harbor expectations. Hunting wolves can track a specific prey for long distances, apparently by maintaining a strong bias in their behavior, often in the face of many potential distractors. All canines can be very persistent in this regard. I have seen my dog maintain a vigilant state for an hour or more, waiting for a particular squirrel or chipmunk to surface from the drainpipe in our backyard. Nothing will easily distract him, and if he is distracted for more than a minute or two, he will always remember to go back to his vigil as soon as the distracting element (most often an invading cat or dog) disappears. His vigil is evidently maintained from within, implying some sort of active working memory process and a selective rejection of other attractors that might distract him.

Another phenomenon that adds support to this line of thinking is the updating process. Memory must be kept up to date. A working memory allowed to drift and left unsupervised would quickly lose touch with the fast-changing environment in which most mammals live. An effective working memory system keeps track of the larger space-time framework within which action takes place and registers the layout of the immediate environment, the presence or absence of crucial rivals, predators, attractions, escape routes, and dangers, even if an urgent task is demanding immediate action. To be useful, this working cognitive map must be updated constantly because it is supposed to provide a useful frame of reference for current action.

Consider the following familiar example of updating. A dog suddenly gets up, bolts out of the room, and returns with his bone. He has gone straight to the place in the house where he left it, without making a wrong turn. Occasionally, if someone moves his bone without his knowledge, he searches the house until he finds it, and he does this very efficiently because he knows the places where he is most likely to find it. To achieve this, he must keep his memory up to date on three important facts: (1) that the bone still exists, half eaten or not; (2) that it is likely to be still where he last left it; and (3) that if it has been moved, it is likely to be in only one of two or three places in the house. This set of contingencies may change, so his memory must be kept up to date about the current state of the house, as well as the current mix of people in it, which can be changeable, because all these things will affect how the house is used. If any key player changes, a clever dog will quickly register the new operating status and behave accordingly.

Ethological studies have demonstrated elaborate memory-updating mechanisms in primates. This is especially true of fruit-eating monkeys. They must keep track of the status of each important species of edible fruit in their region and be there to pick it when it is ready. Otherwise they would miss critical fruiting cycles and possibly starve to death. Primatologist Katharine Milton has discovered that fruit-eating monkeys tend to be almost twice as brainy as their leaf-eating cousins. She surmises that leaves are relatively plentiful and easy to find and that such a diet does not place a heavy demand on spatial memory. Hence leaf eaters don't need a big memory capacity or presumably the large brain that goes with it. Fruit eaters face a more difficult challenge. They must learn the location and timing of each fruit as it ripens because the season is short. They must remember more detail and keep their knowledge current, if they are not to be scooped by competitors. Even if memory capacity is only loosely related to brain size, food-gathering strategy would have conferred reproductive advantages on those monkeys with larger brains, nudging the fruit-eating species toward ever-larger brains. In leaf eaters, neither brain size nor memory would have been so crucial.

But does such a memory capacity constitute evidence for consciousness? It is no accident that the indices of mental autonomy that Hebb and Griffin seized upon are usually identified with conscious processing in humans. We often extrapolate backward from our own capacities to those of animals. This is a reasonable strategy, provided that we know we are using it and that it may introduce bias. Working memory capacity is not unique to humans, nor even to mammals. The emotional and drive systems of primitive mammals descended from equivalent systems in reptiles. Primitive mammals built on what the reptile mind could already do, and newer mammals built on their primitive ancestors. The older systems continued to exist even as new ones evolved, and they undoubtedly remain integral to the new blueprint. There was no qualitative breakpoint where the reptilian mind was suddenly left behind, only a flanking action as the apparatus of memory was gradually adapted to new uses and needs. The mammalian mind emerged slowly, first as an elaboration of its reptilian ancestor and then differentiating gradually into hundreds of variants, adapted to a wide variety of niches. The diversity that now exists should not be allowed to hide the common underlying design.

The most demanding criterion of consciousness is a certain kind of social intelligence that is especially prominent in social mammals and usually not found in herding or flocking creatures. This is the ability to cultivate and remember individual relationships within a working social group. It is difficult to perceive the social events that make and break social allegiances because such events may take a virtually infinite number of concrete forms. Social alliances hinge on objects, past events, and remembered encounters that are not part of the immediate scene in view. They demand considerable memory, since each individual must have a "slot" in the tracker's mind, which must be kept up to date. A socially attuned mind that remembers multiple relationships must keep a current mental model of social structure, a remarkably abstract construct that is never perceptually concrete or obvious. It is a major memory task to follow shifting social alliances such as those of chimpanzees. For those who like elaborate taxonomies of memory, this might be labeled "gang memory." It demands a good updating mechanism. Fail to remember who is currently boss, who is now bonded with whom, or who is no longer allied with whom, and watch out!

A related skill, variously called mindreading, perspective taking, or having a theory of mind, could be taken as another strong criterion of awareness. Mindreading involves understanding that others know things, and this has consequences for how they behave. This is an extremely abstract inference. Some species of monkeys and apes can predict how their colleagues will behave on the basis of what they know or don't know. For example, Kanzi often realizes that his wild-reared friends haven't a clue, and acts accordingly toward them. He sometimes tries to help them out and show them what to do when they are having difficulty understanding what their human caretakers might be asking.

In the wild, apes and monkeys routinely practice little deceptions to achieve personal goals. Scottish primate researchers Andrew Whiten and Richard Byrne have shown that monkeys will sometimes instill a false belief in the mind of another monkey to achieve an immediate objective. For instance, a male might act friendly in order to induce a rival to come close, so he can thrash him. He understands that the other's belief in his friendliness will lead the rival to drop his defenses. Similarly, he might give a false alarm call to drive off dominant rivals and gain access to food. Apes understand social intricacies, change their alliances, remember previous encounters, and adjust their current strategies according to the latest news. It may seem odd to argue that gang warfare and tribalism might indicate a high level of social awareness, but in this context, they do.

A loose hierarchy emerges from these considerations. Bits and pieces of conscious capacity appear in different species. Event perception, short-term memory, flexibility of mind, and mindreading skill might be stronger in one species and weaker in another. But there is a tendency for more of these capabilities to accumulate in certain lines of evolutionary descent, especially in higher mammals, and primates in particular. This loose hierarchy is part of our inheritance as a species. Given the tight fit of mind to brain, human consciousness must have essentially the same combination of deep underlying capacities as that of an ape. Human conscious capacity is the end product of a very long evolution. This idea is incompatible with the idea that humans are the only conscious beings in the terrestrial biosphere. Humans have more of everything. We might be called superconscious. But other species share many component features of our conscious capacity.

BRINGING EXTRA RESOURCES TO BEAR

The fact that consciousness is the product of many components should not distract us from its underlying unity. Above all, conscious capacity must be able to focus, or concentrate, resources. When we speak of animals or humans as having conscious capacity, we usually mean that they can bring extra resources to bear on whatever problem they are currently trying to solve. A fully conscious mind doesn't respond to the world with a canned, stereotyped reaction. It deliberates and experiments, as monkeys do when they are trying to solve a puzzle box. Such a mind does not receive sensory input passively or react in a habitual manner. It explores, actively surveys the environment, organizes its perceptions into coherent chunks or episodes, remembers them selectively, and anticipates the immediate future. It doesn't simply acquire conditioned responses blindly; it notices the incidental details of each episode, calculates the relative probability of different outcomes, and might even use the same conditioned stimuli as cues for other agendas.

The adaptive function served by conscious capacity is thus a very practical one, to provide extra resources when needed. This is a much more energy-efficient solution than designing an animal to go flat out all day. It is also more efficient because it is domain-general and can be used in virtually any situation that a species is likely to encounter. Animals with less conscious capacity are likely to be less adaptive. They are less active, less well motivated, less curious, less vigilant, slower to react, slower to learn, and sometimes oblivious of important cues in the environment. There is a clear fitness gain associated with having a capacity to optimize one's cognitive system, selectively when necessary.

Conscious capacity may be seen as an evolutionary adaptation in its own right, whose various functions have evolved to optimize or boost cognitive processing. Extra capacity is exactly that, extra, not routine. Routine activities are automatic and do not require extra capacity. But when a brain is confronted with a novel, unusually complex, or fast-moving situation, extra capacity may be needed. And it might be needed anywhere. It would be inefficient to evolve extra capacity only in specialized mental modules, for instance, for eye-hand coordination but not for any other purpose. Or to evolve different nontransferable booster systems for, say, spatial intelligence, hearing, or speech. This would constitute a modular solution, and it would be inefficient because of the complexity of the environmental niche filled by higher mammals. Their environmental challenges are inherently cross-modal and require an integrated response. Dangers might be signaled by smells, sounds, or vibrations in the ground. The appropriate response might involve some or all of the animal's repertoire. It is rare for a mammal to be confronted with a neat, modular challenge that can be answered with a neat, modular reply. Birdsong competition might be one example of a narrow, unimodular cognitive challenge that elicits a modular reply. But most of the activities of mammals, whether mating, fighting, fleeing, hunting, or socializing, require flexibility across the board. A nonmodular mechanism is called for under these circumstances.

When conscious capacity is fully deployed, it produces an optimizing effect that corresponds to what we experience subjectively as "concentration" or "mental effort." This capacity is neutral, inasmuch as it can be applied to virtually any kind of behavior or thought, from a tennis game to an intellectual challenge (such as a chess game) or to any number of emotional, social, and aesthetic experiences. Anger and revenge, deception and intrigue, drama and awe: All these can be sharpened and intensified by the allocation of conscious resources. Even a limited amount of conscious capacity will create a more autonomous mental life in a given species, regardless of whether the primary evolutionary result is to give it greater flexibility and independence from the environment.

In the real world a sudden challenge might demand a response from any motor capacity that a creature might possess and perhaps from all of them at once. A porcupine reacting to a wolf might give out alarm signals, run, climb, hide under a low shelf, bite, urinate, or, if close enough, raise his quills; in short, he will do whatever works. He can do this because he has a flexible response and an oversight capacity that can orchestrate this response. It would be fatally limiting for him to be restricted to just making noises or raising his quills. In the competitive world of land mammals, flexibility equals survival. A capacity that is universal in its reach is much more effective because it can coordinate the response of the whole organism to virtually any challenge that might come from the environment. In this sense, conscious ability exists to optimize and coordinate the resources of the brain and to focus available cerebral capacity on the specific problem at hand, whatever that problem may be. The brain systems that provide support for such wide cognitive effects reside outside any specific cognitive domain. In terms of brain architecture, they may be regarded as domain-general.

Domain-general resourcefulness is so common among mammals that it is viewed as ordinary. Consider, for example, the cognitive challenges facing a bear fishing in a stream. Bears may not normally be terribly intense or focused, but fishing is another matter. It is an inherently complex and unpredictable activity. It requires concentration and effort, even for experienced bears. All phases of a fishing expedition demand attention to detail, rapid decisions, and careful coordination. Say the bear wants to locate the best place to fish and to position himself advantageously. This taxes one of his brain's executive functions, his capacity for sustained attention. In this case, he must seek out subtle patterns in the rushing waters of the stream to determine where the fish are swimming. He may also look around for social cues that point to the location of the fish, for instance, in the behavior of other bears or of birds. Social cues change quickly, depending on the specific mix of individuals present, and this places demands on two other executive functions in the bear's brain, his capacity to split his attention (he may be scanning for fish while looking for social cues) and his capacity to update his working memory (the clues are constantly changing).

In addition to gathering information and positioning himself, the bear must get ready for appropriate action. If he sees a shadow in the stream that looks like a fish, he must move quickly to an appropriate location, before rivals do. This sequence engages at least two additional executive operations. First, he has to disengage from what he is already doing (in this case, watching the water for signs) and switch to a different pattern of action. Second, he must initiate the appropriate sequence of actions and keep them flowing in the proper order, directed efficiently toward his goal. To do this, he has to move to his chosen location, monitor that specific fishing locale, strike the fish accurately at precisely the right time, and bat it out of the water and onto the shore. The sequence fails if at any time he loses track of the physical and social context or of his target. It also fails if the order of his actions becomes disturbed.

While doing this, the bear must monitor as well the movements of other bears in the region and other events in the environment itself. None of his actions takes place in a vacuum. Every perception is interconnected with every other element in the scenario, and it can change at any moment, depending on a number of variables. Managing all this complexity and unpredictability requires a mental governor that can keep track of the big picture, while overseeing and switching between several subsidiary mental domains, occasionally altering or recombining components of the scenario to optimize the likelihood of success. The various disconnected bits and pieces of all this would fall apart unless the bear could keep his perceptions and actions organized in a smooth flow. Maintaining direction and flow in a complex behavioral environment is a considerable achievement, well out of the reach of most nervous systems. The bear's dexterity at navigating such complex scenarios is testimony to his brilliant coordination of executive resources.

The bear's fishing expedition is far from unique in the mammalian world. In this, he is typical of all higher mammals: wolves, cats, canines, whales, apes, and us. All these species evolved in complex and demanding ecologies, and they have to be very good at the executive management of their affairs just to survive. They must also be able to inhibit irrelevant tendencies. A bear may feel attracted to a female across the stream and may have an urge to fight the male standing beside her, but being a clever bear, with a prioritizing conscious mind, he will probably eat his fish before taking care of his rival. He can keep his priorities straight and hold to his longer-term direction because he is a highly conscious animal. There are no problems with esoteric definitions here. These are immensely practical considerations, with immediate, and survival-related, consequences. If the bear lost his ability to focus on the task at hand, we would all know instantly, and so would his rivals, not to mention his victims, and meal tickets, the fish.

EMBODIMENT, EGOCENTERS, AND HOMUNCULI

Where in his brain does the bear tie all this together? Is there really a place in his mind that is conscious? Moreover, does this type of proposal reassert the existence of a Cartesian homunculus? Yes, in a sense it does. I think that the idea of a homunculus, or little-man-in-the-head, is closely related to a very old idea in experimental psychology called the perceptual egocenter, which has been a staple of the concept of selfhood since the work of William James and Gordon Allport. The egocenter is not a physical place in the head, although we may have the illusion that it is, because it is the perceived locus of the physical self in most organisms. It is also the deepest reason why William James was right to state, in the opening quotation of Chapter 3, that conscious ideas and images are always *owned*. This owning is highly physical and body-based. It is the farthest thing in the world from a rational or linguistic sense of owning. We own our experiences the way we own our bodies: immediately, urgently, passionately. Most people judge their visual egocenter to be a few centimeters behind the eyes, and nonvisual channels, such as sound, touch, feeling tone, and body sense, seem to converge on the same imaginary location, following the same egocentric coordinates, which are determined in most people by the dominant visual channel of experience. In this, we are behaving like typical vision-dominated primates, like monkeys and apes. The egocenter of a mole, a bat, or a whale probably has a very different raw feeling about it.

The egocenter feeds on the many body maps registered in our brains, and together they form the database for our perceptual homunculus. Each brain map may appear to be dissociated and isolated from the others in the nervous system (there are dozens of such projections in the visual system alone), but the egocenter ultimately ties all these maps together. It is a unifier. It does not tolerate blatantly contradictory messages about the shape or position of the homunculus for very long. A unified personal homunculus is a crucially important point of reference for calculating position, for coordinating movement velocity and direction, and especially for interpreting and directing self-action. To achieve this satisfactorily, the sensory homunculus must convey an accurate impression of our own bodies, located in an objective, three-dimensional space. All conscious experience is referred to that egocenter, not in an abstract or conceptual way but as a direct perceptual given, a building block of raw experience. Even emotions and feeling tones, such as a feeling of dread in the pit of the stomach, are situated with precise reference to our homunculus, which, by the way, includes the entire body, not just a point in visual space. This fact is reflected in expressions such as "I was so proud I felt my heart bursting" and "My feet feel like lead." Our inner landscape is rich, detailed, and very much our own.

This perceptual homunculus is no illusion, but it should not be confused with the Cartesian homunculus that is currently under attack from so many scholarly quarters. That abstract eighteenth-century philosophical category does not do justice to the body-based homunculus I am describing, which is nothing less than the integrated neural footprint of our embodiment, a deeply rooted perceptual and motor phenomenon, and the underpinning of a unified physical selfhood. Our complex egocenter is really a brain model of the physical self and the primary source of self-awareness. Its presence has a huge impact on our thought process. It even shapes our uses of language, as George Lakoff and Mark Johnson have shown. It is undoubtedly very ancient in evolutionary terms. Animals are easily as familiar with their own bodies as they are with their territories. Grooming behavior in birds and mammals and reactions to pain both give away the accuracy of their body maps. These animals can attend to that largely proprioceptive and somatic turf as effectively as they attend to what is outside them. The boundaries between self and nonself are never so clear and basic as when pain is registered, but they also serve as the foundation of our more elaborated aspects of selfhood, and those multilayered Ego worlds that Gordon Allport once lumped together under the umbrella of the proprium, or self-realm.

The egocenter may be nothing more than a self-referential form of perceptual awareness in most mammals, but in humans it encompasses complex external event representations, within which the physical self may be inserted as an integral part. For example, if you hold a rod in your hand, close your eyes, and explore the room you are in by touching it with the rod, you can actually construct a fairly accurate model of what is around you by referring the rod's contacts to your egocenter. This is what a blind person does to explore the world. Elsewhere I have suggested that our capacity to combine an abstract self-representation with a wider representation of events external to the self is surely at the core of our human expressive genius. It allows us, first and foremost, to use our entire bodies as reenactive devices in gesture and pantomime. Ultimately it is the foundation of what we call meaning in the subjective sense. These egocentric body maps define one of the subjective bedrocks of the individual mind, and one of the main obligations of human consciousness is to anchor the base of experience within those personal boundaries. Human self-reference is a product of the most fundamental fact about mammalian consciousness, its grounding in personal embodiment. This creates a loop from our physical selfhood to our representations of the external world.

Meaning is not disembodied from this process. A person-to-person conversation does not float in some realm of pure meaning. The physical "I" talks to the physical "you." The flow of meaning is always physically situated, in that extremely abstract natural habitat called the self or perceptual egocenter. Attention is also situated in the physical self. The physical self settles on what is said, and in the end, it alone chooses to remember whatever it chooses to remember. Shades of meaning are also communicated obliquely by the physical self, through gestures, expressions, winks, masked intentions, lip movements, and so on. These may be incorporated effortlessly into an ongoing stream of awareness by their common relationship to an underlying semantic model, but the model itself is always claimed as "our own" by physical self-reference. This is even true of amnesiacs or people who have temporarily lost their social identities, in so-called fugue states. The physical self remains intact.

Semantic processing is ultimately grounded in embodiment, and this is surely the most elementary fact of existence for most mammals. The so-called brain-in-a-vat thought experiment, on which so many words have been spilled, is only so much word weaving because it proposes the possibility that a brain, disconnected from its body, could develop and maintain a sense of self. There is no reason to believe that this is a possibility. No virtual reality program could adequately simulate the combinatorial explosion of sensory possibilities that characterizes the physical self, but this is a problem only for techies. In the real world, neurological patients with even the slightest partial disconnection of their awareness from body-based feedback systems are greatly disturbed in their consciousness. A consciousness such as our own might actually be more physical than is desirable, in the abstract. It cannot be disconnected from its physical container. As I mentioned in Chapter 3, right hemisphere–lesioned patients, who sometimes develop disconnections from their own bodies, are in many ways more disturbed than those who lose language. This reflects the fact that body-based awareness has very deep evolutionary roots. Embodiment is a basic fact of life for nervous systems. Moreover, awareness, no matter how sophisticated it has become in its current representational gamesmanship, still cannot do without its rock-solid physical selfhood. Without it, consciousness finds itself floating, groundless. It spins out of control and loses its sense of personal reality.

Oddly, our physicality also has a tight relationship with our capacity for symbolic communication, which evolved so recently. This includes language. All expressive systems are ultimately owned in the same way as any other motor system; that is, they are self-rooted. I shall carry this idea one step further in a later chapter. For now, suffice it to say that *language could be the greatest beneficiary, rather than the cause, of the extended human capacity for self-consciousness.* The conscious mind may have reinvented itself and greatly extended its reach in language, but it has never lost its vestigial roots in embodiment. On the contrary, although human consciousness may have had to accommodate itself to the emerging symbolic structures of complex culture, refining, nurturing, and reflecting upon these structures as it expanded its own powers, it has always referred back to its roots in the physical self.

Although its physical basis is not yet understood, the homunculus is psychologically very real. It is the anchor of higher cognition, simultaneously the architect and ultimate expression of selfhood. It guides the self-construction of individual minds as well as cultures, through its capacity to intervene aggressively in the cultural arena. It supervises the countless negotiations that constitute the cognitive basis of human culture. It builds and then governs the human world through its ability to coordinate the actions of the executive brain.

THE EXECUTIVE SUITE: DEFINING THE PRIMATE "ZONE OF PROXIMAL EVOLUTION"

Direct evidence on the relative conscious capacity of the human brain can come only from comparing human executive abilities with those of apes and monkeys, with roughly similar paradigms. There are several paradigms that provide some evidence on this issue. Some measure metacognitive skills, such as the ability to evaluate one's successes or failures. Others test self-representation or self-regulation. Still others require subjects to take a third-party perspective on their own behavior. Some have also been developed to compare the attentional abilities and nonverbal communication skills of humans, apes, and monkeys.

The most informative paradigms are those that are positioned right on the edge of what primates can achieve. They allow us to map out something akin to what Russian psychologist Lev Vygotsky called the "zone of proximal development." This is a theoretical space intended for studying child development and optimizing pedagogy. Parents and teachers can optimize a child's learning by intuiting exactly what he or she is ready to learn. They must estimate exactly what skills fall just outside the child's immediate ability. That zone, the space just beyond the child's current competence, is called the zone of proximal development. When the teacher judges this successfully, teaching and learning become very efficient. The reasoning behind this is simple but compelling. There is no point in trying to teach the child something that is far beyond that zone because it is too far out of reach and the child won't catch on. Anything less demanding has already been mastered, and the child will be bored. But teaching just inside the zone works well.

In principle, the same dynamics apply to cognitive evolution. The functions that are most vulnerable to selection pressure would be those that fall just on the edge of a species' capacity, so that only a few members of the species have some ability in that direction. We might call this the "zone of proximal evolution." This is where evolutionary shifts are most likely to occur. If certain individuals gain some reproductive advantage from having an ability that falls within that zone, the gene pool will drift inexorably in that direction. We usually estimate, as a matter of convenience, that our most distant ancestors, the Miocene apes, were roughly similar in appearance, brain size, and cognitive ability to modern chimpanzees. On this admittedly large assumption, their zone of proximal evolution would probably have corresponded fairly well to the adaptive possibilities that exist in modern chimpanzees and bonobos. Thus it is important and potentially crucial to demarcate the areas in which the basic executive skills of humans and apes diverge. This might give us a feasible starting point for the emergence of a distinctively human conscious capacity. It might also show us the ancestral zone of proximal evolution from which humanity emerged.

Conscious capacity involves many brain subsystems, some of which evolved independently of one another. However, there is reason to think that in early hominids, all these systems might have changed together, in a fairly synchronized manner, as a sort of evolutionary cluster. When a cluster of traits evolves more or less simultaneously as a piece, it is sometimes called a suite of adaptations. I have called the cluster of skills underlying human conscious capacity, the hominid Executive Suite. We don't yet have a complete or detailed picture of the Executive Suite, but I am confident that the abilities shown in Table 4.1 must be included on the list.

TABLE 4.1

The Executive Suite: Domain-general skill clusters that fall within the primate zone of proximal evolution and are very highly evolved in humans

	Monkeys	Wild Apes	Enculturated Apes	Humans
Self-monitoring	Yes	Yes	Yes	Yes
Divided attention	No	Maybe	Some	Yes
Self-reminding	No	Maybe	Maybe	Yes
Autocuing	No	No	Yes	Yes
Self-recognition	No	Yes	Yes	Yes
Rehearsal and review	No	Maybe	Yes, but limited	Yes
Whole-body imitation	No	Partial	Yes, but limited	Full
Mindreading	Minimal	Minimal	Yes	Yes
Pedagogy	No	Maybe	Yes	Yes
Gesture	No	Doubtful	Some	Yes
Symbolic invention	No	No	Protogesture	Yes
Complex skill hierarchies	No	No	Some	Yes

1. *Self-monitoring of success or failure.* This is sometimes called metacognition, although I tend to prefer a wider definition of that term. It is the ability to monitor one's own success and failure. It has been measured in primates using a paradigm developed by animal behaviorist David Smith, called the bailout option. Monkeys are trained to respond to visual stimuli with three possible key presses: yes, no, or bailout. Pressing the bailout key automatically skips the trial and advances the monkeys to the next one. This allows them to escape punishment for a wrong response. If they estimate that they will fail on a given trial, their best strategy is to press the bailout key and skip it. Monkeys can master this task easily. This implies that they are aware of their own likelihood of success or failure and are capable of self-monitoring. Rats and cats apparently cannot perform this task. Humans have a highly developed ability to do this, of course. This is an essential capacity for making rapid progress in acquiring any complex skill. The old stock market rule is based on this ability: Run your wins, cut your losses.

2. *Divided attention.* There are many ways to measure split attention, but the ones that highlight human-primate differences are those that involve more than just attending to several sources of input. The most telling studies focus on the production side, where monkeys and apes are asked to perform two tasks simultaneously. They have great difficulty with such tasks. Humans can do them fairly easily, and there is a large human experimental

literature on speed and accuracy trade-offs, which require the subject to do two or more things at the same time. In a management context this ability is sometimes called multitasking. Sue Savage-Rumbaugh has suggested that chimpanzees and bonobos have most of the cognitive components for language, with the major exception of multitasking ability, which she believes is the main cognitive difference between apes and humans. She has found it difficult to train even her most brilliant bonobo pupil, Kanzi, to do more than one thing at a time. Her favorite example is the game of basketball. Apes cannot learn to play basketball, she argues, because they cannot handle so many things at once. This difference between them and us is not qualitative, only quantitative. With a great deal of time and training, apes may learn to manage more than one thing at a time. We simply have more of this ability. It is worth pointing out that the difference is mostly in the domain of action. Many higher mammals can attend to several things at once, as in the case of the bear fishing, provided they don't actually have to perform more than one action at a time. Humans, on the other hand, can manage several things simultaneously. As we saw in the previous chapter, this is crucial in language, where speakers must keep track of both their own behavior and that of others, splitting their attention among several channels, often at very high speeds.

3. *Self-reminding.* This is the mnemonic use of action, to keep a longer sequence of action going. It is not the same thing as explicit memory but is undoubtedly a close relative. To acquire or maintain any complex skill that involves contingencies and rulelike linkages between various component actions, we must be able to repeat part of a long sequence in order to remind ourselves of the next part. This is very helpful in maintaining the longer sequences of action that have to be remembered. An example of this might be making a simple tool, such as a stone ax. There is a fairly long sequence in making the ax that involves locating and acquiring the necessary materials, modifying those materials, and shaping the tool. Each of these actions can be broken down into smaller ones, and each part of the overall sequence must be completed roughly in order before final assembly can begin. To maintain such a sequence in the absence of constant verbal rehearsal, humans usually stop and repeat sections of the sequence, as if to self-remind. Children often do this, long before they can speak well enough to remind themselves verbally. The result is a series of short repetitions, assembled into longer sequences. Apes do not seem able to self-remind, at least not systematically and certainly not as effectively as human beings. Psychologist Franz de Waal has claimed that self-reminding actions do occur in bonobos, and the enculturated apes of Savage-Rumbaugh have been reported to leave

markers in the forest as reminders of where they have traveled. Although controversial, these reports suggest that apes may be on the verge of self-reminding ability.

4. *Autocuing.* This is a more explicit self-triggering of memory than self-reminding. It is the basis of recall and achieved entirely from internal cues. It is the foundation of the human capacity for explicit recall in language. Apes are capable of some autocuing but seem relatively poor at it. The most convincing examples of autocuing in apes are the sign language performances of enculturated ones, such as the Gardners' chimpanzees, which mastered a significant amount of manual sign language. The active production of sign language requires each message to be triggered internally. But in the wild there is no solid evidence for explicit recall in nonhuman species, and this is the main reason for the huge gap between human and animal research on memory. Usually animals cannot be tested by being asked to recall material from memory, and they have to be tested indirectly for implicit, rather than explicit, recall. This limits the kinds of symbol systems that apes can learn. They are much better at using prepackaged symbol boards than they are at signing, because symbol-boards require fewer internally triggered responses. Humans could never have acquired or used a vocabulary of thousands of words without the ability to search rapidly among the thousands of word forms in their memories and find the ones they needed to build a sentence. This is a precondition for any kind of language, even the most primitive. Although rare in primates, it is just inside the zone of proximal evolution, for both bonobos and chimpanzees.

5. *Self-recognition.* This is the ability to "objectify" consciously the perception of the physical self. The standard test for this ability in apes is the mirror test pioneered by G. G. Gallup. An animal is left in a room to become familiarized with a mirror. Secretly a dot is painted on its forehead, and the mirror is brought back. Most animals will take no notice of the dot and continue to treat the image in the mirror as if it were another animal. But certain ape subjects instantly recognize themselves in the mirror and touch their foreheads as if they knew that (a) the forehead in question was their own and (b) they didn't normally have a dot on it. Monkeys and other mammals do not behave this way. They do not see themselves in the mirror image. This suggests that some apes have a more objectivized image of themselves than do other animals. They can decipher the rotations and transformations of their images in a mirror and see the connections between their self-initiated actions and those in the mirror. Needless to say, it hasn't been necessary to make a "control" demonstration of this ability in humans (a visit to any

clothing store would provide more than enough evidence). Humans are generally very good at self-recognition, from an early age. Actors and athletes use objective self-visualization as a part of their training regimens, using mirrors, videotapes, and so on. As in the previous example, this ability is just inside the zone we are interested in.

6. *Rehearsal and review of action.* Apes can acquire some fairly complex skills. They can learn to master a joystick to solve computer mazes, for example. But they do not typically rehearse those actions to improve their performance. One of the most essential components of advanced skill is the capacity to rehearse and review one's own actions. This complex self-supervisory capacity is unique to humans, but apes are quite close to having it. Human children effortlessly and endlessly rehearse and refine their own actions, from a very young age, often without any obvious reinforcement. For example, they may practice throwing stones at a tree or balancing on one foot, for no apparent reason. Although all primates play repetitive games, they do not appear to rehearse and refine them spontaneously. Thus baboons throw projectiles but never practice to improve their skill. The rehearse and review loop is not in evidence, yet its components seem to be present. Rehearsal and review must have evolved by combining many of the skills listed above, such as self-monitoring, autocuing, visual and kinetic imagination, and advanced self-representation. All these exist in limited form in primates. In a domain-general executive system, a large increase in capacity by itself might allow such combinations of components. On the same logic, purposive rehearsal might also have been the archaic form of intentional representation, in which an act is used to represent itself (a simple form of self-demonstration).

7. *Whole-body imitation.* Imitation is a very active field of research, especially in the area of ape-human comparisons, and there is no room here to review in detail this immense literature. Suffice it to say that in the wild, several species of apes and monkeys are capable of a limited form of imitation, sometimes called emulation. Enculturated apes are much better at this, but their imitative skills are not anywhere close to those of human children. Apes find the precise, detailed reproduction of actions particularly difficult, perhaps because they do not always understand the objective of the action. Orangutans and gorillas in the wild routinely perform "program-level" imitations. Researcher Richard Byrne has shown that young gorillas use this to learn how to prepare foods that are difficult to access, such as nettles, by acquiring hierarchical programs of action. Anne Russon, working in Indonesia, has indicated that orangutans imitate their human neighbors—by trying to untie and paddle a boat, for example. They can reproduce the

overall shapes of the acts without actually doing them well, perhaps because they are not inclined to rehearse the acts or to review their performances in a self-critical way. They seem unable to keep focused on, and remember, the specifics of the imitated sequences and often fail to grasp the connection between the details of the actions they are trying to copy and the result of those actions. Nevertheless, they get the "program" right. On the other hand, the enculturated bonobo Kanzi can master such tasks easily. He has even learned, through imitation, to manufacture simple stone tools and use them purposefully. He has learned many other skills without explicit instruction, through imitation. The difference between enculturated and wild apes in this regard suggests that imitation falls within the zone of proximal primate evolution. Surely it could have been subjected to selection pressure early in hominid evolution, once the environment made such skills relevant to fitness.

8. *Mindreading.* As already described, this is the capacity to understand that other minds know things and that this knowledge predicts behavior. This is central to the repertoire of social skills needed in complex societies. Some researchers, such as autism specialist Simon Baron-Cohen, believe that mindreading skill is a module, isolated from other mental abilities. However, this idea does not stand up to analysis. Although mindreading is clinically dissociable from other skills, such as speech, it does not follow that it is a true cognitive module. It is inherently a domain-general skill that shows no modality specificity, either in humans or in apes. Cues that can help us understand another's mind ordinarily come from vision or hearing, but they may also originate in other modalities (touch can be particularly useful in this regard). Mindreading can almost be called a classic executive task, a divided-attention problem, in which observers usually follow the behavior of others without losing sight of their own actions. It is also a working-memory problem since it is contingent on following the behavior of a single individual for some time. We cannot rule out the possibility that mindreading has a modular component, perhaps in the realm of emotion and empathy, but its overriding cognitive demand is domain-general. Mindreading ability is present, in a limited way, in both monkeys and apes, as testified by the evidence of deception in social interactions, produced by Richard Byrne, Andrew Whiten, and many others. Although this conclusion is still controversial, it seems safe to allow that primates are able to deceive one another in small, concrete ways. This ability seems to have two principal cognitive components: (1) a very abstract event-perceptual capacity, which enables the perception of social events and interactions, and (2) metacognitive skill, which can calculate the relationship between what others do and one's own actions.

The perceiver's mind is somehow able to gain perspective on its own operations, so that it can generalize specific actions to particular circumstances in the behavior of others and remember those relationships. Empathy alone could never explain how a monkey might successfully use deception (reflecting the feelings of others is a more limited skill), but with a metacognitive overview of one's own behavior to use as a template, empathy might make it more likely to perceive the internal states of others. The existence of deception in monkeys and apes suggests that this ability also falls within the zone of proximal evolution of the primate mind.

9. *Pedagogy*. This is an elaboration of mindreading. It involves two-way mindreading, by both teacher and pupil. It might better be called mind-sharing. In effective pedagogy, one person consciously regulates the learning process in another, while the learner tracks the teacher's intent. This is the basis of Vygotsky's pedagogical zone. From the earlier years of infancy, the teacher-learner relationship is central to human culture. It involves not only understanding the intentionality of others but also engaging in intentional exchanges and understanding how the mind of the other has been affected by the communicative act. This capacity exists in people who lack language, such as deaf nonsigners. It is also crucial for language acquisition and thus takes priority over language, in evolution. The eminent behaviorist David Premack has stated that it is generally lacking in wild-reared apes, although some signs of it have been reported. There seems to be a quasi-pedagogical relationship between mother and infant in the common chimpanzee, in the transmission of such skills as the use of rocks as hammers and anvils in nut cracking. However, in this case, the infants might be emulating the behavior of their mothers and subsequently being rewarded by their own success or failure at getting the nuts. There is no evidence for systematic instruction by apes in the wild. However, enculturated apes have bona fide pedagogical skills. Once again the Rumbaughs' evidence on bonobos is the clearest in this regard. Kanzi and Panbanisha, their most highly enculturated bonobos, both have tried to show their wild-reared cousins how to do things, on many occasions, usually without success. Thus, even though pedagogy may not exist in wild apes, some of them have the potential for it, and culture can bring this out. Natural selection might also act on this potential. The leading edge of the ape's potential for pedagogy was evidently close enough to the point of evolutionary combustion to be a possible factor in hominid evolution. It falls within their zone of proximal evolution.

10. *Gesture*. This is sometimes called iconicity in expression. Gesture creates a visual metaphor of something else, usually another action. This

kind of expression is common in human children during their first year, when they can easily learn simple gestural signals, such as representing "fish" by puckering their lips or "give" by reaching out and turning their hands. Apes don't do this, but once again they come close. Although there are some reports in wild apes of pseudogestures, such as pointing to a desired object or capturing attention in another by deliberate movements, these have been observed unequivocally in enculturated ones. Once again, the executive fundamentals of gesture are not all that different from those of imitation, pedagogy, or self-rehearsal. Gesture requires splitting attention, remembering relationships, self-supervision, rehearsal, and metacognitive review. Enculturated apes have developed these skills more highly than unenculturated ones. It is all a question of degree and early experience. Any gestural skills latent in primates, or in the early predecessors of humans, such as the australopithecines, may well have been subjected to selection pressure.

11. *Symbolic invention.* This is the spontaneous, unsolicited creation of novel expressions. It has never been convincingly demonstrated in apes, even enculturated ones, although pockets of evidence suggest that they may have some innovative ability. There have been a few reports of gestural invention by primates, but most of them have, to my knowledge, been discounted. There was a published claim that the hand-signaling chimpanzee Washoe could sometimes combine two hand signs to create a new word, the most famous one being the combination "water-bird" for swan. However, in a critical review Herbert Terrace, of Columbia University, showed that this could have been a chance combination and that it was never adopted as a word in continuous use. Terrace's criticism has been rejected by psychologist Roger Fouts and linguist Philip Lieberman, who have supported the Gardners' claims for their chimp, but the evidence at this point is still somewhat equivocal. Primatologist Richard Wrangham has argued that there are many minor variants on existing primate expressive signals, and although they cannot be accepted as fully developed gestures, these might appear to be "proto-gestural." In strong contrast with the difficulty of establishing gestural invention in apes, it is easy to see that human children are highly inventive from a young age, both in signaling their intentions and in innovative dyadic interactions, such as games. However, even though it has not been convincingly demonstrated, in my opinion, simple proto-symbolic invention is not far from the limit of what apes can do in the realm of gesture. Imitation, skilled rehearsal, gesture, and gestural innovation form a continuum, and there is no reason to think that on a gestural level symbolic invention would involve different mechanisms. It is a matter of degree. Whether

or not it was already present, this capacity was probably not far removed from the capacities of Miocene apes, given their similarities to modern apes.

12. *Assembling complex skill hierarchies.* The final level of nonverbal mastery that characterizes human beings and that is almost or partly present in apes is the ability to self-assemble complex skills. Humans build skill upon skill, creating very complex contingent hierarchies, as in driving or piano playing. A driver must learn a whole range of somewhat independent actions for driving, such as starting, turning, backing up, steering, accelerating, braking, checking the mirror, shifting gears, monitoring traffic, reading road signs, maintaining speed, keeping track of directions and street names, and so on. These subskills are usually self-taught, self-rehearsed, and self-evaluated, with some overall guidance. The result is an amazingly complex chain of habit systems, or demons, each with its own executive demands, which must eventually be integrated into a massive metasystem that coordinates all of them (otherwise we might find it difficult to negotiate a crowded expressway on a Friday evening). No other primate can assemble hierarchies that are anything close to this complex. Our capacity for self-installing skill systems of this magnitude serves as the foundation of culturally transmitted custom. It involves most of the components of the Executive Suite, working in cooperation with one another. These hierarchies, once learned, become automatic, going underground into the unconscious, and only their on-line control remains conscious. Although the ability to self-supervise the assembly of elaborate skills is special to humans, apes are able to acquire simple skill hierarchies. This is especially clear in the Rumbaughs' enculturated apes, who have self-assembled all the interconnected skills required for a limited form of symbolic communication, in a naturalistic setting, without explicit training. Once again, the differences between us and them are quantitative, not qualitative, and natural selection had ample opportunity to increase hominid capacity in this area.

In sum, there are many domain-general skills that distinguish humans from apes, independently of language capacity. These functions are all reliant upon the executive brain. It is reasonable to conclude that the whole executive complex, including its anatomy and function, evolved as an adaptation, or a series of adaptations, in the hominid line. A more detailed archaeological record of the expansion of the hominid brain might provide us with a rough index of when and where the Executive Suite evolved.

Although it does not in itself encompass language, the Executive Suite provides a foundation for the evolution of language, the most complex of all human skill hierarchies. Not even protolanguage could have evolved before the hominid executive brain had reached a critical point in its executive capacity because language in any form demands a great deal of executive management. The moment-to-moment difficulty of acquiring and managing language in the real world is very high, because the cultural environment that encompasses language is constantly changing, creating a need for a much more powerful acquisition process. This rapid rate of change may not have been true in the earliest days of its evolution, but it is certainly true today. Human speakers often carry out several complex operations at once, in several modalities, simultaneously maintaining parity with multiple recipients of their communications. This demands tremendous attentional control, to a degree unheard of in other species.

Because of its complexity and speed and its simultaneous demands on attention and working memory, when compared with other highly skilled tasks, language is unique in the sheer magnitude of its executive demands. The architecture of language is really a gigantic metahierarchy of acquired skills, each requiring the executive coordination of an entire hierarchy of subtasks and sub-subtasks, all regulated from working memory. Language routinely programs and engages executive ability at all levels, including the most mundane subtask, and builds these programs right into the brain's language system. Human language is also characterized by rapid, effortless autocuing; that is, improved access to memory. Gaining voluntary access to our own memory systems is a singularly powerful skill. This is not unique to language. An actor can carry a repertoire of hundreds of conventional gestures, postures, sounds, facial expressions, and attitudes. However, this skill pales in comparison to language, where the explicit retrieval of words and idioms must be achieved much faster and more accurately. This is no small matter when a person must find a single word from a vocabulary of tens of thousands.

Split attention is also necessary for language. Speakers hold in their working memories a rough record of their own past communications with others, as well as what those participants have said, tagged for source. Otherwise a two- or three-person conversation would soon cease to make sense. This multichannel updating skill cannot be modular in any meaningful sense of the term because it must remain open to all sorts of inputs, including nonlanguage ones. Moreover, the working model must be immediately available for constant updating; one might say that the model *is* the speaker's effective working memory. It guides the overall contour of communication for considerable periods of time. This kind of continuous model building and memory updating is a major component of human executive capacity.

The inevitable "other face" of language is also an executive ability, our capacity to automatize the entire linguistic skill hierarchy. The acquisition of any skill, including language, depends on executive control. But its efficient use depends even more on automatization. Conscious capacity must remain available to deal with urgent temporary matters, the novel, the complex, and the unexpected. The basic operations underlying language, such as searching for a word, assembling sentences, and generating the correct sounds when speaking, must become largely automatic, so that the speaker is free to take a complex communicative context into account. By increasing our executive ability, including an expanded capacity for automatization, humans have effectively become self-programmable organisms, able to install nearly any cultural algorithm in our brains and make it automatic. We can build our own demons, consciously and deliberately, and the biggest demon of all is the language hierarchy itself.

We stand at the far end of a long process of evolution. The material origins of consciousness started in specific kinds of nervous systems. Conscious capacity evolved, with its various component systems, in parallel, in many species, in a series of slow-moving adaptations of the vertebrate brain, each for its own local reasons. The earliest conscious functions were focused on achieving basic perceptual unity, and more recently evolved ones on gaining a better fix on short-term events in the environment, as well as achieving more effective control over behavior in the intermediate term. The core brain systems of primates remain the foundation of human conscious capacity, but they have been greatly expanded in the human brain, through the neocortex.

The Executive Suite is a uniquely human configuration of conscious capacities. It stands at the peak of a hierarchy of executive capacities that have a long evolutionary history. There could have been many other neuropsychological solutions to designing a powerful executive system. Our particular system reflects its primate inheritance. It also has a mirror image in the brain itself, which brings us to the next chapter.

5

Three Levels of Basic Awareness

Thinking in terms of physiological processes is extremely dangerous in connection with the clarification of conceptual problems in psychology. . . . [It] deludes us sometimes with false difficulties, sometimes with false solutions.

—LUDWIG WITTGENSTEIN

It is, I suppose, no accident that Descartes, who has affected the ways of thought of the world more than most people, was an anatomist and physiologist, and was deeply concerned about how to speak about the brain.

—J. Z. YOUNG

Whatever amount of conscious capacity we may concede to animals, there is no question that during the course of our evolution, our species crossed a great divide, a cognitive Rubicon that no other species on this planet has crossed. How can we usefully conceptualize that crossing? And what connection might it have with consciousness?

THE MYTH OF THE ISOLATED MIND

I would suggest, only half in jest, that human evolution might be reconceived as the Great Hominid Escape from the Nervous System. The main difference between apes and us is culture, or more specifically symbolic culture, which is largely outside, not inside, the brain box. Culture distributes cognitive activity across many brains and dominates the minds of its members. Despite this, cognitive science studies the mind as if it were confined entirely within a single brain. Culture is not usually included in cognitive theory, except as part of the environment. The consequence of this is that cognitive science has an "isolated mind" bias, in which the cognitive system is always treated as a self-contained entity or monad. This mind-in-its-box belief is accepted by Cartesians, Behaviorists, and Cognitivists alike.

The isolated mind model works well in the study of nonhuman animals. Animal cognition rarely escapes the boundaries of its own embodiment, and whatever there is of animal culture, it plays a small role in forming the individual mind. Even highly sophisticated animals, such as apes, have no choice but to approach the world solipsistically because they cannot share ideas and thoughts in any detail. Each ape learns only what it learns for itself. Every generation starts afresh because the old die with their wisdom sealed forever in their brains. An older ape cannot tell the younger, "There is an orchard beyond that ridge," or "Avoid that tree; I once got very sick while eating its fruit." If the young one wants to learn that fact, it must do so either by observing others get sick or by getting sick itself. There are no shortcuts for an isolated mind. As a result of this isolation, the rate of cultural knowledge accumulation is very slow, and there is much less cultural variation between troupes of apes than there is between groups of humans. The tiny variations that exist between ape populations, such as those documented by primatologist Richard Wrangham and his colleagues, consist of slight differences in tool use, hunting strategies, and social life, many of which are attributable to significant differences between their habitats, rather than to culture itself.

In contrast with this, humans link with a vast and diverse cultural matrix in early infancy and profit from the rich storehouses of knowledge and skill that have accumulated in our cultural memory over many millennia. Since cultural knowledge accumulates rapidly, there can be dramatic differences between human cultures. Our dependency on culture is very deep and extends to the very existence of certain kinds of symbolic representation and thought. Socially isolated humans do not develop language or any form of symbolic thought and have no true symbols of any kind. In fact, the isolated human brain does not act like a symbolizing organ, any more than an ape brain does. It is apparently unable to generate symbolic representations on its own. It does so only through intensive enculturation. This is consistent with the standard definition of language as an arbitrary and conventional system of expression. A century ago Ferdinand de Saussure observed that languages are always the product of circular interactions between two or more brains, rather than of the operations of a single brain. Translated into a neural context, this implies that symbols evolved to mediate transactions between brains, rather than to serve as an internal thought code for individual brains.

Once learned, languages can also be used by brains in isolation. As a result, we can think silently to ourselves in the language we acquired from our culture. But our brains are not equipped to invent language in the first place. Isolated brains can formulate categories and concepts and think in an analogue manner, entirely on their own, without enculturation. But they cannot create languages or explicit symbols. This raises questions. If language and symbols are taken from culture, where could symbolic culture have come from in the first place? How could a vestigial primate brain, slightly altered, dig its way out of its solipsistic box and generate a shared cognitive universe?

Interesting questions, you are thinking, but irrelevant. Obviously consciousness stands above this debate, in magnificent isolation, because it is the one aspect of cognition that absolutely cannot be shared. Consciousness is owned by and locked within each of us. Surely the role of culture is virtually nonexistent in the case of consciousness. But is it? Awareness may be physically confined by its embodiment, but it dissolves, on the one hand, into the infinite inner spaces of the brain and, on the other, into a rapidly expanding cultural universe. Modern human awareness is shot through with cultural influences, and regardless of its physical boundaries, it is in the firm grip of the cultural web. It is woven as seamlessly into those collective networks as it is into the submolecular depths of the physical universe. Connected minds prosper in proportion to the richness of their links with culture. Minds grow with the collectivity, and isolated minds wither. Everything that is specifically human about our mode of awareness is a product of our long-standing symbiosis with culture. Symbols, our notions of selfhood, and the basics of autobiographical memory originate outside the monad, in culture. The unique character of human awareness cannot be understood if it is kept in its container, as if all explanations should lie there. We are what we are because of enculturation, plain and simple. This is not true of any other species.

Let me clarify this statement before I am misunderstood. There are certainly many other species that have social minds, especially if we include insects in our discussion. The actions of a bee can change tremendously as a function of what the whole swarm is doing, and there is no way to predict the behavior of the swarm from the capacities of any single bee. A prime example is the choice of a queen, whose selection is unpredictable and whose role and physiology change dramatically once she has been chosen. The same principle is evident in fire ants. The collective intelligence of this species is truly terrifying. Fire ants are able to adapt to changing circumstances as a group, and the single ant is reduced to a replaceable component in a social machine. E. O. Wilson referred to fire ant societies as biological superorganisms because their behavior cannot be understood at the individual level. In this case, the collectivity is much more than the sum of its parts. It has a mind, if that is the appropriate word, of its own.

Another example may be drawn from research on the construction of termite nests. These are complex three-dimensional structures, comparable in their relative height to human skyscrapers, with exquisite architectural structure. How could a beast so simpleminded as a termite build things of such complexity and beauty? The answer cannot be found in the capacity of any individual termite. There are no plans for nest building in any termite's brain, no designs or models of the nest in some Chief Architect's memory system. It is even debatable whether any termite has the ability to perceive its own nest as a category. Nevertheless, termites build these fantastically elaborate structures. Psychologist James Gibson explained this with an "ecological" theory of vision, which saw the organism and environment as integral to the same system. The answer, he suggested, lies more out there in a distributed cognitive system than it does inside the individual nervous system. The distributed system includes the information contained in the larger social setting and in the interactions between the termite colony and the changing physical environment. It is the synergy between these forces that produces these architectural masterpieces. In a sense, the nest system builds itself. It is the result of myriads of interactions that are ultimately mediated by the simple reflexes of individual termites. A large-scale system, distributed between the termite nervous systems and the signals given off by the changing environment, guides the construction.

Insect superorganisms are distributed across many rudimentary minds. There are no conscious planners here. Rather the insect mind-in-its-environment constitutes a social-cognitive system with its own structure. Such structures do not make heavy demands on individual brains, and no one would seriously suggest that insect societies are an appropriate model for human culture, let alone for consciousness. The point is that these might be called cognitive systems but could never be called conscious ones. Insect societies remind us of how much cognitive complexity can exist without demanding either consciousness or intelligence from individuals. In contrast, the achievements of human culture are contingent on conscious individual minds working together.

Human societies and their institutions are products of conscious minds in collision. This is true whether we are speaking of the crew of a sailing ship, the members of Plato's Academy, or the directors of a corporation. They depend greatly on the creative confrontation among social complexity, the physical environment, and individual consciousness. The result is a revolutionary new phenomenon, distributed symbolic systems, whose properties cannot be predicted from what individual brains can do in isolation. There is a parallel revolution inside the individual brain, whose functional organization is changed by those systems, in equally unpredictable ways. Viewed in itself, an expansion of the hominid executive brain could not have foretold the radical cognitive innovations that appeared in hominids. Proportionately similar increases in the ape brain, relative to monkeys, resulted in no equivalent change. In humans, even after our expanded brain is factored in, something remains that cannot be accounted for by innate properties. That additional element is enculturation. The specific form of the modern mind has been determined largely by culture. The creative collision between the conscious mind and distributed cultural systems has altered the very form of human cognition. It has also changed the tools with which we think. Language, art, and all our symbolic technologies have emerged from this collective enterprise.

As a neuroscientist who accepts the need to push materialism as far as it can go these examples leave me with one chastening and potentially liberating thought: Some aspects of human cognition and even of consciousness may not have to be explained entirely in terms of individual brain capacity. But wait, you say, weren't you about to hang all our achievements as a species on the innate conscious capacities built into our brains? Well, sort of. This brings us back to what brains, isolated or not, can and cannot do.

THE GREAT COMPUTATIONAL DIVIDE

This discussion might become more focused if we revert to one of our most overused but powerful metaphors, computation. The great divide is really a computational divide. Humans use a mode of computation, symbolic computation, that is unique in the biological world. We use the symbolic mode in common discourse, in everything from street talk to filling out income tax forms. We also use it to program computers and to create expert artificial systems, such as chess-playing computers. This is epitomized by the symbolic programming strategy used in Artificial Intelligence, which spells out everything a computer must do, and in what order, in thousands of lines of commands. Each apparently simple step, no matter how trivial, must be explicitly spelled out in symbols and fixed precisely in relation to all the other steps in the algorithm.

The word "algorithm" itself gives away the computational strategy used by AI. In its strict definition, algorithms (named after the Persian mathematician al-Khuwarizmi) are notations that capture the solution to a mathematical problem. Strictly speaking, there are no algorithms in nature. They exist only in our descriptions of nature; that is, in language. Causal chains in the natural world are commonly referred to as algorithms, but this can lead to a misleading use of language. It confounds symbols with what they represent. The etymology of the term is clear enough, and algorithms are symbolic inventions, plain and simple. AI programs are entirely symbol-dependent. They ultimately control the assembly languages that run the hardware. Once it receives a sequence of instructions, the computer doesn't need any understanding to produce correct results. Trigger the dominoes, and they fall exactly as they have been arranged. Moreover, it doesn't matter what specific hardware implements an algorithm. It might be executed by a computer, a human being, a well-trained pigeon, or all three in tandem, as long as the program is correctly followed. When we carry out addition and subtraction, unreflectively and automatically, as we learned in school, we get correct results whether or not we understand the underlying logic. In doing so, we become computers, unbelievably slow, lumbering, unreliable, but computers nevertheless. All that is needed for an algorithm to do its work is a physical device to implement its instructions.

In the natural world, nervous systems are normally locked into a computational style that is very different. They don't use symbols, such as numbers or words. Real brains work by fuzzy recognition, using a nonsymbolic, holistic, or analogue style, a bit like the popular concept of "right-brained" thinking. This mode of cognition can be simulated in a very limited way on digital computers, using an associative or connectionist strategy. The brainchild of this approach, the field of neural net computing, tries to emulate the principles by which nervous systems operate. It may seem ironic that neural net simulations must be run on digital computers, rather than analogue ones, but we don't know much about analogue computing, which is closer to the workings of real nervous systems. Neural net simulations are implemented in virtual nervous systems that can learn on their own, without instructions, by a process of pattern recognition. They are networks of randomly wired virtual neurons, whose connections change with experience. They contain "hidden layers" of neurons that store memories without using any explicit symbols to define them. They just form and store impressions.

Human beings maintain a precarious balance between these two computational modes. This is the most fundamental sense in which we might be called hybrid minds. We are part analogue and part symbolic in our mode of operation. Because we are living animals, with conventional nervous systems, it is a safe bet that most of our early experience is gained in a nonsymbolic manner, built around our nervous system's basic habit of forming impressions. But we are also capable of constructing languages and symbols, such as those found in storytelling, art, and mathematics. These typify the symbolic mode. We bequeathed a specific branch of that mode of thought to our greatest twentieth-century invention, the digital computer, which became a direct technological extension of our symbolizing powers. When Donald Griffin wrote in defense of animal consciousness, he was not suggesting that animals have symbolizing powers, like humans. Rather he was referring to the analogue mode of thought that is common to most cognizing creatures. Basic animal awareness intuits the mysteries of the world directly, allowing the universe to carve out its own image in the mind. This is a largely receptive mode of knowing, and we share it with our animal cousins. In contrast, the symbolizing side of our mind is more aggressive in its approach. It creates a sharply defined, abstract universe that is largely of its own invention.

The main difference between these two modes is reflected in the two main branches of computational studies of cognition. The artificial minds built by these two branches differ in the way they obtain and manufacture representations of reality. Artificial neural nets acquire experience by changing their connection patterns after repeated exposures to the environment. They form impressions in essentially the same way a time-exposed astronomical photograph does, by passively gathering data over time. Several exposures to an object allow neural networks to extract consistencies in the world that relate to the object. These are stored in stable connection patterns. In effect, the neural net classifies the world, without preconceptions about what the major classes ought to be. In contrast, symbolic computation takes in the world in prepackaged categories. It is given the major classes of experience in advance. This difference can be seen in the way artificial neural nets and symbolic systems learn to recognize a face. The neural net is simply shown a set of faces, over and over, until it forms an impression. The symbolic system is provided in advance with a definition of the category called face, with lists of features, identifiable parts, and acceptable proportions. It might use a metric to assess the relative sizes of the parts, such as eyes and noses, of various faces. For this, it needs a reference system, which allows it to compare the faces it receives with a standard template. The neural net system is obviously much simpler and more direct. After a number of exposures, with simple corrective feedback (the simulators' equivalent of reward and punishment), it will learn to classify each face. Recognition, in this case, consists of familiarity and consistency of response. It does not depend on symbolic capture, or remembering rules of thumb, or looking up a list of features.

There is one important insight to carry away from this comparison: Simulated neural nets are, in principle, autonomous systems. They acquire their own experience, like living creatures. As we saw in Chapter 4, attributing consciousness to animals has something to do with their degree of cognitive autonomy from the environment. A conscious animal carries around a representation of the world that is unique to its brain and its history. In this sense, it is a very Jamesian creature: Its experiences are owned. Real nervous systems vary in the degree to which they can achieve this, and this determines how much conscious capacity we are willing to attribute to them. Then why not apply this idea to artificial systems? Someday neural nets might have autonomous experiences of some complexity. They construct the world in roughly the same manner as real nervous systems, and it is probably just a matter of time and technology before they actually have such experiences. At that point we shall have to grant them some degree of consciousness, even if it is only the most primitive kind.

In contrast, purely symbolic machines, such as AI systems, can never have such autonomy, no matter how sophisticated they become, as long as they remain purely symbolic in their mode of operation. Such a system, even in principle, cannot own its experiences in the same manner. By definition, it receives all its codes from the surrounding culture. Everything it does is given meaning by culture. The programmer, serving as the rather godlike interface of the machine with its culture, defines its cognitive universe, down to the last detail. In the case of our own minds, our cultural gods may be less visible to us because we are so dominated by them. But culture infuses the symbolic side of our minds with meaning. What, then, about our claim of cognitive autonomy? The answer is, if we are autonomous, it is primarily because of that deep nonsymbolic engine. Autonomy lies in the richness of our raw conscious experience. Human awareness has its deepest roots in the indigenous, impression-forming, presymbolic side of mind. Symbols have such a crystallizing impact on how experience feels that they create the illusion that they are the true source of that experience in the first place. This is wrong. They are never the source. They exist only by the good graces of our ancient analogue minds. Our claims for autonomy should be adjusted accordingly. Symbolically encoded thoughts seem autonomous, but in reality, they acquire their autonomy only from their nonsymbolic engine.

When it is fully conscious—that is, fully up and running—the conscious human mind is a frighteningly clever beast. Its intuitive power is astonishing, and even so, its symbolic creations seldom do justice to the richness of its experience. If human beings seem hugely brainy, to the point of being overbuilt, it could be because our analogizing minds had to be very clever indeed to invent symbols in the first place. Our semantic engine had to be more powerful, and much more clever, than any of the languages it has invented or any of the technologies that it has conceived. Symbols themselves are devoid of any inherent meaning. Their meaning has its anchor elsewhere, down below, and without that deeper foundation our symbol worlds would float in an empty ether. This applies even to the neural support systems that make symbolic processing possible, such as the speech brain, the literacy brain, and many of the external symbolic systems that exist outside the brain, in libraries, archives, the Web, and so on. Symbols can be viewed as a go-between, a fantastically elaborate technological network that ultimately mediates only one thing: the connection of two conscious minds, one sending, the other receiving, across time and space.

When we observe computers in their natural habitat, we can see that they are merely extensions of our cultures, like all previous symbolic technologies, including everything from megalithic observatories to writing, clocks, sextants, and calculating machines. Oddly, the same statement applies to the symbolic tools that are installed in our minds, such as languages, which exist solely because of a virtual invasion of our brains by external symbolic systems. They are all extensions of culture. If we have stories to tell and archetypes to paint, it is only because we have been properly equipped and fitted out to have such things. Part of the equipment is outside, and part inside, our heads. Culture leads, and mind follows. In other ways, mind leads. Symbolic culture is nothing without the human mind, while the human mind, without culture, remains locked in incoherently upon itself.

THE DAWN OF HYBRID MIND: ACCESS TO MEMORY

Humans thus bridge two worlds. We are hybrids, half analogizers, with direct experience of the world, and half symbolizers, embedded in a cultural web. During our evolution we somehow supplemented the analogue capacities built into our brains over hundreds of millions of years with a symbolic loop through culture.

I have suggested elsewhere that the earliest emergence of symbolic culture must have depended on a radical change in the hominid memory system, specifically on our gaining voluntary access to it. Difficult as it is to comprehend, most animals have no voluntary access to their own memory banks. This limitation can be visualized with the aid of an image (a time-saving concession to our archaic analogue mode).

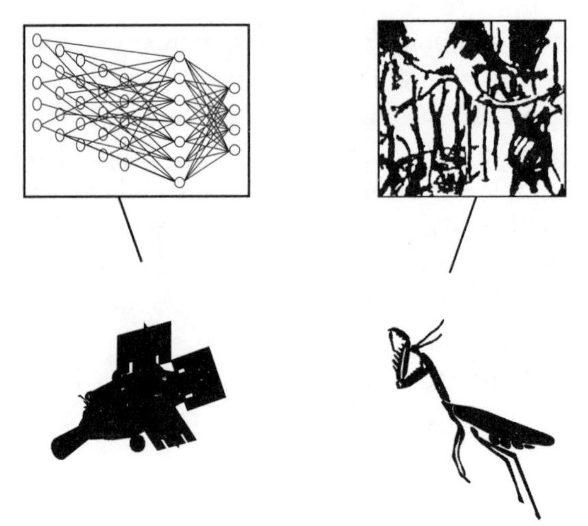

Figure 5.1

In Figure 5.1 the drawing on the left shows a hypothetical satellite pattern recognition system built with artificial neural nets. It has learned to recognize certain terrestrial objects, perhaps continents, or smaller things, such as specific lakes, after many exposures to their silhouettes. On the right there is a drawing of a creature from the natural world, a *Mantis religiosa*, or praying mantis. It recognizes a fly by means of a process that, in principle at least, resembles simple perceptual learning in a neural-net system. It responds to the stimulus by striking and eating. It is able to produce the appropriate response, from a very small number of built-in options, simply by refining its impressions of the world and reacting to what it sees. It is locked into this impression-forming strategy.

Impression-forming, analogue minds are ubiquitous, and as a result, most organisms are reactors, rather than actors. Some of them are very smart, by any standard. However, they need huge nervous systems if they are to achieve any amount of intelligence. They don't benefit from the efficiencies of symbolic systems. They rely on a massive pattern recognition capacity mediated by neural clusters, whose cell bodies are usually gathered in one place. The Roman physician Galen called such agglomerations ganglia, or walnuts, presumably because to his unaided eye, they actually looked like walnuts. In a very simple nervous system, such as a spider's, there might be one ganglion receiving an input from each sensory path into the nervous system. Such ganglia typically connect to only one eye, leg, or antenna, but they are wired into a larger network that carries signals back and forth between them and between larger groups of central ganglia. Some insects have a head ganglion, which is somewhat similar in function to a vertebrate brain. Integrative functions, such as directed locomotion, are quartered there, because they require integrating inputs from several subsidiary ganglia.

Some ganglionic nervous systems are very large. They are also architecturally complex. The octopus brain fits this description. It carries the invertebrate nervous system to its ultimate expression, having somewhere in the neighborhood of eight hundred million neurons. The octopus lives a timid, rather Kafkaesque existence in the shadows of the deep. It has remarkable perceptual intelligence, explores reefs and caves in a slow, cautious manner, and is equipped with an exquisite sense of touch and very good vision. But it doesn't have very many options as far as action is concerned. It can either approach and eat or run away and hide. When it approaches a potential food source, it moves very carefully, in a style that we would call stealthy. It uses vision, taste, and a delicate system of touch receptors that cover its tentacles to decide whether to approach or flee. This is apparently a computationally difficult task, and the octopus uses up to 80 percent of its central neurons to arrive at such decisions. It also uses its formidable neural resources to solve tricky instrumental challenges. For instance, it might have to open shells or explore spaces with its tentacles in order to eat. On present evidence, it manages all this with endless pattern crunching and no voluntary recall. Like all ganglionic creatures, it has several specialized neural systems, each of which constitutes a module, one for each major skill system the species has inherited. Its actions are ruled by gigantic networks of neurons, dedicated to specialized functions. The physical medium of memory storage seems to lie in the patterning of connections in, and between, these large networks. There is no voluntary access route into such a system. Memories are triggered only passively, by associative cues. Thus the creature is a slave of its environment, unable to conceive of an agenda of its own.

Figure 5.2

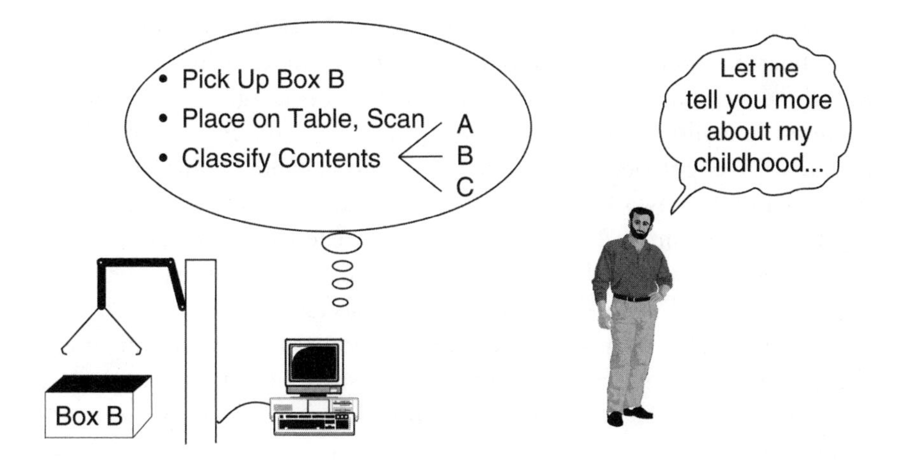

Figure 5.2 might help convey the striking contrast between this kind of mind and a symbolizing mind with an addressable memory system. The left image shows a prototypical symbol-using robot, which has been given a language, an operating system, and specific instructions for exploring and classifying the contents of a box. The robot scans the box, using a program that tells it exactly how to proceed, down to the meaning of the last pixel. The right image shows a man, trying to recall something from his childhood, using a symbolic search strategy. The man is a very different creature, of course, but in one crucial respect he resembles the robot. Both man and robot have direct access to their memory banks. They can go to a specific item, look it up, and retrieve it. The computer does this unconsciously and automatically. When it checks out a list of features or looks up a number, we do not claim it is aware. The human usually (but not always) performs the recall operation in awareness; that is, consciously and deliberately. The man directs, evaluates, and triggers his search consciously, even if some memories might pop up unconsciously. In this, the man is different from the computer. However, this should not obscure the similarities between man and computer, especially the fact that the vast majority of existing natural minds cannot perform this recall operation at all. The computer has been programmed by humans to do this, of course. *But so has the human.* In both cases, cultural programming established symbolic skills that were not given, in fact not even hinted at, in their basic design.

We know how computers got their memory retrieval skills (we invented them in our own image), but where could voluntary access to memory have arisen from in human evolution? This issue requires a digression on memory in the nervous system. Memories are stored in networks. As a rule, single neurons don't do much on their own. There are some exceptions to this, in very primitive creatures. For example, the worm *Myxicola* has only one neuron, which controls the muscles that draw it into its tube. A small thing perhaps, but vital. Some small squids have only two neurons, which control the sac whose jet action propels them through the water. Again, vital, but not cognitive. Cognition, even of the most primitive kind, seems invariably to require large populations of neurons wired into networks. Thus recall must be a function of networks, rather than single neurons. Could access to memory have arisen simply from an evolutionary redesign of neural networks themselves? It is not at all clear how this could have happened. Such networks do not lend themselves to easy internal access.

Large neural networks evolved initially to serve simple motor functions. The nematode worm *Ascaris* (a parasite) has a small neural net, exactly 162 neurons, whose main function is to control the worm's wriggling movements. Its rhythmic movements result from waves of electrical activation that go through the neural networks, triggering muscle contractions. In such simple creatures the linkages between neural nets and behavior are easy to grasp. Memory doesn't even come up. However, the relationship between more complex neural networks and cognition is less easy to understand. Cognitive networks, such as those of the cerebellum, hippocampus, and cerebral cortex in humans, often have exquisite internal structures, with unique wiring diagrams. The same applies to advanced sensory systems, such as pattern vision, which depends upon many pools of neurons, organized into several distinct circuits, each analyzing different aspects of the visual image and feeding their calculations back to the primary sensory region. Memory storage seems to occur in such nets, but it is perceptual memory, not subject to voluntary recall. There is no evidence that we have explicit voluntary recall of anything that is stored solely in the sensory regions.

Could voluntary access to memory have originated in the cross-modal linkages between different sources of sensation? The combination of one sensory source, such as visual input, with other sources of information, such as touch and balance, takes place in specialized networks that combine incoming data from the three modalities. Nervous systems must learn how to combine these into a unified perception of space. But this kind of perceptual learning is common to many species, and such memories are not accessible to consciousness except, perhaps, as a feeling of familiarity. Complex pattern recognition is also a cross-modal operation. A dogfight, for example, combines vision, sound, taste, and pain sensation with many other possible inputs. The cross-modal memories formed from such experiences are very abstract, but even they cannot necessarily be recalled at will. Some artificial systems, such as the famous Carnegie Mellon robotic insects, can emulate the actions of real-world cross-modal perception with uncanny accuracy without needing an addressable memory system. Robotic millipedes are particularly convincing, as they crawl over three-dimensional obstacles like real insects, on the basis of their inherent mechanics and neural net wiring. They can do this without any symbolic programming or any possibility of memory recall. We can be sure that cross-modal integration of this sort does not need an explicit memory address mechanism and is very unlikely to be the evolutionary root of voluntary recall. Most nervous systems, including many that have formidable perceptual skills, simply do not have any ability to recall. They lack the basic capacity that I have called autocuing; that is, the ability to self-trigger recall from memory.

How could a hierarchy of connectionist neural networks, dedicated to various specialized sensory and motor operations, have evolved autocuing capacity? In theory, there are only a few ways to do this. It is conceivable, if improbable, that a nervous system might evolve the same strategy that engineers use to build hard disks, by preassigning memory addresses. In that case, every instruction, input, and output would need a preassigned location in a searchable medium. This would allow each specific memory item to be tagged as it was stored and to be located later, when needed. Computers usually look up tables, such as file allocation tables, to find things. A nervous system might have evolved a similar strategy to locate memory items. A neural time tag has often been mentioned as a possible candidate for this, but such a tag could never give us the kind of cross-indexing of categories that is needed for full-blown symbolic thought.

The only potential physical mechanism that I can think of for a built-in memory address system would be the biochemical gradients that guide growth and development. Imagine that the human nervous system somehow evolved a way to convert these diverse molecular gradients into an address system, using coordinates based on them. Memories might then be systematically located and recalled at will, but only provided that there was some intelligent, or potentially intelligent, mechanism of retrieval that mapped stored meanings onto these gradients in a fixed manner. To evolve a system that could penetrate the maze of neural net circuits we have inherited, our preexisting gradients wouldn't have served anyway, since they are fully occupied with the traditional roles assigned to them, such as controlling the growth of the neural matrix itself. So new ones would have been needed, and if such pathways existed, they would presumably be visible. But the pathways aren't there. On present evidence no such things evolved.

There is no point in extending this speculation any further since it is only meant as an example of how not to approach the problem. Human evolution didn't take this course. The human brain does not have file allocation tables, innate allocation mechanisms, or anything remotely like them. There are chemical gradients in our brains, but even this process is somewhat unpredictable because it only guides the growth of axons toward their targets, under constantly changing conditions. These gradients cannot fix in advance how the memory matrix will be connected and cannot, even in principle, preallocate memory space. Moreover, there is no known candidate for an intelligent allocation mechanism. Because infant brains are so plastic in their course of development, memories could end up virtually anywhere in the labyrinths of the brain. There aren't even any reliable neural equivalents of the cruder memory partitions used by computers, such as labeled disk sectors. It is true that there are broad regional subdivisions in the brain that seem to sort out types of memory storage, such as the rough partitioning of memory between right and left brain. But there is no hard evidence for fixed localization of explicitly addressable memories in the brain. The hippocampus plays a role in the fixation of such memories, but it is, at best, a consolidation device that helps build retrieval paths at the time a memory is acquired. It does not contain the addresses, the memories themselves, or any essential part of the retrieval pathways. This is proved by one clear and dramatic fact: Long-term memories are retained and recalled even after complete destruction of the hippocampus. However, new explicit memories cannot be acquired. Thus the hippocampus helps us set up new explicit memories, but once they have been consolidated, it is no longer needed to retrieve them.

The cortex is often touted as the location of explicit memories. But this is too general a statement to be helpful. Antonio Damasio has narrowed this down by suggesting that different classes of memories might have distinctive locations within the cortex. Within language, for instance, nouns and action verbs seem to engage broadly different regions of the left hemisphere. However, this evidence seems to reflect only the rather boring fact that memories involving visual sensation must rely to some extent on the visual cortex, motor memories on the motor cortex, and so on, hardly a surprise. This doesn't help locate the really substantive parts of the human memory system, which must store entire scenarios and narratives, such as the latest gossip and news, and complex social events, such as scientific meetings, weddings, and funerals. Such complex memories are held in vast interconnected subclusters of subsidiary items (the bride's dress, the family's grief, the flowers, old friendships renewed, and so on). They also involve a mix of ancillary sensations and emotions, which engage other areas of the brain. The best hypothesis about how the brain can do this is the notion of neural reentrance, which holds that the same brain circuits can be reused for a variety of purposes. The cortex is designed so that reentrance is inherent to its circuits. We should be in awe that such tangled memories can be smoothly and accurately dragged out of unconscious storage, intact, into the light of day, many years after the event.

How extraordinary that such memories are so easily found when we need them! When we want to recall a really good bit of gossip, the third act of last night's play, or our feelings of nausea after losing a crucial game somewhere in the distant past, we can, despite the fact that the brain does not, indeed cannot, have any built-in address system for these memories. Those memories might be passively activated, of course, by simple association. But this doesn't count, and it doesn't explain very much about explicit recall. Passive recall is a simple matter of cued association, but precise explicit recall is a highly ordered process that depends heavily on having a symbolic system in place in the brain. Some symbols can amplify the autocuing process by giving us a consciously accessible shorthand for triggering prepackaged memories. This capability cannot have been built into the brain entirely, since it is derived from culture, yet it could not function without some sort of innate autocuing capability. Again, our dependence on such a system, which in its neuronal instantiation must be both arbitrary and unique in every individual, reflects the tremendous interdependence of brain and culture. Culture not only guides us in encoding our memories but gives us an entire system of addresses and memory management skills that allows us to recall them voluntarily. Access to memory is the key to our symbolic cultural system. It is the foundation, as well as the product, of that system. But it cannot be accounted for entirely by cultural programming. Its possible basis in the human brain will be discussed in Chapter 7.

Culture provides more than mere programming for explicit recall. In its deeper aspects, it provides something that is perhaps more akin to an operating system. It gives us a shared structural framework for public, conventional systems of thought. As such it could not have come into existence within the closed boundaries of single nervous systems. The invention of symbols does not occur in socially isolated children, even those born with the neural capacity to speak. Deaf children who are not exposed to language do not even suspect that such a thing might be possible. A solipsistic system of self-generated symbols might be useful to them since it would convey some of the advantages of symbolic thinking. But such systems never come into existence. Brains must be programmed by culture first.

Both animals and simulated neural nets can operate without culture. They accumulate their own experiential histories. Their archrivals, symbolic machines, have no equivalent of personal experience, but they have plenty of symbols, indeed nothing but. Humans are unique in having both modes available to us, and we play them off one against the other. We have hybrid minds. Like the monsters of Greek mythology, we are two creatures struggling within a single body. We are capable of operating entirely within that same fuzzy analogue mode that constituted the whole of the cognitive universe for our ancestors, while another part of us operates like the symbolic machines we have made. But mostly we muddle through with various patched-together hybridized modes of thinking and feeling. Our conscious experience reflects this. These two sides of our being are engaged in a constant struggle for the ownership of awareness.

MODELS OF MODELS AND THE TERTIARY REGIONS

The key to crossing the great divide was a physical change in the brain. The human brain got a great deal larger, but its expansion was not an indiscriminate, across-the-board increase in size. It expanded in a very specific way, led by the most abstract regions of the cortex, the so-called tertiary, or association, areas. These areas expanded more than any other during our evolution. They receive inputs from all the major sensory and motor fields and are in a unique position to perform abstract processing. They are also the main control structures for executive functioning.

Figure 5.3

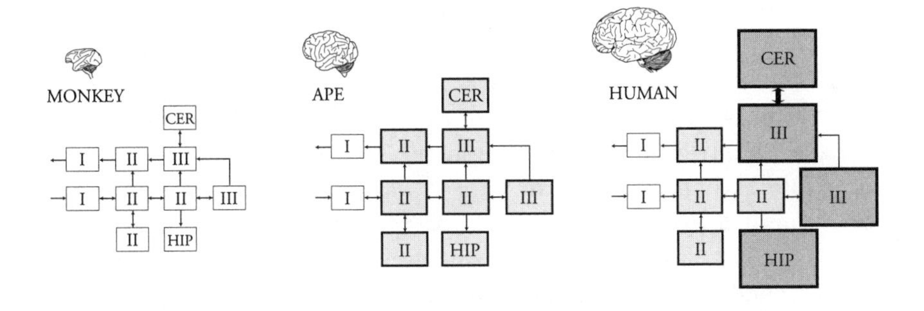

Nonexpert readers may become impatient with this section, but I ask their indulgence because these details cannot be left out. The tertiary regions of the cortex have a very systematic pattern of anatomical connections with other areas of the brain. This pattern is repeated for all the tertiary regions, whether sensory or motor. The vastly oversimplified schema shown in Figure 5.3 shows some of these connections and conveys a general impression of the relative size of the three major classes of cortex: primary (I), secondary (II), and tertiary (III). These are shown for monkeys, apes, and humans, from left to right respectively. This is a generic diagram that applies to the regions for vision, audition, and touch. Note how the relative balance of power within the cortex shifted over the course of primate brain evolution. Primary cortical areas remained unchanged in relative size. Secondary and tertiary areas became relatively larger in the brains of apes, as they evolved from a common monkey ancestor. Tertiary areas became especially large in humans. They are the largest structures in the human brain. This reflects our pattern of encephalization, or cortical expansion.

This pattern of change has major consequences for human cognition because these three kinds of cortex do very different things. Primary sensory areas are, by definition, connected directly to the peripheral nervous system.

They receive signals from eye, ear, and skin, via sensory pathways, and extract the basic features of sensation, such as color, location, contrast, loudness, pitch, and brightness. Primary motor areas send commands directly to the motor neurons that control muscles. They can control the actions of individual limbs directly. Thus the primary regions are located at the interface of brain and world. They are subject to a great deal of influence from the peripheral nerves. A creature dominated by the primary cortex has a sensorimotor mind, focused on perception and immediate pragmatic reactions.

The secondary areas are located farther from the effects of the periphery. There are secondary sensory areas for each of the major sense modalities, visual, auditory, and somatic. These receive their major inputs from the primary sensory cortex, rather than from the sensory pathways themselves. They also send signals back to the primary cortex, in very systematic patterns of reciprocal connections. As might be expected, the secondary sensory areas carry out high-level integration of the data forwarded to them from the primary cortex. In humans they perform advanced perceptual functions, such as object and face recognition, and the resolution of sound patterns, such as those of musical themes and spoken sentences. The secondary motor (so-called premotor) cortex forwards signals to the primary motor cortex and supervises the execution of motor programs and plans of action, rather than moving specific muscles. It also couples action to context. Secondary regions are much larger, in relative terms, in apes than in monkeys. This suggests that apes should have more abstract event perceptions and finer motor control than monkeys, and this is generally consistent with observational data.

The tertiary regions of the cortex are superabstract. They have no connections to either the peripheral nerves or the primary cortex. Thus they are completely removed from direct influences from outside the brain. They connect only to the secondary cortex and other tertiary regions, as well as to various important integrative subcortical parts of the brain, such as the cerebellum. The tertiary regions are devoted to even more abstract cognitive processing than secondary areas and are truly amodal, or supramodal; that is, domain-general. They receive major inputs from all secondary sensory and motor regions and return those connections. They are thus perfectly located to influence and supervise cognitive processing and perform executive functions, such as working memory, selective attention, metacognition, and planning.

The tertiary cortex is more highly evolved in apes than in monkeys, but it has become a gigantic, dominant presence in the human brain. Our massive tertiary regions are the signature structure of our brains, accounting for most of their evolutionary expansion, in terms of absolute volume. We have large tertiary cortical regions in each of the frontal, temporal, and parietal lobes. They receive inputs from all secondary cortical regions and send outputs to the secondary motor regions, the limbic system, the hippocampus, and the neocerebellum. Damage to tertiary areas does not usually result in sensory loss or motor paralysis. It does not even result in specific perceptual distortions or disturbances. Instead it causes disorders of a very abstract nature, in language, planning, self-representation, metacognition, reasoning, and the long-term supervision of action. This gigantic tertiary-based system is the core of the executive brain system.

From a connectionist perspective, the tertiary regions lie at the deepest point inside the nervous system, the place in the brain that is the greatest number of connections removed from the outside. This area is the logical governor of the entire cognitive hierarchy. While the primary sensory regions might be regarded as designed for sensation, and the secondary ideally designed for event perception and voluntary action, the tertiary regions are relatively free of sensorimotor constraints. They have what neurologist Kurt Goldstein once called an "abstract attitude" (cortex with attitude no less!). The tertiary regions played an especially important role in the expansion of the human executive brain. In fact, the special executive features of the human Executive Suite map precisely onto the evolutionary expansion of the hominid tertiary cortex.

The ultimate result of having so much tertiary cortex is our ability to build mental models on a very abstract level. Mental models are the most deliberate, conscious productions of the mind. The ultimate model of models, the human self-in-its-environment, is the most frequently and intensively rehearsed of our mental constructs. It is developed and rehearsed through play, reflection, and self-correction, with the advice and contributions of parents, siblings, friends, enemies, institutions, and society. It is consciously evaluated, reevaluated, and updated, over and over, throughout every individual's lifetime. Although we cannot speak to the details of how this is done, we know from clinical neurology that the right tertiary parietal regions seem to provide the vast amodal resource we need for this kind of self-modeling. This is the source of the perceptual egocenter, including our updated body image, discussed in Chapter 4. It is also the place where the brain achieves the integration of emotion and cognition. When this area is damaged, we have difficulty understanding gestures, facial expressions, and humor, all of which require us to connect on an emotional level. The semantic meanings that lie deep beneath the surface of language are vulnerable to injury in the equivalent tertiary regions on the left side. Our correct use of grammar can fall victim to injuries of the tertiary regions surrounding the Sylvian fissure, also on the left side. Similarly, other abstract functions, such as long-term planning, self-evaluation, social judgment, and intermediate-term working memory, may be affected by injury to the tertiary regions of the prefrontal cortex, on both sides. Thus the tertiary cortex is implicated in our most wide-reaching mental functions. It controls the highest functions of consciousness.

CHASING PHANTOMS

If our uniquely hybridized conscious capacity was built up in several evolutionary stages, as argued in Chapter 4, it follows that even our most basic conscious capacities might have several vestigial layers hidden beneath the surface. But how can we establish this? It is difficult to verify this idea on a purely psychological level because conscious processes are so complex and interconnected. If we want to gain some insight into the deeper structure of awareness, the most promising route to follow is to explore the evolution of its brain mechanisms. But before we can excavate its ancient brain substrates, we need a conceptual hook, a physical fix on its presence. Admittedly, this looks like an exercise for masochists. When I suggested to a colleague a few years ago that I was about to write a book on consciousness, his only comment was that I must really like pain. Looking for consciousness in the brain is a bit like one of those old silent comedies in which everyone is chasing everyone else, all at once. After all, you might say, anyone actually audacious enough to try to track down the Ghost in the Machine deserves exactly what he will get: nothing. However, I gained some solace from one of my favorite H. G. Wells stories, *The Invisible Man*. It is about a scientist who makes himself invisible. He is accused of crimes, but the police can't see their suspect. Still, his body leaves traces, and so he is eventually found, even though he remains invisible. He gives himself away, in his clothing, his effect on the environment, and traces of his outline. Like a sandworm that displaces the surface of the sand, he leaves a mark. We need an equivalent means of tracing conscious processing in the brain, by watching its indirect effects on a physical medium.

We have one good candidate for that medium, electricity. The brain generates several kinds of electrical signals that can be identified with aspects of conscious processing. One might claim that conscious processing causes peculiar electrical events in the brain or that causality always runs the other way, from electrical states to conscious experience, but this engages us in unprofitable word games. Experience and the brain states corresponding to experience are presumably two sides of the same coin. We are simply seeking a tracking device, some method of finding the coin in the first place. Our main clue consists of certain electrical wavelets that occur in the brain only during conscious processing. In some cases, these wavelets seem to correspond to consciousness fairly closely, almost on a moment-to-moment basis, thus giving us a way to track its trail. The first empirical case for this was built more than sixty years ago, by the physiologists Giuseppe Moruzzi and Horace Magoun. They showed a very close correlation between consciousness and brain wave patterns. At the time their discovery was seen as a revolutionary breakthrough. Brain waves are ionic currents that have their origins in the chemistry of neurons. Neurons are metabolically active and generate currents that wax and wane with global changes in our mental states. Brain waves change with many variables, including excitement, depression, curiosity, surprise, emotion, and concentration. The mere fact of their existence is astonishing and should have discouraged even the most fervid anti-Materialists. (For a while it did seem to throw them off their agenda a bit, but they are a resilient breed.)

This early work was followed by a second wave of studies, led by pioneers of postwar electrophysiology, such as Herbert Jasper and Donald Lindsley, who expanded our knowledge of the systems that regulate human consciousness. Their generation showed that conscious activation is regulated by a hierarchy of structures that exist in the ancient core brains of all mammals. When these systems are blocked, awareness suddenly ceases. When they are stimulated, awareness is instantly restored, even from deep sleep. A sleeping cat could be made fully awake within seconds after its brainstem arousal system was stimulated with a very mild current. When the stimulation stopped, the cat immediately collapsed into a deep sleep, as if a light switch had been turned off. It was that simple. Turn the switch on, and the cat is fully awake, standing up, reacting vigorously to the world, and generating a waking brain wave pattern. Turn it off, and the same cat plummets into a coma, dropping like a stone, its brain pumping out delta waves, the signature of deep sleep. This was a powerful demonstration. However, it was a crude effect and did not show any fine-grained correlation with any specific aspects of consciousness.

We have discovered since then that much finer correlations can be observed in human subjects who suffer from nonconvulsive status epilepticus. This rare condition causes continuous low-level epileptic seizures in some patients. These are usually mild spells restricted to a limited area that occur without the weakness and coma that accompany grand mal (big, bad) seizures. Status patients do not close their eyes, fall down, bite their tongues, or lose muscle tone during their seizures. However, they cease responding to the environment and simply stop being aware for a few seconds or so. Their spells are time-outs, rather like the Roman Saturnalia, the seven days each year that didn't count on the calendar. But unlike the Romans, these people can never celebrate their psychic saturnalias. They cease to exist as conscious beings for these brief slices of time. Then they return. Over and over. It is often difficult to evaluate these patients psychologically because they flicker abruptly in and out of awareness, from second to second, even while being tested.

Insidiously, their condition can be mistaken for an intellectual disorder. To rule this out, tests of their mental capacities are sometimes accompanied by simultaneous recordings of their brain waves. These can verify whether their hesitations and apparent absences may be attributed to seizures. In this procedure, their brain waves are viewed on a split-screen monitor, with their test performance displayed on the upper half of the screen, and their brain waves below. The split screen can reveal a startling second-to-second correlation between awareness and brain waves. I can remember very clearly one such session, recorded more than two decades ago, in New Haven, Connecticut. The patient, a young man, had just been presented with a set of numbers to memorize and repeat. His brain waves betrayed a pattern of almost continuous epileptic seizures in the temporal lobes, most of them less than a second in duration. They showed up on the monitor as intermittent bursts of rhythmic spikes, while his surface behavior didn't show any serious disturbance other than that he hesitated and paused on each trial. He did this so smoothly that we could easily have been fooled into thinking that he couldn't remember the numbers he had been given.

The time correlation between his awareness and his brain waves was perfect. When the spikes were present on the screen, his speech and his actions would halt abruptly for a moment, as if pausing. For example, if this happened while he was reciting a list of digits, he would hold his position, and just as the epileptiform spikes stopped, he would pick up right where he had left off, completing the list he had started, without noticing that he had stopped. From his subjective reports, his experience of time was never interrupted during the session. But to the outside observer, it was. Brief slices seemed to have been excised from his conscious stream. Yet from the inside, he perceived that his experience continued to flow smoothly. This is relativity, with a vengeance! His consciousness was thrown into suspended animation by those brief seizures and returned instantly when they stopped. Somehow, the electrical state of the brain was reflecting the course of his awareness, down to the second. His brain waves revealed his state of consciousness indirectly, as the shifting of sand reveals the sandworm.

This gives us a whiff of the phantom. But can we get any closer? Fifty years ago Donald Hebb speculated that we should be able to find a much tighter interdependence between awareness and brain electricity. He proposed an electrical model of awareness, the notion that the basis of short-term memory is the electrical activity of neurons. Hebb was thinking of a different kind of electrical signature from the one we see in gross recordings, such as those used to monitor seizures. He was speaking of very local activity in specific groups of neurons he called "cell assemblies." Each assembly consisted of hundreds of neurons, connected together by synapses, formed into tiny functional units. To dominate consciousness, all the cells in a given assembly would have to become simultaneously active, so that the entire circuit was engaged. Hebb predicted that cell assemblies might also be embedded in longer chains that could regulate longer, more complex behavior and stay active for longer periods of time. He called these "phase sequences." Research has confirmed many aspects of his theory.

Cell assemblies, or neural nets that behave very much like them, definitely exist in the brain. Under some conditions, the electrical activity of groups of neurons seems to correspond to awareness. One of the clearest proofs of this idea is the phenomenon of retinal rivalry. To produce this effect, each eye must receive a different image at the same time so that the two images compete (thus the rivalry). Show a full-screen image with no blank areas—say, a landscape—to the right eye, and using a mirror, reflect a different image—say, a city street full of people—to the left. As long as there are no obvious blank spaces in either image, the subjects will see only one image at a time, either the right or the left. Intuitively we might expect them to see both images at once, jumbled together in a sort of collage, and this sometimes happens. But more often than not, the two images tend to alternate in awareness. They do this every few seconds, with one image completely dominating awareness, then the other. We know that this reflects a struggle between right- and left-eye bands of neurons. These are wired together in the brain so that they expect to register roughly the same image in both eyes. If the eyes are very discrepant, they inhibit each other's images, and usually one eye or the other will win the struggle, temporarily. This victory is short-lived, however, since the winning eye's neurons become fatigued, and those of the other eye then gain ascendancy for a while, allowing the first to recover. And so on and on. The two eye images will alternate their dominance indefinitely, if the viewers can stand it.

The electrical model of awareness rings truer now than it did fifty years ago, when it was regarded by many as somewhat radical. However, some aspects of awareness are much slower-moving than the electrical discharges of neurons, and need a different kind of explanation. An interesting variation on the electrical model was made in the late sixties by Karl Pribram, who proposed that the physical mechanism of awareness might also involve slower electrochemical events in the brain. These are longer-lasting gradients that reflect chemical shifts in the nooks and crannies of the cortex, especially the so-called dendritic tangles that are found in the upper layers of the cortex. Those shifts can be very complex. Neurons can talk to one another in at least eleven different ways, and complex interactions between different forces can create slow activity patterns that are extremely intricate and subtle.

Slowness and complexity were the very features that interested Pribram. He called these tangled cortical networks the "junctional microstructure" of the nervous system, best imagined as a neuronal jungle packed with clusters of interwoven cell bodies and dendrites. Even a small jungle of this sort would involve millions of connection points, each ruled by particular chemical and electrical conditions. Taken together, these may even generate stable, slow-moving patterns. Pribram proposed that every consciously registered event might be based on one of these microstructural patterns. While this idea has not been fully explored, it is true that complex interactive patterns might endure over time, much longer than the electrical activity cycle of any single neuron, and might better account for the stable, lasting aspects of experience. Slow waves of activity might explain things that Hebb's short-term electrical theory cannot, such as the persistence of intermediate-term awareness, but this idea has proved hard to test, because of limitations in our recording technology. On the face of it, slow, complex electrical fields could maintain the continuity of experience during a nonconvulsive status epilepticus seizure, in which the patient's perceptual world remains fairly stable despite brief bursts of interference from seizure activity. Of course, we would still have to account for the rapid-fire switching of awareness, off and on, with each miniseizure, and for this, faster activity would provide a better explanation. Perhaps the latter provides the vivid foreground, while the former provides the seamless background, of perception.

These two classes of theory have a common grounding in brain electrical activity, and both could be partly right. The traces that dominate short-term awareness might be initiated by short bursts of self-propagating neural networks, but held together over the longer run by slow activity in junctional microstructures. This view would make awareness into a tournament of electrical knights, where the highest-placed, longest-lasting combatant carries the day. Inevitably, this still leaves a great deal out. Neither class of theory can explain why one active neural network should ever gain a competitive advantage in awareness over another equally strong one or why we couldn't become aware of more than one of them at the same time. We have already mentioned the subjective tunnel of awareness, the bottleneck that allows only a few things through at any given time. This is a well-documented limit to our capacity. But why does this limit exist? Why can't we have an infinity of electrical traces active in the brain, all present in awareness at once? This seems to happen sometimes, but only as the effect of hallucinogenic drugs or damage to the brain. This chaotic, free-running state of mind is very rare, and extremely undesirable, in normal consciousness. This fact needs explaining.

These two theories also have difficulty accounting for our ability to become aware of very weak stimuli, in the presence of stronger ones. Weak stimuli almost invariably produce fainter bursting in neurons. Thus the neuronal assemblies they trigger should be weak and easily overwhelmed by any strong stimulus. But this doesn't necessarily happen. Under the right circumstances, faint stimuli can be highly compelling and even dominant. Two decades ago neuroscientists Terry Picton and Steven Hillyard showed that an omitted stimulus in a train of stimuli can trigger as robust a brain response as can an actual one. Their results have been replicated many times. This might be called the clock-not-ticking effect and is at the extreme end of the continuum, but it is common to experience the dominance of a faint stimulus. This effect can be huge. Take the standard adolescent fantasy of an unexpected amorous message, whispered into a receptive ear in a dark theater. This faint whisper, once registered, will overpower any sound the theater can produce! The content of the message apparently turns up the volume of the brain's response, even though the physical magnitude of the sound is small. This can happen only because there is a powerful internal selection mechanism inside the brain, tinkering with its ionic ebbs and tides.

Many laboratories are seeking the neural mechanism of this kind of selection, this fine-tuning of awareness. If such a mechanism could be linked directly to working memory, our model of conscious awareness would start to get really interesting because it might explain how the brain chooses certain experiences over others, in the short term. If this same circuitry could be brought under the direct control of a symbolizing process, the model might even go some distance toward explaining the longer-term, and more esoteric, features of awareness. There is already some good electrical evidence on the selection process. This has come from studies that are much finer-grained than the ones already described. These experiments follow the effects of conscious selection further into the system, closer to the source of awareness itself. The technique used, averaged evoked brain potentials, takes advantage of the fact that every sensory stimulus leaves a particular signature trace in the brain. This trace is an ionic disturbance created by the stimulus as it travels through the brain. In this respect, it is rather like the wake left behind by a motorboat as it plows across a lake. As the event travels through the brain's networks, a chain of local electrical storms branches out from it into a number of ganglia and nuclei, culminating in a complex pattern of activity in the cerebral cortex. These traces reflect what people experience and how they behave with remarkable fidelity. Even traces that never reach awareness and extinguish soon after they enter the brain leave a faint trace. These are the casualties of experience. But those that reach consciousness leave a characteristic pattern of waves, the signature of awareness.

The ear provides a good model for demonstrating this effect, and it has been studied widely. When a sound hits the ear, it produces several wavelets of electrical activity, first in the organ of hearing, the cochlea, and then in the brainstem, as the central nervous system begins to register the sound. The trace then diverges, traveling to several places in the brain concurrently, including the auditory cortex. This happens even when we are deeply asleep. During sleep the trace is fleeting and weak, restricted mostly to a few regions of the lower brainstem and midbrain. When we are fully awake, the same sound elicits more than this, including a measurable cortical trace, even if it is not consciously registered. If, however, the sound reaches consciousness, its trace is larger and different in shape. Under some conditions, a stimulus will have a tremendous impact on several cortical areas simultaneously. Attention controls this process and allows the brain to magnify or diminish it, as it chooses.

Steven Hillyard and his associates in San Diego were the first to prove this conclusively, about twenty-five years ago. They observed amplitude differences of about 100 percent in the brain's electrical response to sound that could be attributed to the voluntary shifting of attention. I tinkered with the parameters of that same experiment until the effect was maximized and found some remarkably robust attention effects, which revealed large discrepancies between wavelets evoked inside and outside the attentional focus. Some differences were in the order of 500 percent! In other words, competing sounds of equal magnitude can produce very different results in the conscious brain, depending on where attention is directed. Some of these wavelets never occur at all, outside of awareness. Further research by Marta Kutas, Emanuel Donchin, and their colleague Greg McCarthy confirmed the ties between some of these wavelets and a different aspect of executive function, stimulus evaluation. These wavelets were linked, within a few dozen milliseconds, to the time at which events had been perceptually registered. Taken together, these studies provide evidence of fairly tight temporal linkages between conscious processing and certain kinds of brain electrical activity.

This points to a neural gate that controls access to awareness, editing experience at high speeds, quite near the time point when sensory inputs enter the cortex. It has proved very difficult to locate the exact source of many of these electrical wavelets, and for many years this haunted those who worked in the field. In the heyday of localization theory it seemed to be a crushing limitation not to be able to pin these wavelets down to one particular place in the brain. More recently, however, this apparent weakness has turned into a strength. We now realize that awareness engages many brain regions simultaneously and that most of the networks of conscious regulation are distributed widely across several brain areas. Thus we should not expect to localize every wavelet to a fixed position in the brain. More likely, most complex experiences share common neural resources. The nervous system may use the same bit of tissue to handle many different challenges, embedding it into many different functional networks. In this case, we should not expect highly localized activation during complex performance. Instead we should be trying to identify and follow the wider networks that support conscious processing. This will require a paradigm shift. Cognitive psychologist Michael Posner has called for such a shift, suggesting that the same basic set of executive brain structures may be engaged in many different mental tasks.

The most promising direction for future research in this field will be tracing entire functional architectures, or networks, in the nervous system, as they are carrying out their duties. Rather than look for localized modules, we should be using brain-imaging techniques, such as functional magnetic resonance imaging (fMRI), to identify entire networks. Unfortunately, these technologies are not yet fully developed. Moreover, existing methods are not always carefully used. In the interest of easy results (and publicity), experimenters sometimes create the illusion that highly localized brain regions perform such complex functions as reading and speaking. Images of cognitive "hot spots" in the brain abound in the popular literature. These brain images are obtained by means of elaborate data laundering. This involves adjusting thresholds and subtracting control from experimental conditions.

This seems reasonable, but it exposes the results to a serious risk of misinterpretation. For example, we might try to localize reading in the brain by subtracting the brain image obtained while the subject looks at meaningless letter strings from the image obtained while he reads sentences. The difference between the two images would supposedly represent additional processing associated with meaning. The technique assumes that if we subtract the brain activity for viewing letters from the averaged reading response, we would be left only with brain activity for reading. This is a questionable assumption because it takes for granted that seeing and reading are done by two distinct processes, each independent of the other. However, this seems rarely to be the case. Additivity (the idea that each distinct operation adds a new brain circuit to the pattern of activity) is a highly questionable assumption because many brain circuits are reentrant in nature. This means that the same network could, in theory, be used and reused for a number of different cognitive agendas. This is devastating for the additivity assumption. The brain might read with the very same networks it uses for many other visual tasks, perhaps with a few extra neurons thrown in to increase its power. In that case, most of those spectacular brain images that appear in the newspapers every day, supposedly localizing such functions as reading, remembering, and thinking, are probably artifacts of subtraction. They are impossible to interpret at this stage.

There is one finding in this literature that is very solid, however. The harder we concentrate on a task, the brighter the brain image burns, regardless of the task. Effort equals more metabolic activity, involving a wider set of circuits. It is not obvious why this should be the case. What does this extra activity achieve? If I play the same piano piece, once with great conscious effort and again absentmindedly on automatic, it is the same performance after all. Embarrassingly, it may even sound better without so much effort, with the fingers executing the same actions. So why is the brain image so different with conscious effort? What does the concentration of effort do for us?

The answers to these questions depend on how we conceptualize the role of the executive brain system. Its wide interconnectivity would explain how it can activate all the regions of the cortex simultaneously. When we consciously register something in the world, the central attention system can light up virtually the whole brain, like a Christmas tree. It can also make the same areas go suddenly dark when we deliberately suppress the outside world and direct attention inward. Significantly, however, the same brain areas are activated, over and over again, when conscious capacity is engaged, no matter where it is directed. Brain scans taken during effortful tasks, such as listening to the news, doing math, or reading the paper, can show activation of roughly the same regions used when reciting poetry or mentally rehearsing a golf swing. This is due to their common dependence on the executive brain. They might also show a task-specific pattern that remains active after performance becomes automatic. The executive component will usually drop out after enough practice, when less effort is expended. But difficult tasks, whatever their nature, engage the brain's executive system until they become easy. Then the task-specific components will take over.

Conscious effort is more than simply increased metabolic activity or diffuse activation of more brain circuitry. These ideas were tested long ago, and it is clear that the mechanism of cerebral effort is much more specific than either of these very general principles would allow. It seems to provide the extra computational resources needed while we are closely evaluating and reviewing our perceptions, memories, and performances. It also directs this review of performance, focusing on the task at hand, whether this involves forming a better perceptual image, coordinating action, solving a conceptual problem, or fixing a memory in place. Concentration, with all its curious stable of eccentric habits, such as furrowed brows, worried frowns, pacing, clenched fists, and so on, actually changes the way the brain does business and gives us a clue to what parts of neural anatomy should be included in the executive brain.

This returns us to the core question about the functional role of conscious capacity. The most important effect of cerebral effort is that it establishes temporary functional networks in the brain, in order to solve various immediate problems. These networks can be remembered, and recalled, and can effectively rearrange the geography of processing in the nervous system. The brain that solves your math problems on a calculus exam is a different brain from the one that negotiates a social network, inasmuch as it is temporarily set up very differently. This may explain why it is often so difficult to switch gears from one mind-set to another. When we exert conscious effort, we narrow our focus, which means that we organize our brains around a single project for a long period of time. Once we have done this, it is not easy to switch. When we do switch, we have to reassess the new status of the world and "wire ourselves up" accordingly, in a new way, to face a new task.

New York researcher Michael Posner has studied this process of temporary network rearranging during performance, using brain imaging to identify the subcomponents of conscious processing. He has found that subjects rearrange their mental circuitry in distinct ways for different classes of performance. Reading requires the participation of certain regions, hooked up in a certain sequence. Writing may rearrange the same circuitry, using roughly the same components. Speaking out loud triggers a somewhat different cerebral pattern, with some new components and some old ones, because it taps some of the same components used in reading. This applies to many types of mental work. During a period of effortful switching, as we shift from task to task, the brain will actually generate a temporary operational system appropriate to each performance. This implies that cerebral effort involves a great deal more than simple activation. Its effects can also be structural.

Somehow, in ways that we do not yet understand, conscious effort helps set up these temporary functional networks and assembles a strategy for dealing with the problem at hand. The brain self-organizes toward achieving these goals, and conscious capacity is important in allowing it to do that. This may sound a bit like simply opening up a new software program to do a particular job, but it is not so open-ended. The brain must carry out its functional rearrangements within the strict confines of its built-in anatomy. For this reason, what it does is probably less like opening software and more like temporarily rewiring its own hardware. This process occurs at many levels in the nervous system, affecting, at one end of the continuum, the smallest minicolumns of a few hundred neurons and, at the other, those vast cross-hemispheric networks of millions of neurons that we need for semantic memory.

This may not give us a tight causal chain from atom to neuron to awareness, but it is a start. The electrical activity of the brain is ineluctably tied to consciousness, and conscious effort is the single most reliable predictor of the patterns of brain activity. Metabolic imaging confirms that conscious processing plays a central role in shaping the brain's activity patterns. It also reveals the existence of a system whose job seems to include setting up those patterns to cope with the day-to-day tasks we face.

However, there is one compulsory caveat here. The conscious brain can never become aware of itself, and we should not allow ourselves to be drawn into pointless debate about why the boundaries of awareness shouldn't encompass the brain's own activity. There is simply *no* direct awareness of the brain's activity and no possibility of achieving it. Brain activity is the end of the line. It is the source, never the object, of direct experience. Asking why these strange ionic ripples can make us aware of everything but themselves is pushing the question too far. There are limits to science. Although it seems to evolve and expand endlessly, at any given moment it is finite, like space. We must respect that limit, wherever it may fall, in each generation, even though we may be impatient for a complete theory. At the moment we still don't know why there is a strong gravitational force or why certain macromolecules come alive. For now we must accept that there just is and they just do. Brains that pulse with certain patterns of electrical activity are conscious. Why? They just are.

LEVEL-1 AWARENESS: SELECTIVE BINDING

Have brain waves helped us determine the underlying structure of the nervous system that generates it? In truth, they have helped enormously. Those electrical phantoms have led us directly to a new theoretical universe in our understanding of consciousness. The most interesting new theories have combined the results of several other fields, including cognitive science and computing, with electrical data. This is a turbulent, creative literature right now, full of piss and vinegar, as it should be, promising to deliver better and better theories. It would be presumptuous, indeed impossible, to give any kind of balanced or comprehensive review here. It would also be premature since these theories have a long way to go. Nevertheless, I shall single out certain important ideas and try to place them into a unified evolutionary framework (when all else fails, chronology is a great organizer). I shall cluster our knowledge of the basic mechanisms of consciousness around three levels of brain evolution, corresponding to three levels of capacity. These levels emerged in a certain evolutionary sequence, establishing the underlying structure of human awareness. They make up the physical half of our unique brain-culture symbiosis. They determine the shape of human conscious cognition and form the core of the brain adaptation that enabled the Executive Suite to evolve, with its ensuing cultural revolutions.

The first component of this hierarchy is a phenomenon known as binding. Binding is the theoretical basis of object perception or, more accurately, the neural means of attaining perceptual unity. The mechanism of binding is ultimately responsible for our ability to perceive complex things, such as objects and events. Since events are the very stuff of experience as we know it, we might take the binding problem as a paradigm for pondering the larger question of how we extract such exquisite order from the seeming chaos of light rays scattered about the environment and the myriads of sound waves accosting our ears. Since the days of the Gestalt theorists, we have known that perception is an abstract cognitive achievement and not a simple matter of interpreting an already crystal-clear physical stimulus. Objects and events are not given directly to the eyes and ears, as are color, loudness, and brightness. They must be sought out and derived from a very noisy barrage of physical energy. Before any species can become sensitive to such abstract aspects of the world, it must evolve the brainpower to detect the existence of things that hide behind raw sensation. It must be able to find patterns in space and time that reveal the existence of objects and events.

The capacity to do this is not present in most living creatures. Most primitive nervous systems are attuned only to the crudest, most global dimensions of sensation. A slug, for instance, is sensitive only to the pull of gravity and the push of light. It tends to climb up and hide from light. Place it on a tilted board, half in shadow, and you can predict exactly where it will go. Increase the tilt, or move the shadow, and it moves accordingly. But it doesn't care about the finer-grained aspects of its world. It does not have the neural means to discriminate complex things, such as people and trees. It only has simple tropisms and the sensory detectors it needs to drive them. Slugs can't perceive flowers, walls, or rocks as such. They can't even perceive other slugs.

Objects can only be perceived by combining discrete features into a unified percept. To construct a percept of the red apple sitting on my desk, I must be able to bind together its color and form, which are processed in different parts of the cortex. I must be able to combine simultaneous sensory inputs in two different brain regions just to perceive the red apple. In reality, this is an oversimplification. To be really sure of its identity, I would probably have to bind much more information than that. I might have to adjust for its size (it might not be a real fruit, like the giant apple sculpture standing on a highway near my home), hardness (it might be a decoy, made of plastic), texture (the lighting of my study may distort its appearance), color, and so on.

I shall assume that the raw feeling definition of awareness is an appropriate description of the most basic level of awareness. Given this assumption, binding is the logical grounding for most raw feelings. It is the basis for all the more abstract forms of conscious event perception. Put simply, without a binding mechanism, awareness would have no structure. Objects could not be perceived as coherent entities that remain constant under a multitude of different viewing conditions and contexts. A binding mechanism is essential for any sophisticated perceptual experience, especially event perception. Events that endure for several seconds are not easy to synthesize into single, bound percepts. For example, I perceive the event "mosquito biting my hand" as a unit, with natural boundaries and an internal structure that is easily differentiated from other concurrent happenings, such as my attempt to slap the same mosquito. Thanks to the design of our brains, incredibly complex event percepts such as these are delivered to our awareness in neatly structured packages. Our mammalian genius for parsing events must have started with a simple binding mechanism and might be an extension of binding.

Could binding be unconscious and automatic? This is possible. Our best experimental data on humans and other mammals suggest that even the most elementary kinds of perceptual binding demand attention. But some kinds of perceptual binding may not require it. One instance where there could be some form of unconscious binding is event perception in insects. They do manage such things, believe it or not. In the tail-wagging dance of the bee, the visual stimulus is complex and multidimensional and must be integrated over time. A bee observing another dance must recognize (1) that the dancer is a bee; (2) that the choreography is impeccable; and (3) that it is pointing in a certain direction. This is not a simple perception, and it requires a binding process of some kind. The bee's dance has a certain spatiotemporal shape, and determining its underlying vector is a very different perceptual challenge from figuring out the species of the dancer or the direction in which it is pointing. Three kinds of information must be combined into a unified percept. If this isn't binding, I don't know what is. Yet few of us would argue that a bee is conscious in the sense that mammals and birds are conscious. Bees are fixed, narrow, and obviously unable to perceive many events that any mammal could easily perceive. They seem to bind certain patterns automatically, without any attentional control over the process.

The origins of the mammalian binding mechanism seem to be related to the evolution of attention. In fact, the mechanisms of selective attention may have evolved initially to enable the brain to synthesize more complex objects in a more flexible way in a variety of environments. Many researchers have seized on this relationship between binding and attention as a possible fast track to solving the neural basis of consciousness. In the literature on brain electricity, the term "binding" often refers to a hypothetical integration process, driven by attention, that ties the bits and pieces of information the brain receives from the environment into a unified circuit. Sensory physiology has shown in some detail how the brain initially sorts out various aspects of the sensory world. Different areas of the visual cortex become active simultaneously, extracting the form, color, self-motion, object motion, orientation, and eye of origin of an object in different places with different electrical codes. This presents the brain with a complex circuit to bind together. If we see a green grasshopper jumping in the grass, the larger textural frame of the image (the lawn) is analyzed in one area of the visual cortex, while the finer-grained foveal pattern (the grasshopper) is analyzed elsewhere. The colors and patterns of movement are extracted in other brain regions, while the sounds associated with the grasshopper, the smell of the grass, and the feelings experienced by the observer are registered in others. Somewhere, somehow, all this electrical activity is bound together, and the result is an integral percept. Remarkably, the brain does this very quickly, usually within a fraction of a second. Computational simulations of events such as this are notoriously bad at making similar judgments. They are slow, and even if they could achieve such integration, they would probably have to crunch numbers for several hours, after which they would print out something like the following message: "There is a 75 percent probability that there is a grasshopper on the lawn, a 60 percent probability that it is a praying mantis on a carpet, and a 30 percent probability that it is a backhoe excavating a distant swamp." So they would go, on and on, listing other possibilities. Yet we see the event instantly for what it is. Binding is fast, clear, sharp.

This seems to occur in two stages, a fast, "early" stage of seeing that instantly extracts features and the basic shape of the scene, and a slower "late" stage, taking half a second or more, that extracts objects and events. We cannot yet explain in detail how the brain achieves either of these stages, but as MIT vision scientist David Marr observed long ago, it is obviously designed to deliver bound percepts as its end product, rather than just provide us with disconnected fragments. It is clear that binding itself occurs in the primary and secondary regions of the sensory cortex, under the control of the thalamus, as well as in some regions of the tertiary cortex. One theory holds that the right and left hemispheres serve complementary binding roles in objects.

Given that different parts of an event activate different parts of the visual brain, how is such a circuit unified? One popular theory holds that perceptual unity is mediated by relatively high-frequency brain waves that are rhythmic; that is, have semiregular oscillations. The oscillations occur in the high gamma range and are called 40-Hertz rhythms because they oscillate at a rate of forty times per second or more. This oscillation occurs only as long as the object remains in awareness. This shared oscillatory frequency seems to give the neural circuit registering the stimulus a temporary "coherence." This makes the entire circuit stand out from any competing neurons, which tend to fire away on their own without becoming part of an oscillating cell assembly. Many researchers now believe that these bound circuits could conceivably be the "atoms" of conscious experience.

Francis Crick, the codiscoverer of DNA, and his colleague Christof Koch have pursued the 40-Hz, or gamma, hypothesis and made the absolutely crucial claim that certain classes of perceptual binding are dependent on attention. They concede that the simplest forms of binding, such as those involved in defining a boundary between two points in space, might occur without attention. But they believe that even the simplest act of object recognition requires active attention. Attention thus directs any complex binding process. The most basic function of the attentional searchlight is to bind the disparate elements present in a complex stimulus. This idea converges nicely with psychologist Anne Treisman's earlier suggestion that attention is needed whenever we try to conjoin two or more features in object recognition and with linguist Ray Jackendoff's notion that consciousness intervenes in perception at an intermediate level; that is, as soon as several features must be interrelated. This might be a good candidate for the ancient adaptive function that might have led to the evolution of attention in the first place. Since attention is at the very heart of conscious capacity, this may be the starting point we are looking for if we want to develop an evolutionary theory of conscious capacity. *Conscious capacity might have appeared originally as an adaptation to increase the power of perception.* Or if we concede the bees their smidgen of automatic binding, we might prefer to set the bar higher and link attention, binding, and perceptual learning. Selective binding is essential for perceptual learning, and most laboratory tests of binding make use of this fact.

It follows that selective binding should be included on our list of basic executive functions, as a by-product of attention. It may even be the more elemental of the two. It is also obviously an adaptive process, one that would confer major advantages on any organism that evolved it. A capacity to perceive objects and abstract events is useful in virtually any environment, for all those typically Darwinian challenges: detecting food sources, noticing predators, understanding complex topography, monitoring the environment, and so on. It is also a long stretch to see this as the sole physical basis of all complex perceptual experience. But science is full of surprises, and this could be another instance of theory taking us into realms that were formerly unimaginable.

But binding theory still has to overcome some major obstacles. For one thing, it seems to fall back on the same old conceptual trick that I have already criticized. Conscious circuits are supposedly built simply by linking unconscious elements, active neurons and cortical columns, by making them fire, or oscillate, together. But why would such oscillations, in certain neural circuits, generate awareness? What is it about these oscillating circuits that might pluck them from unawareness? Hackneyed though it may seem, I find myself falling back on John Searle's famous attack on AI, with an allegory he called the Chinese Room, about which an inordinate amount of ink has been spilled. His fictional room appears to understand messages sent into it, regardless of the fact that it contains only one person, who understands nothing of the messages because he doesn't speak Chinese. However, he has an instruction manual telling him how to "answer" various written ideograms as they appear by sending back certain other ideograms, which he doesn't understand either. He does this so faithfully that he creates the impression on the outside that the room is intelligent. This, says Searle, is all computers can do. When cleverly programmed, they create the illusion that they are intelligent. The philosopher Ned Block pushed that allegory even further, by proposing a room in which there were teams of such people, none of whom understood the language of the messages, but whose collective skill at following instructions created an even better illusion of intelligence. This he called the Chinese Gym.

We may be tempted to snicker at the presumptive consciousness-raising powers of the gymnasts in some putative neuronal Chinese Gym, tuned up with a rhythm or two, with their endoskeletons aligned by a few proteins. Suddenly their collective singing elevates the social superorganism to consciousness. Somehow, a bound cluster of uncomprehending hordes becomes aware as a group, despite the unconsciousness of the units that form the group. This is a megastretch, and has understandably been scoffed at. However, many theories in the physical sciences push our credulity to the breaking point. The idea of a bound neuronal circuit is surely no stranger than relativity, black holes, event horizons, or the bacterial theory of disease must have appeared when they were first proposed. Nobody said our expedition into the darkest corner of the mind would be easy, familiar, or reassuring. Haldane probably understated the case in saying that the universe is queerer than we can imagine. Its queerness will undoubtedly exceed all expectations when it comes to our own minds.

Neural binding, as presently understood, seems to be a small local phenomenon and the present form of binding theory may prove inadequate to account for the broader geographical effects needed in the real world of event perception, where we might have to bind together all the sensory material needed, say, to follow prey running across a field and into a specific burrow. Bound percepts regularly occur on the fringes of consciousness, in automated tasks such as driving and talking, in which we routinely execute very complex operations without allocating much, if any, attention. These tasks run smoothly because our automatic brain mechanisms can assume the unconscious registration of fairly complex events, such as spoken words, grammatical correctness, and localized traffic signals. If such things can be registered unconsciously, how can we maintain that binding is the exclusive product of attention? This can be answered if binding can be embedded into an automatized skill, whereby it will not always demand capacity. Any paradigm that demands a conscious response, rather than an automatic one, will link binding and attention. But attention may be needed only for initial learning. After this, we might be able to achieve the bound registration of a familiar stimulus unconsciously in a routine task, such as reading.

In sum, the primordial emergence of consciousness took the form of selective binding, guided by attention. This could not in itself have predicted its more evolved forms, which have had such a determining influence on the shape of human cognition. But cognitive evolution is a matter of building upon existing strengths. Conscious capacity, which amounts to the domain-general realm of cognition, had to start somewhere, and key elements of conscious capacity seem to be present in simple binding. The dependence of binding on selective attention in mammals suggests that fairly early in its evolution, two crucial ingredients of the executive system, selection and amodal integration, were already in place.

LEVEL-2 AWARENESS: SHORT-TERM CONTROL

The second component of this tripartite model is short-term working memory. Simple binding is restricted in its application to the raw feeling definition of awareness. It does not address the issues of governance and control that dominate research on mammalian cognition. As we saw in Chapter 4, the attribution of conscious capacity to higher mammals hinges on short-term working memory, which is where controlled processing, as we know it in humans, probably begins. Perceptual binding, in itself, is stimulus-bound and concrete. Short-term working memory grants a species autonomy from the tyranny of the immediate environment. Although the evolutionary momentum for this was probably to extend the mechanisms of binding into a wider time frame, it effectively changed the nature of awareness because the mental models it was able to generate extended perception into the domain of simple events. It also introduced the possibility of more than one active focus. Awareness became able to include one focus of sensation and a concurrent one for working memory. Working memory is best regarded as the second functional level of conscious capacity because it follows logically from the first and introduces some new functions that are ancillary to binding itself.

Moreover, our knowledge of comparative ethology confirms that working memory appeared later in evolution than did simple binding. An elementary binding mechanism can only synthesize and parse whatever the immediate environment presents to the brain. Level-2 conscious capacity goes one step farther. This can be seen clearly in the emergence of a capacity for delayed response, which is found in many mammals, including rats. Species with this ability can hold an image or memory in consciousness, for some period of time, independently of what is going on around them. They "carry around" their images and thoughts. Short-term storage also allows the perceptual world to expand along a temporal dimension. Longer perceptual events and scenarios can only be bound into unified percepts with the help of a working memory system that can hold the various components of an event, which are experienced over time. These are the kinds of perceptual problems in which pieces of a puzzle must be kept in mind while the solution is sought. A large landscape is a good example. You cannot fixate the entire landscape at once. Say you are looking at a waterfall, a mountain, and a grove of trees. To resolve this scene into a unity, all these elements must be examined in a series of visual fixations, and these must be held in mind somehow before the pieces can fit together. The final resolution of the scene is not simply stored away in short-term memory, however. It is a direct perceptual resolution, an image that seems stable but is never wholly present at any one time. The landscape has a feeling of solidity, even though it cannot be seen all at once. This principle also applies to complexity within a single frame, which can require many fixations to resolve. Look at a village nestled in a river valley from above, and there is usually no way to resolve it, even though it can be seen in a single glance. Many fixations are necessary within the visible scene, and these must be assembled into a larger percept, just as the landscape was. These multiple-scan percepts place demands on short-term storage.

Social events are even more demanding, if they take any time to unfold. An extended social event, such as a long courtship between a male and female lion, takes considerable time and cannot be perceived without integrating a number of apparently disjointed component segments. Such events could not be perceived accurately from the vantage point of any single set of frames because the meaning of such frames is not always clear without reference to the overall scenario. To unify something as perceptually challenging as a courting ritual into a perceptual unit, the brain needs a storage mechanism that can temporarily hold, or buffer, the various frames that make up the sequence, so that the sequence itself can be broken down and parsed.

In itself, this evolutionary development did not necessarily represent a qualitative break with the deep past. Even a primitive binding mechanism contains within it the germ of short-term memory. Crick and Koch acknowledged this, pointing out that even the most rudimentary forms of object binding require some form of very brief storage. Moreover, in his original cell assembly proposal, Hebb argued that the same neuronal assemblies that could serve for perception must also serve as the basis of short-term storage. Thus binding and short-term memory must be closely related. But this relationship shifted, between level-1 and level-2. In the latter the balance of power shifts from binding to short-term memory. Working memory is active storage, involving evaluation and problem solving. The short-term store in this sense is the place in the mind where active mental modeling takes place. This is epitomized by the delayed response paradigm, alluded to in Chapter 4. In that paradigm, a stimulus is presented, then taken away. After a brief delay memory is tested. If the animal can perform the correct response after a delay, it is said to have short-term memory.

In its simplest form this kind of storage amounts to sustaining the perceptual trace in the absence of the stimulus. But in a more demanding delay paradigm the animal might have to remember how to respond despite the presence of conflicting stimuli. This involves more than merely sustaining the existing pattern of brain activity because something other than the binding mechanism itself would have to decide where to attend, which memories to sustain, and how to apply them to the task at hand. In this sense, working memory is equivalent to the older concept of the central processor, with most of the basic executive capacities we identify with advanced mammalian controlled processing. These include evaluation, selection, problem solving, and response choice.

Theories of consciousness tend to combine research on selective attention and short-term working memory. On level-2, attention is really only the gatekeeper aspect of short-term working memory. This applies not only to experience but to short-term memory storage itself, which is filtered by attention. This opens up a whole new set of questions and calls for a larger anatomical circuit than the one needed for binding. The brain structures involved in short-term memory have been widely studied, and most theoretical models emerging from this literature focus on those parts of the mammalian brain that play a role in executive capacity. They say little about the crucial question of what exactly is going on in each of these brain regions or of how neurons interact microscopically to regulate most aspects of cognition, including perceptual binding. But they specify the overall wiring diagram, or the gross anatomy, of the executive brain.

Michael Posner and his colleagues have used metabolic imaging to record how these different regions of the brain are activated during the performance of different tasks. Their main interest has been precisely those short-term functions I identify with level-2 conscious capacity. They have proposed that the typical laboratory task can be broken down into a series of simple operations, each of which can be localized within a different part of the executive system. Posner's executive network contains several subsystems. One of these is a spatial attention network that regulates eye movements. Eye movements are so crucial for primates, including humans, that they tie up at least three of our twelve cranial nerves and are controlled from at least three different cortical regions. They involve several discrete executive operations, each with a different cerebral substrate. The first of Posner's eye movement operations is called Disengage. It allows the viewer to break with the current fixation point. This is not as easy as one might think, and neurological patients who suffer from a disorder of this system often report that the things they are looking at have a "magnetic" attraction that they cannot resist, and they cannot look away from their current fixation. Disengagement appears to be controlled from the right parietal lobe. The second operation is the command Move. It initiates movement and guides the eye to a new location. It is controlled from a midbrain structure called the superior colliculus. The third is simply called Enhance. It sharpens the image where the eye lands. It is like the focusing action that might follow a panning shot with a movie camera, except that it is not so much a matter of focusing the eye as of increasing the internal contrast settings within the brain. It is controlled from the pulvinar nucleus of the thalamus.

These examples show how complex the elementary suboperations of the executive system have become in primates. Every voluntary eye movement, no matter what we are looking at, is driven by this standard sequence, disengage-move-enhance. The sequence is programmed by three successive bursts of activity, in the parietal lobe, superior colliculus, and thalamus. This has been verified with electrical recordings, positron emission tomography (PET), and magnetic resonance imaging (MRI). Although this sequence undoubtedly has deep roots in our primate past, the executive control of gaze might be somewhat distinctive in humans, because we exert so much voluntary control over its use in fine-grained skills, such as toolmaking and reading.

We can break down any learned visual skill into a series of component operations, controlled by the brain's executive networks, which regulate information flow between various specialized modules. This produces a formidable flowchart of control for even the simplest visual tasks, as demonstrated by neuroscientist A. R. McIntosh of the Rotman Institute in Toronto, in his dynamic analysis of brain images. The drawings in Figure 5.4, derived from PET scans, show very different flow patterns between the same right brain regions (each marked with a conventional Brodmann number tag) when subjects are locating, as opposed to identifying, objects. This figure shows that the pattern of activity in the right hemisphere changes with the nature of the task. Solid lines indicate positive flow, and dotted lines indicate negative flow, or inhibition. The width of lines reflects strength of influence. Note how the temporary flow between the same areas differs dramatically with the task. For example, during object recognition, area 46 (in the prefrontal cortex) has a negative influence on area 19 (in the secondary visual cortex), whereas during spatial vision it has a strong positive influence.

If all this seems too complex to believe, keep in mind that real-life tasks we routinely undertake—driving a car, playing tennis, carrying on a conversation, watching TV, cooking a meal—could not be performed without equally complex arrangements of elementary operations. At first glance, we

Figure 5.4

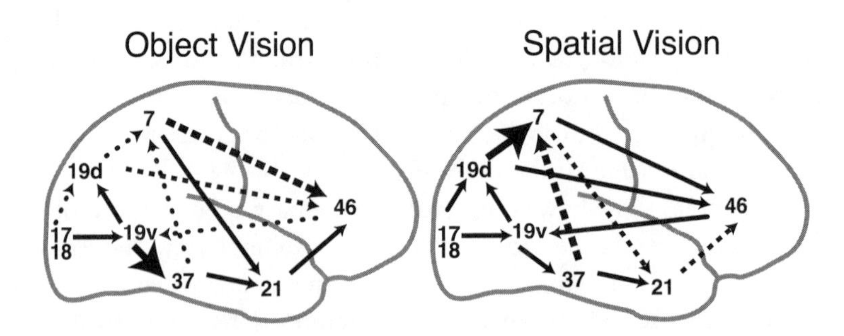

may appear to engage so many areas of the brain that most tasks may seem to be using all of it. Of course, this isn't true. Each task has its own specific circuit pattern. But it is a fine-grained pattern and cannot be contained in a single anatomical module. To describe such circuits clearly, we need to recognize that conscious mental activity can alter temporary flow patterns in the brain and reshuffles the same basic set of neural cards in accordance with the task.

The key to short-term executive governance is thus temporary network rearrangement, or the construction of temporary functional architectures. The brain must set itself up one way for reading and another way for listening. Another arrangement is needed for playing cards and so on. In the short run, this works very much like the short-lived divisions of large corporations, established for, say, designing a new aircraft or carrying out a construction project. They are set up, used, and then set up in a different way, after they have performed their function. The executive brain is constantly involved in any number of different tasks, and it has to reshuffle the temporary network architectures needed for each of them, using and reusing the same central control functions, over and over, in various combinations.

To return to our eye-scanning example, reading is not the only task that requires a highly programmed series of eye movements. All detailed visual work, from engraving to watch repair, from botanic classification to computer design, from tennis to flyswatting, places demands on roughly the same system, but each demands a different sequence of programmed control reactions, with vastly different parameters. Eye movement control is not just devoted to a single specialized set of skills or operational tasks but may be used to set up and optimize many skills. The executive brain seems to behave very much like a cerebral CEO in all this, guiding an elaborate managerial apparatus that sets priorities, switches between programs, and maximizes the flexible use of limited cerebral resources for a wide variety of operations. This level of control is necessary only in species that have significant learning capacity. Massive memory demands massive management down on the ground in the short term, and working memory must contain whatever knowledge is relevant to the situation.

This kind of executive management also includes a major suppressive component. It is important to inhibit irrelevant or undesirable emotional, attentional, and other impulses. Inhibition of emotion is certainly one of the most difficult aspects of effective executive control. It is crucial to be able to suppress strong emotions that might compromise one's agenda. This is epitomized in the Spy Who Came In From the Cold syndrome. In John Le Carré's novel, the spy assumes a false identity, lives a life of elaborate deception, and then, in the end, desperate to return to normalcy, tries to shuck the acquired identity, with tragic results. The key to success in this is suppressive control. Assuming a false, deliberately deceiving identity is a very difficult conscious act (ask any good actor). It requires incredibly, some might say unbearably, sustained attention to every detail of one's actions, and in the case of espionage, this vigilance has to be sustained for years at a time. Immigrants and refugees also create new identities, although in their case it is not to deceive but to fit in. To achieve this metamorphosis, they have to attend to the tiniest details of life and, more important, suppress their habitual responses. For instance, they might no longer wish to respond to the sounds of their real names or to a host of other giveaways (their spouses' names, national symbols, music, facial expressions, and gestures, to name a few). They must respond instead to new names and identities and associations that are much less firmly established than deep childhood memories. This is an extremely difficult task.

Because it is essential for emotional deception, the suppression of deep expressive habits might even have been one of the original adaptive functions driving expanded executive capacity in humans, one of our specialties as a species. Emotional deception, in the form of temporarily suppressing what is the most obvious, or appropriate, response to a situation while cold-bloodedly manufacturing a false one, may not always be socially correct because it is perceived by everyone else as dangerous. However, it is a very effective social weapon. Suppressive control may be the best proof of which is really in charge of your actions, the unconscious or the conscious. It is a bit like acting, but with real consequences. If you act poorly in a drama, you usually don't die for it (or if you die, you die metaphorically, not literally). But overcontrol is also dangerous. Eventually the assumed identity may dominate the original one. Then, as in the case of our spy and not a few actors in real life, you will face an existential crisis of the first magnitude. Too much self-control can be emotionally counterproductive; we need a close connection with our deep feelings, just to recognize ourselves.

Inhibition may be important and possibly primordial in short-term control, but another executive function, selective attention, is even more critical. Without it we could not navigate our way through an hour's worth of social activity in a society as complex as ours. Selective attention allows you to focus on one dimension of the world, say, the violin section of the orchestra, to the virtual exclusion of the others. It is the bedrock of all complex perception because this involves choices between many alternatives. Say you are confronted with a large garden. The visual image is initially so complex that it allows any number of different attentional approaches. If you were Monet, you would focus on the colors or perhaps the shadows; if a horticulturalist, on the rarest species; an entomologist, on the insect life; an ecologist, on the entire ecosystem; a child, on trees to climb and places to hide; a cat, on the birds; a dog, on the cats; a typical husband, on the amount of maintenance required; and so on. On a different day you might be actively looking for something or somebody and see it all completely differently. Yet the objective scene would be the same. Attentional selection is at work, with its powerful transformative effect on what you perceive.

One of the most important areas for attention control is the thalamus, an egg-size cluster of nuclei buried deep in the middle of the brain, whose influence radiates out over the whole cortex. One thalamic structure in particular, the reticular nucleus (so named because it is wired up like a reticule, or net) is closely involved in the conscious switching of attention and is ultimately controlled from the prefrontal cortex. The reticular nucleus may be involved in mediating the filtering influence of the prefrontal cortex on the sensory relay paths that pass through the thalamus. One interesting hypothesis about the thalamus is that it also serves as a neuroelectric amplifier. David LaBerge has recently proposed that it is the key to setting up the temporary triangular circuits that control the attentional searchlight. These are temporary architectures that engage three crucial brain areas simultaneously. The two most common areas engaged in this kind of circuit are the prefrontal and the sensory cortices. A third area, the midline thalamus, seems to be critical for completing any triangular circuit because it can amplify the activity of its nonsensory components. Risking oversimplification, we might say that LaBerge has proposed that the thalamus energizes the attentional "searchlight" because it can amplify, or highlight, a specific brain network.

Why would such an amplifier be needed? LaBerge has made the exciting, if still unproved, suggestion that the brain's internal activation level cannot compete in intensity because the outside world is a very powerful attractor. Since our thoughts are usually activated internally, they are much less intense than sensations. Most brains are dominated by images from the environment and cannot gain any degree of autonomy from sensation. To become autonomous, our brains must dominate their sensations. Humans have inherited a powerful thalamic amplifier that can level the playing field by amplifying nonsensory activity. The thalamus allows the brain to fight against the overweening influence of external stimulation and create internal foci of activation that can compete. Other subcortical regions also seem to be involved in attentional selection, but they play different roles. Two limbic structures, the hippocampus and the amygdala, form part of an internal selection mechanism that regulates future access to memory storage. The neocerebellum is also crucial, especially in the deliberate acquisition of new skills, because it can help automatize the attention mechanism in certain contexts.

The general pattern of interrelationships between these cortical and subcortical structures has been reviewed extensively by Bernard Baars, who has summarized a wealth of empirical data on the machinery of short-term consciousness. Baars and his late colleague John Newman have built a more comprehensive model of short-term conscious control than either Posner or LaBerge. It integrates most of our current knowledge about the executive brain. In their model, executive functions are controlled by an elaborate neural network called ERTAS, an agonizing acronym for a hypothetical entity, the Extended Reticulo-Thalamic Activating System. This system includes most of the anatomical structures already discussed, coordinated from the prefrontal cortex. It absorbs most of Posner's ideas into its vast maw and goes well beyond it, summarizing three decades of data on conscious processing.

ERTAS is the gatekeeper of Baars's unique central theoretical entity, called the Global Workspace. This is a great chunk of neocortex, mainly but not exclusively tertiary cortex, that serves as a common computational resource for many mental operations. The Global Workspace is like a neural blackboard on which many different brain systems can write. Awareness may be dominated at any given moment by a particular message written on the Global Workspace, which is in effect the place where all our difficult cognitive work is carried out. Although Baars and Newman are vague about the location of the Global Workspace, if it exists, it must lie somewhere in our vast stretches of tertiary association cortex. These form a lending bank, a sort of cerebral International Monetary Fund, that provides space for the intensive neural activity that sustains short-term working memory.

Experiences can become conscious only if they are projected to the Global Workspace. Specific conditions have to be met before an event—that is, a potential experience—can trigger full awareness. First, there must be global broadcasting of the event to the work space. Once broadcast, the experience receives wider and deeper extra processing, while events that do not receive such processing remain unconscious. Second, the event must have a minimum duration of global activation to reach consciousness. Very brief broadcasting doesn't work. The minimum effective period appears to be between a tenth and a quarter of a second. Less durable events are only weakly felt or not felt at all. Third, the event must have internal coherence—that is, make sense—or it will not register properly in awareness. An incoherent event may actually present itself briefly in consciousness as a distracting moment of confusion. Fourth, the conscious event must fit into whatever context one is thinking about at the moment. If it doesn't, and is still strong enough to persist into awareness, it will trigger a major startle response, a kind of wake-up call for the brain to update its working memory records. To achieve this, even more conscious resources have to be called in. Thus the executive brain surveys and reclassifies the environment and then reorients itself, by recalling appropriate new contextual knowledge from long-term memory.

The fifth condition for the achievement of full, blue-chip awareness seems special, if not unique, to humans. To be made fully conscious, an experience must be accessible by the self-system. This amounts to being integrated into the elaborate autobiographical fabric that we hold in long-term memory. Continuity of self and even continuity of the physical world with reference to the self are essential for the effective projection of an event into consciousness. This integrated self-system is based mostly on body awareness and a stable model of the environment, but it also depends on long-term, particularly autobiographical memory. The self-system encompasses much of what was said in Chapter 4 about egocenters and homunculi.

The great virtue of the Global Workspace idea is that it offers a testable model of how the executive brain might actually work. By controlling access to something like a Global Workspace, the executive brain could in theory amplify, optimize, and integrate the localized patterns of cerebral activity into a broader contextual representation. If I have any criticism of this idea, it is that the Global Workspace has been conceptualized as too passive an entity. It is used mostly by outside players (the unconscious modules of the brain), rather than serving as an active player in its own right. This places it in the same class as some of the models that have come from a purely cognitive perspective and makes it vulnerable to the criticisms I leveled at the Hardliner approach, including the fact that it overlooks the intermediate term. However, this model is compatible with my own evolutionary perspective on level-2 conscious capacity. Baars and Newman did not specifically address the question of evolution, nor did they explain how the appearance of uniquely human capacities, such as language, might be related to the organization of executive function. But, unlike many other theories of human consciousness, their theory was built on solid neuropsychological evidence, and they acknowledged the debt we owe to our mammalian heritage.

More important for my own agenda, they have made a very convincing case for the interconnectedness of the entire executive system. This has implications for evolution, because it follows that *the entire executive brain could have evolved as a unit* and been subjected to the kinds of selective evolutionary pressures that shape cognitive functions. The potential for significant fitness gains by evolving a more flexible executive system would have been very great because evolutionary improvements to the executive brain would have created an amplifier effect on existing cognitive capacities. Major modifications to executive functions affect preexisting innate behaviors, even those that were formerly outside the range of conscious control, because simply by evolving more wide-reaching executive capacity, any given species can get better at what it already does. Together with certain primitive structures in the brainstem, the executive network forms the skeleton of an intricate "parallel brain" that infiltrates, connects, and surrounds all the sensory and motor modules, allowing the brain to switch among them, select within them, and integrate their activities, while regulating storage in working memory. This elaborate system mediates all our short-term executive functions, including the management of working memory and the setting and maintaining of immediate priorities. Level-2 conscious capacity also oversees the unconscious modular domains of cognition, touching on virtually all cognitive domains.

There is one remaining key idea that needs treatment under the heading of short-term executive processing. This is the notion that automatization is a corollary process of consciousness and a major by-product of executive processing. One of the primary objectives of conscious processing is to eliminate the need for engaging it in the future by making learned skills as automatic as possible. The key to this seems to be a neural structure that is a relative newcomer to this debate, the neocerebellum. For many years this massive subcortical structure was thought to be completely outside the executive system. As recently as a decade ago the cerebellum, including one of its most recently evolved parts, the dentate nucleus, was regarded as the very antithesis of the conscious brain. It was seen as a specialized region dedicated to the execution of automatic motor skills, normally too fast for detailed conscious oversight. Examples of such skills include highly overpracticed routines, such as playing an arpeggio in a difficult Chopin piece, and other fast-moving skills, such as shifting gears on a standard-shift automobile during a race. The cerebellum was once regarded as the brain's fast-time auxiliary computer, in charge of relieving the conscious mind of the burden of managing rapid, on-line performance. It was supposedly our storehouse for automatized skills, the place where skills were sent once they had graduated, as it were, and no longer needed conscious oversight.

That theory has been drastically revised during the past five years. It is now clear, primarily from brain-imaging studies, that the cerebellum is one of the most needed neural structures when conscious capacity is engaged. It is very active while learning new skills, but surprisingly, it seems to tune out during automatized performance, unlike the motor cortex, which remains highly activated. This is the opposite of what we might have predicted. However, it actually fits in well with something we have known for decades: that the motor cortex is crucial for the on-line monitoring of action. It cannot be turned off when skills are automatized because each act must be adjusted to the specific conditions at hand. Thus, if I want to write my name on a sheet of paper with my foot, I must change the muscles I would normally use. This is where the cortex is crucial, in taking context into account and adjusting action patterns to contemporary conditions. Since the specific context is never precisely the same, the motor cortex can never rest. But the cerebellum apparently can. This seems to contradict the old rule-of-thumb principle of cerebellar function, that the cortex was the seat of consciousness, while the cerebellum was its opposite, an unconscious slave, a habit machine that smooths out the bumps. However, it seems that the cerebellum is implicated in some important aspects of conscious self-regulation, while some areas of the cortex are involved in the unconscious, and automatic, guidance of movement. The precise extent of the neocerebellum's participation in executive function is not yet fully established, but Posner regards it as an integral part of the executive attentional system.

The take-home message is simple: Level-2 conscious capacity is based on an elaborate brain system that supports short-term working memory. It is a well-documented reality in many advanced species, including all mammals. This system broadens the reach of binding, by enabling the brain to extend its perceptual framework over several seconds, allowing the perceptual and conceptual world to expand. It also increases the complexity and length of the automatized routines that it can learn and off-load from consciousness.

LEVEL-3 AWARENESS: INTERMEDIATE- AND LONG-TERM GOVERNANCE

This leads us to level-3 of our exotic Dantean underworld of consciousness. This level represents a massive expansion of level-2 powers and completes the basic primate brain design. Table 5.1 summarizes the properties of the three levels proposed in this model.

TABLE 5.1

Three levels of basic conscious capacity

	Level-1	Level-2	Level-3
Anatomy	Sensory cortex	Secondary cortex	Tertiary cortex
Sensitivity	Objects, simple events	Complex events	Long episodes
Time span	Milliseconds	Seconds	Minutes, hours
Selectivity	Simple bias	Complex selection	Multifocal selection
Operations	Selective binding	Short-term awareness	Extended awareness

Level-3 is concerned with the intermediate- and longer-term regulation of thought and behavior, something that human beings do uniquely well and that chimpanzees can manage to some extent, as we saw in Chapter 4. Level-3 capacity builds upon the first two, further expanding the range of experience. But it adds two new elements, both related to self-evaluation. The first is the extension of awareness into the domain of voluntary movement, or self-initiated action. The second is a supervisory, or evaluative, dimension within which level-2 executive function itself can come under scrutiny within a wider world, in which time is perceived as flowing over a much larger landscape, and the perception of space extends well beyond the perceptual horizon into imagination. These new elements are a fairly straightforward evolutionary expansion of some powers of the level-2 system, inasmuch as a level-3 model of models is still an event percept, albeit on a Wagnerian scale. This type of awareness encompasses a vision of oneself, acting in a three-dimensional world, and has considerable breadth and width to it.

The key change is the extension of voluntary attentional control into the domain of action, especially the control of limbs and the vocal tract. This change might qualify as the central defining characteristic of human conscious capacity. It not only made our perceived world larger, since it placed us squarely in the center if it, but revolutionized skill. The perceptual ego-center became an action-based control device, exercising considerable discretion over the shape of action. This was a paradigm shift of the first magnitude. In other mammals, with the possible exception of some apes, consciousness seems dominated by the perception of everything but one's own actions. Awareness, in that context, is about objects, opportunities, paths through space, pain, threat, and bodily states. But level-2 conscious capacity cannot manage much control over action, especially over the long run. It cannot even gesticulate. Level-3 awareness changed all that. Prior to it, the self-conscious actor did not exist. But-but-but, you protest, surely you aren't suggesting that we climbed all those Darwinian mountains, just to evolve a race of ham actors? To which I reply, never underestimate nonverbal representation. Once the physical self entered the conscious world-model, a whole new world of possibilities was opened up.

The principal brain adaptations involved in level-3 conscious capacity are the great hominid expansion and differentiation of the prefrontal tertiary cortex and the parallel growth of other regions of the tertiary cortex, the hippocampus, and the neocerebellum. The tertiary cortex almost quadrupled, and the neocerebellum tripled in size during human evolution. More significant perhaps, the interconnections between the prefrontal cortex and the neocerebellum increased by a factor of at least five, which points to the increased traffic between these structures, presumably caused by the hominid revolution in skill. The level-3 system followed roughly the same expansion pattern, building on existing strength, except for a hugely enlarged cerebellar component, which seems to be a quantitative, if not qualitative, break with the past. Just as the proposed locus of level-2 capacity is in the cortical expansion of a level-1 framework into the secondary cortex, the level-3 system seems to have depended upon expansions of the latter into the tertiary cortex, as Figure 5.3 (p. 165) implies. One result of this was a great increase in the number and complexity of reentrant circuits in the brain, on a tertiary level. This in itself could have extended the potential time span over which neural activity could be maintained.

In the process of hominid expansion, the prefrontal cortex invaded some new territory, captured control of several strategic subcortical motor nuclei, and radiated into new secondary cortical regions. The expansion of the prefrontal tertiary cortex made us aware of self-action. It also gave us direct access to the levers of governance over action, resulting in our remarkable capacity for conscious, deliberate rehearsal, review, and refinement of action. In addition, a major new element was introduced. The greatly enlarged anatomical loop between the cerebellum and the prefrontal cortex contributed to our expanded capacity for automatized skills. This level of involvement in the conscious control of movement seems to be a distinctively human adaptation, based largely in the newest parts of the cerebellum. Physiologist Henrietta Leiner has suggested that the neocerebellum is a cerebral optimizer, whose role is to amplify, and make more effective, other functional systems. Under the guidance of the prefrontal cortex, this creates a powerful mechanism. The importance of its role is reflected in the rich pathway between the frontal tertiary cortex and the main cerebellar receiving areas in the brainstem, called the pontine nuclei. This pathway contains over 20 million nerve fibers on each side, which makes it one of the largest, if not the largest, pathways in the human brain. By comparison, each optic nerve has 1.5 million fibers, and the entire pyramidal motor pathway has only 1 million. The fact that these pathways are connected to high-level cognitive regions places the cerebellum in a strategic location, right in the core of the highest level-3 control system. The overwhelming size of this connection to the prefrontal areas suggests an important executive role, probably in the generation of automated programs of attentional control. It is not yet clear how this system works in detail, but the human cerebellum seems to be in a perfect position to mediate between the high-end frontal system and the lower-end nuclei that dominate the actual programming of movement.

The governing role of the frontal lobes has been well documented. Almost two decades ago Tim Shallice and Donald Norman suggested that the human prefrontal cortex is the core of the mind's supervisory system.

Donald Stuss and Frank Benson came to a similar conclusion, based on their massive review of frontal lobectomy research. To judge by the nature of the cognitive deficit in frontal lobe patients, the prefrontal lobes may be essential for sustaining many high-level metacognitive operations, such as self-evaluation, long-term planning, prioritizing values, maintaining fluency, and the production of appropriate social behavior. The important role of the prefrontal regions in Ericsson and Kintsch's long-term working memory paradigms has been confirmed by researcher R. W. Engle and his colleagues. However, we know that there is considerable differentiation of frontal lobe function, and there seem to be subregions in the human prefrontal cortex that have distinct functions. These have resisted a cut-and-dried, one-size-fits-all theory of their functions perhaps because there may be significant differences between individuals in the organization of this region. As we become better at the functional imaging of individual brains, these questions may be resolved.

The clinical neurological literature shows that the prefrontal tertiary cortex is a key player in the top-level coordination of many disparate executive functions. This has led researchers to coin such terms as "supervisory system," "prioritizing function," "hierarchy formation," and "working memory management" to describe the functions of the human prefrontal cortex. But these functions are all interrelated on the metacognitive level. This vast cerebral resource can reflect on the cognitive process itself. It judges the appropriateness of one's thoughts and actions, evaluates whether a change of tactic might be profitable, and so on. This is the ultimate requirement for achieving a significant degree of autonomy from the environment.

This expansion of capacity may explain many things about the speed, inclusiveness, abstraction, and overall power of conscious processing, but how would it extend conscious control into the intermediate term? This raises questions about how neural activity might be sustained into a longer time range. One reason why psychology has been so locked into the short-term range is simply that our earlier models of neural activation would not allow such long time parameters. When Hebb wrote his book on cell assembly theory, he was more concerned with explaining the length of short-term memory, which seemed longer than neurons could possibly sustain. At that time, it was believed that neurons could be active for only very brief periods, amounting to fractions of a second. That problem was eventually solved, as our knowledge of neurons improved. But we still find it difficult to conceptualize a continuing activation that lasts as long as the kinds of working memory models needed to describe our uses of language in conversation, for example. These mental models remain in place for hours on end.

The long-term memory feeder systems proposed by Ericsson and

Kintsch may solve the problem by making long-term working memory just a primed subset of long-term storage. This would require a very rapid activation mechanism, so that long-term memory traces could be called up in fractions of a second. This is possible. But I am not comfortable with that idea because it seems to violate our usual notions about long-term memory, and in particular, it doesn't deal with the governing role of intermediate-term awareness. My point is that longer-term metacognition endures *in awareness.* It doesn't just pop out of an unconscious nowhere every once in a while, as if we were waking from a dream. The cognitive framework of a conversation, for instance, is always hovering in the background. It corrects, evaluates, and updates itself continuously. It doesn't behave at all like a long-term memory trace, sleeping peacefully in cerebral hibernation, while the waking mind continues to play its games unsupervised. On the contrary, it is integral to the supervisory process, present and accounted for, front and center.

This raises another possibility, which would explain how the nervous system could maintain a slower-moving, more stable, intermediate-term activation of working memory. Intermediate-term regulation might be provided not by conventionally primed long-term traces but by an active trace that moved more slowly than the short-term one. For too long we have accepted that short-term memory relies on rather unstable short-term electrical activation. We have never outgrown that assumption, but perhaps we should. Slower processes based primarily on the activation of slower chemical and dendritic interactions may allow the nervous system to support an enduring intermediate-term memory trace. Such a trace would, in theory, be mediated by waves of slow postsynaptic activity that would not necessarily involve much axonal transmission but would rest in an active background state in a large population of neurons. It could even involve changes in the state of the extracellular space, which could exert a major influence on local transmission. Such a process, interacting with third-order reentrant circuits, might maintain a number of active foci for long periods of time.

This remains to be proved, but we have had good electrical evidence for the existence of very slow shifts in brain activity for many years and have not known how to interpret them. Some of these shifts, even within the limitations of our recording equipment, can last as long as half an hour. Shifts of ten seconds or more occur frequently, in many different structures. In the late sixties, when I was recording brain activity using direct current amplification, I could sometimes follow a set of slow shifts for an hour or so. I routinely observed very slow shifts that could last for several minutes before stabilizing. Many animal researchers have observed very slow shifts inside the brain, but their significance has not been well understood either. For reasons of method, we have not pursued these slow events as feverishly as we have chased the faster activity (a sign of impatience, perhaps, but more probably because we have had no theory on which we could hang our studies). The brain's slow activity should be reexamined using better recording and imaging technology, to search for the mechanisms behind a slower-moving class of working memory.

It is important to remind ourselves that the activity of the executive brain is not equivalent to consciousness itself. Consciousness is the product of the executive brain's activity. But it amounts to much more than the sum of its neuroanatomical parts, just as any living cell is much more than the sum of its parts. The executive brain is a physical resource that determines both the quality and the efficacy of consciousness. It supports and modulates consciousness, sets limits, and regulates which mind states might dominate at any given moment. But its activity is not awareness itself. Nevertheless, the executive brain is an important index of conscious capacity. Its presence in a species, even in part, suggests strongly that the species has some of the same aspects of conscious experience that humans have and share the same executive components. It is reasonable to link conscious capacity to anatomy, even if we acknowledge that the neural reentrance principle might actually change the functions of component structures as evolution proceeds.

One final note. Since the executive brain system is supposedly dedicated to various nonmodular or amodal cognitive functions, it may appear ironic to find that it is componential; that is, made up of distinct structures that look and behave suspiciously like modules. But this should not be misconstrued as a problem because it is not. Conscious capacity is domain-general on a cognitive level. This is inherent in its definition but does not lock us into any specific neural model. Each of the various components of the executive brain has a unique evolutionary history. These components are present in various degrees in different species, whose specific executive capacities can vary enormously. Some might be better at certain kinds of attention, and others might have better working memory in some sensory domain, because the modules feeding the executive system impose constraints. But all executive systems adjudicate between modalities and influence more than one cognitive domain.

EPISODIC AWARENESS

The three-level hierarchy of the executive brain is, to my knowledge, complete only in primates and fully developed only in humans. Lacking the normal human enculturation process, this brain system forms a distinct kind of awareness I have called episodic. Episodic awareness is defined primarily by elaborate event representations. For an episodically competent mind, experience is not normally remembered as a confusion of objects, actions, colors, or raw sensations but as a series of events. These events are the givens, the raw data, of its memory, which consists of experience that has been segmented into and remembered as a sequence of discrete episodes. All mammals do this to some extent. A typical mammalian episode might be eating food in a certain place, or marking a territorial boundary, or battling a rival, or being chased by a farmer. Each episode is composed of hundreds of simpler bound percepts, which are batched together into coherent chunks. This batching process can be regarded as a kind of metabinding. Its mechanism is not yet understood and constitutes perhaps the greatest single challenge to cognitive theory today.

The complexity of this kind of cognition can be illustrated by analyzing a very routine matter for a typical mammal, a remembered fight. For example, a dog engaged in a fight receives many disconnected fragments of experience—sounds, smells, tastes, pain, visual images, and so on—in a chaotic, helter-skelter way. If these impressions were strung out in serial order, as a film director might, frame by frame, with each sense modality following its own vector in time, we would see a bewildering chaos, a sequence in which visual images might be superficially out of synch with the sounds, smells, and tastes of the episode. Yet this does not happen. Such events are stored as single, unified episodes, and future behavior is affected by memories of such episodes. This memory for specific, coherent, detailed events is the essence of episodic cognition.

There are varying degrees of this ability. To have episodic knowledge, animals must integrate experience across many sources, including internal body feelings, and organize them into large-scale scenes that will be remembered as such. Thus "fighting with x" or "mating with y" or "grooming with z" will be effortlessly parsed and remembered. To achieve this, the brain must attend selectively not only to the outside world but also to its memories of similar episodes. Episodic competence demands considerable executive skill. The parsing of episodes is after all a matter of registering events selectively and interpretatively. But above all, it demands an efficient working memory system. Various facts and implications inherent in episodes must be kept in mind when negotiating, for instance, the subtleties of chimpanzee social behavior. These facts might include one's place in the dominance hierarchy, territorial interests, and past interactions with other players, as well as one's immediate objective (for instance, to be groomed by a particular individual). These facts need constant updating because social relationships can change suddenly.

In humans, this kind of awareness was the foundation stone and starting point of the mind before it began its long cultural journey. Alone, this capacity cannot explain language or symbolizing capacity. But in the right cultural environment, it provides the raw cognitive potential for those things to prosper. Human uniqueness lies in our radically improved conscious capacity. This may have evolved for many reasons. It would have made primates better at most things they were already doing. The extraordinary intellectual heights to which humans have climbed are partly attributable to this raw executive capacity. This should not confuse us into thinking that humans are not specialized for any particular niche. Of course we are. We are specialized for executive functioning. Our niche, broad as it is, makes huge demands on executive processing. It requires a larger perspective on time and space and a more flexible mentality. We are specialists, like all other mammals, but our particular specialization is for high-level executive control.

ENTERTAINING A RADICAL POSSIBILITY

Is this the end of the line then? No, because we have not yet escaped from the nervous system. We have not accounted for our symbolizing capacity or our complex cultures. To return to our opening theme, symbolic minds are not self-sufficient neural devices, like eyes. They are hybrid products of a brain-culture symbiosis. Expanded conscious capacity provides raw capacity, but it is unprogrammed capacity. Without culture, we could never have become full-fledged symbolizing organisms. We might have become something else, perhaps, with very powerful perceptual-motor systems, like those of a superprimate, but we would be only powerfully episodic, not symbolic. To understand consciousness fully, the generation of culture must be explained.

Enculturation has been neglected as a possible formative process in its own right, but we have no alternative other than to give it pride of place in any evolutionary theory. True, culture follows, to some degree, its own evolutionary trajectory. True, the brain seems to have followed a different one. But the two have been linked together for a very long time. Perhaps our neglect of enculturation is due to our great difficulty in objectifying our own intellectual dependence on culture, embedded as we (especially scientists) are in it. Given this bias, it is not surprising that the most striking evidence of the power of enculturation has come not from studying ourselves but from studying another species, the chimpanzee.

Some very strong claims have been made about a hypothetical ape language capacity, and I don't agree with many of them. But I cannot avoid making at least one major concession. Enculturated apes can do things that they are not supposed to be able to do. Moreover, this is entirely the product of a special cultural environment. We have not tampered with their biological inheritance. They have exactly the same brain design as their siblings in the wild. Yet in some respects they behave like a different species. Many of their remarkable skills are completely absent in wild-reared apes. Their capacity to use symbols and understand some spoken language could not have evolved directly in a species that lacked these skills in the wild. They were implanted in their brains by the cultural influence of another species, our own. This demonstrates convincingly that the enculturation process can successfully uncover, and make good use of, cognitive potential that sat unexploited for millions of years.

A symbol-using culture can become a formidable force in reshaping the primate mind. Some enculturated apes have learned to use visual symbols to communicate with one another. This was first shown by the Rumbaughs two decades ago, when they demonstrated symbol-coordinated tool use in two chimpanzees named Sherman and Austin, who had earlier learned to use half a dozen tools, including keys, and to name them using symbol boards. They were allowed visual contact with each other through a window and could communicate using their symbol boards. They were also able to "mail" objects back and forth, from one room to the other, through a drawer that could be pulled through the wall. They shared a food locker, which was situated in one room, while the key to the locker was kept in the other room. When one of them wanted to eat, he had to flash a request, using the visual symbols, to the other chimp, who was the temporary caretaker of the locker. The latter would read the message, get the key, and send it through the porthole. Then the first chimp would open the locker, get the food, and share it with his colleague. They managed to coordinate their behavior in various scenarios, using several different tools, even though they were never explicitly trained to perform these specific exchanges. This was an impressive demonstration of symbolic communication between two members of a species that supposedly lacks any capacity for doing such things.

These achievements are regarded by many as a clever illusion because they don't occur naturally in the wild. Let's accept this at face value. In some respects, these symbolic skills are undoubtedly an illusion. In fact, Sherman and Austin are from outer space, as far as most other chimpanzees are concerned. But this is not a substantive criticism of the research. Rather, it has very serious implications for how we should see our own species. Certainly, to our archaic forerunners, most of us would also appear to be strangers in a strange land. Our modern intellectual armamentarium has been constructed on the scaffolding of many previous cultures, and we have traveled a great distance. Just like Sherman and Austin, we are illusory creatures, products of a cultural revolution that has kept raising our intellectual bar, higher and higher, pushing us toward heights we weren't really designed to reach.

There might be a hint about what we are as a species in the failure of enculturated apes to generate actual cultures. Despite the brilliant efforts of modern primate researchers, apes still use symbols for traditional primate agendas. If symbols alone had been enough, many apes would presumably be much farther down the human road by now, moving rapidly toward collective representational systems, like the denizens of *Planet of the Apes.* But they are not. While individuals have made great advances, collectively they have never been inclined to construct their own symbolic cultures, not even on the small scale of a small working or family group. Sherman and Austin never extended their use of symbols to hold anything resembling a conversation. Not even on the level of, say, "Hey, Sherm, how's it goin'?" Nor is there any evidence that they modified their traditional social behaviors, using symbols. In their case, competence in the use of symbols was not sufficient to generate a cultural revolution, not even a small one. This negates any notion that symbols have any immediate, transformative power to generate culture. Lacking the necessary expanded level-3 capacity, Sherman and Austin were unable to push beyond the limits of their capacity for making mental models, which cannot be changed by culture.

There is a great deal more to generating culture than a sprinkling of words and a smattering of grammar. As I said, this leaves us with a tricky chicken-and-egg question when examining our own origins. Which came first in human emergence, symbolic skill or culture itself? Obviously, many of our symbolic skills are due to enculturation. The most indisputable examples of this are the skill sets associated with literacy and mathematics. Given the recency of their invention, they cannot be vested in evolved modules. There has been no time for such evolution.

Unthinkable as it may seem, this allows us to entertain a radical possibility: If the exquisitely intricate networks of the literacy brain have to be acquired in culture, why not language itself? The truth is that we are not certain that spoken language is vested in a dedicated brain module. The actual sound system of language, phonology, may be modular, but not the higher intellectual drivers that give language its great representational power. We cannot dismiss the possibility that language itself, especially in some of its most esoteric semantic and grammatical features, might be just another product of culture, driven by our expanded level-3 conscious capacities.

6

Condillac's Statue

> We cannot recollect the ignorance in which we were born. Reflective memory, which makes us conscious of the passage of one cognition into another, cannot go back to our beginnings. To say that we have learned to see, to hear, to taste, to smell, to touch, appears a strange paradox.
>
> —CONDILLAC

It is one thing to state that expanded conscious capacity equips human beings to cope with complex cultures. It is quite another to describe the mechanism that links culture to an expanded consciousness. It is far from obvious why increased conscious capacity should in itself make human beings able to cope with cultural complexity. In this chapter I shall examine in more detail the linkages among early development, cultural learning, and conscious capacity.

This draws us into the field of cognitive development, especially attentional development. This field is too vast to cover comprehensively; thus I intend to borrow from it very selectively and only where it clarifies the functional role of consciousness in dealing with the cultural environment. Much of the discussion will focus on attention. During early infancy, cultural influence rests chiefly with certain figures, such as the mother, father, and other close family members. These are powerful forces in the mental life of infants because they influence attention. They do more than dominate attention; they also train infants to share attention with them. Perhaps the most important lesson they teach their infants during the first year of life consists of the basic rules of attention sharing. Once this process is well established, it serves very well as a fast-track social learning instrument, in a variety of situations. Joint attention develops into a primary cultural guidance device. It allows children to follow cultural signals that will become increasingly more abstract as they expand their horizons.

Joint attention, or more accurately, the interlocking of two or more conscious minds, is the cognitive foundation of pedagogy. Human cultures are powerful pedagogues because their members regulate one another's attention, through a maze of cultural conventions. Our developmental symbiosis with culture culminates, as we shall see in a later chapter, in formal symbolic thought, which emerges from a hierarchy of culturally programmed experiences. This kind of pedagogy is not always deliberate, but it always engages conscious capacity because by its very nature, it demands the sharing of attention. Joint attention can influence everything, from the perceptual binding of objects to the symbolic capture of abstract ideas. But it is not the whole story.

In fact, attention is only one aspect of the three-level model of human consciousness proposed in Chapter 5. It is also the gateway to the executive brain. By controlling attention, adults also influence and structure the development of the child's working memory. Children generalize very quickly and learn to anticipate where attention should be directed without explicit cuing. They soon learn shortcuts that allow them to align their working memory systems and their habits of thought and memory with those of their adult guides. The same applies to habits of self-supervision, rehearsal, recall, and a number of other metacognitive routines included in the Executive Suite. These are shaped by culture, and taken together, they can wire a child's executive brain into a powerful social learning device and a nimble cultural navigator, adapted to a specific historical context. Not only does each culture teach children how to find out where to direct attention in a given situation, but it also teaches them what to remember and how to learn. These basic interactive routines become the foundation stones of more complex cultural learning as children grow older.

Children's minds are thus self-assembling systems, to a degree. As we shall see, this important idea, first identified with philosophical Constructivism, is enjoying a renaissance. Children acquire many of their basic mental structures through experience. They show remarkable adaptability, or plasticity, in this. Human developmental plasticity is revealed most clearly in extreme cases of early deprivation, which expose the crucial role of consciousness in cultural learning. This topic, as well as the complex attentional machinery that interlocks our mental development with culture, will be discussed at some length in this chapter. But first, we need to review a few facts about the developing brain.

MINDS IN MOTION

Brains are often made to appear as static entities. Depictions of brains in anatomy texts freeze them in time and project a rigid immobility, akin to, say, family portraits of Victorian royalty. They convey a certain prim stolidity, where Princess Amygdala poses self-consciously alongside Prince Hippocampus, with his handlebar mustache and medals, together with their great-grandchildren, the mammillary bodies, all dressed up in sailor suits for the occasion. The anatomist forbids movement, and his subjects obey, stiff-necked and stock-still. This conceals their true dynamism. In this motionless, structured world nothing ever changes, time stops, and the sun seemingly never sets.

Visual images such as this are in everyday use, and although we should know better, they affect our language when we talk about brains. Out of habit, and occasionally out of necessity, we coin phrases like "axonal transport mechanism," "language region," and "executive apparatus," as if brains and their workings were unchanging, solid things, like steam engines, the motherboards of computers, or the Precambrian shield. But they are not. The images are a clever touch of legerdemain. They impose order, or rather a temporary restraining order, on the unceasing blizzard of activity that is the true moment-to-moment reality of the brain. Those frozen icons are all we have, and they go as far as our direct images of microanatomy can travel at the moment.

But it would be closer to the truth to say that brains and minds are born as a bagful of possibilities in a sea of chaos. Instead of Victorian royalty, their cellular inhabitants are more like swarms of cybernewts, fighting for survival in some hostile ocean on an alien moon. They never stand still, and the more we turn up the power of the microscope, the more frenzied and crowded their patterns appear. Neurons are embedded in systems of molecular exchange, and like the world's stock exchanges, they are active twenty-four hours a day, seven days a week. They turn over their material parts rapidly and retain none of them for very long. In their own way the brains of living systems are just as immaterial, and just as slippery to grasp, as economies or languages. Only the blueprint of the brain outlives the elementary particles of which it is composed and endures throughout an entire lifetime. By the time we are adults, we do not have a single atom of our childhood left. Yet the physical "we" still exists.

The cognitive entity growing at the center of all this activity seems even less solid. It is a ghost, in the truest sense of that High German word *Geist*, which means "soul" or "spirit." Starting as a faint glimmer of light in the fetal darkness, it will grow into an ethereal being (perhaps one of us), as it emerges from its primordial dawning. From a cognitive vantage point, there isn't much there to start. Gene kits may provide the basic instructions needed to get a brain going, but the catch is that there is no one around to assemble it. It has to assemble itself, using only a blueprint, a few tools, and a built-in, or innate, dynamism that comes from below, rather than from above. With the right technical tools, we might be able to manufacture better images of this unfolding inner universe. Such images might help us see the process more clearly if our brains were able to comprehend their own complexity, which remains to be seen.

Evolution and development share a common interest in this process of unfolding, or genesis. The word "genesis" means "the coming into being of something." "Phylogenesis" refers to the evolution of groups, or phyla, of genetically related organisms. "Ontogenesis" refers to the development and growth of single individuals. In both cases, time is a frame within which the pattern of change can be visualized. In phylogeny, time is measured in multiples of many lifetimes, whereas in ontogeny, it is measured in fractions of a single lifetime. Emergence in both is usually broken into coherent stages, or periods of transition. In the standard phylogenetic chronology, chimpanzeelike ancestors metamorphose into australopithecines, who become archaic hominids, who become us. In the standard ontogenetic chronology, newborns metamorphose into children, who become adolescents, who become us. The second chronology is never just a recapitulation of the first. To put it crudely, we do not go through a "chimp" stage, followed by an "australopithecine" stage, followed by a "human" stage as we grow up.

Nevertheless, theories of evolution and development must agree on basic principles. We are describing the same beast, after all, and evolution can affect developmental variables. A dispassionate judge would probably say that humans are just chimpanzees with a slightly different developmental bioprogram. A human fetus starts with essentially the same brain design as a chimpanzee fetus but develops the blueprint in a different way because of evolutionary changes in the regulatory genes that control relative growth. Different parts of a developing body can be made to grow faster, slower, or longer by such genes, and this can have a dramatic effect on the final shape of the organism. A human skull, for instance, is a geometric transformation of an ape skull, in which the same plates of bone have been programmed to grow at slightly different rates, for longer or shorter times, from a larger or smaller number of source cells. Altering these variables has transformed the shape of the adult human skull. It can do the same to any other developing structure, including the brain.

SUPERPLASTICITY

This picture is complicated by an interesting evolutionary conundrum, the superplasticity of the human brain. Neural plasticity simply implies that there is some flexibility in the way a nervous system unfolds in reaction to the environment. When cognitive researchers speak of neural plasticity in development, we usually mean that the physical side of the equation has a quirky, somewhat unpredictable relationship to the mental side. This often translates into uncertain localization within the nervous system. For example, depending on a host of variables, such as early brain injury or hormonal development, the neural control of speech might be localized primarily in the right, the left, the frontal, or the posterior regions of the cortex. Moreover, the brain's memory systems, which rely on a high degree of neural plasticity for their very existence, might store their records of experience in a virtual infinitude of different networks, depending on the specific life histories they chronicle.

Neural plasticity is not unique to humans. It is present in some degree in most nervous systems, and because of it, accidents of birth, dietary and hormonal influences, disease, injury, and experience all can affect how the brain is wired up in the adult. Some readers might be shocked to know that even that paragon of apparent fixedness the visual cortex is not restricted to any preordained region of the cortex or preset in its functional organization. The "visual" cortex is defined by its role in vision, rather than by a strictly anatomical criterion. It is the consequence of a developmental collision between the growing optic nerve and some part of the growing cortical mantle. Once the nerve collides with the cortex, wherever that collision may occur, the patterns of stimulation mediated by an active eye-brain linkage will generate a ripple effect on the cortex, and the area of this interface will gradually become transformed into a visual region, expanding to recruit many other regions of the brain into its activities. Thus, although its location is on average very similar in most of us, this is mostly a result of the fact that our optic nerve has a standard location and a standard rate of growth. The internal organization of the visual brain reflects accidents of development that can change the standard pattern. If, for some reason, the growing cortex receives inputs from three eyes, it will be programmed for three eyes. If there are no eyes, there will be no visual cortex at all because that same region will be cannibalized by other input regions, such as those for hearing or haptic touch, in the battle for connectivity that takes place during development. During the critical periods when various parts of the cortex mature, the rule is that as the pattern of use goes, so goes the function of those areas. As a colleague once put it, "The cortex is what it eats," to which another replied, "Perhaps, but this just shows that there is more to vision than meets the eye."

Once again, this reinforces the notion that static metaphors don't work for the brain. The brain's real geography is closer to that of plate tectonics, or drifting continents, than to the fixed Earth of ancient times. This is true even in adults. Physiologist Michael Merzenich has shown that the sensory areas of the cortex can expand and recede in the adult brain, reflecting their current level of use. In principle, if a blind man learns braille, the brain areas representing his fingertips will actually expand, recruiting adjacent cortical columns, as his sensitivity improves. This implies that the brain is constantly making adjustments to accommodate its current pattern of use.

Plasticity is a design feature of the brain that can change in evolution, and as a consequence, it varies greatly between species. Humans have superplastic brains. A developing human brain is a sort of snowballing cognitive leviathan that adapts to everything and anything close to it. Learning is one aspect of extreme plasticity, and creativity is another. Any species that can do such things as play with the world, imagine it, remember it, and expand its circles of experience, while constantly increasing its purchase on the surrounding social universe, will ultimately start to experiment with its own fate. It will try out and evaluate many different strategies for survival. It will exchange and share its discoveries. Most species do not have this ability because their cognitive strategies are very specialized and fixed in their genes, and they can generate new strategies only at the agonizingly slow rate of physical evolution. But a superplastic species, such as ours, can generate new options at a rapid rate, in fractions of a single lifetime. Most of these innovations will not improve fitness, but some will, and natural selection will seek out and select those genes that nurture the most successful innovations. In this way, cerebral plasticity speeds up the rate of cognitive evolution. This is the so-called Baldwin effect, an invisible positive feedback loop that welds phylogeny to ontogeny in certain fast-learning species, producing a multiplier effect on the rate of evolution. Since plasticity itself is subject to natural selection, the Baldwin effect can also amplify itself. Many believe that this is the main reason for our "runaway brain," as biologist Christopher Wills has called the extremely fast expansion of the human brain during the past half million years.

As the human brain accelerated toward its superplastic state, the ancient slow-moving stereotypy of the mammalian mind was left behind. We have developed enormous variety in our expressive behavior, and this has generated ever more variable cultures, which have demanded ever more neural plasticity. Generation after generation, the human nervous system has drifted inexorably in this direction, evolving us into the neural chameleons of the biosphere. Our neural plasticity has led to a direct cultural consequence: The range of variability between human cultures and between individual members of those cultures is now vast. One of a pair of identical twins could, in theory, have been raised in a preliterate Gravettian Stone Age society in the Dordogne Valley thirty-five thousand years ago, and its cloned sibling in twentieth-century central Paris by a member of the Académie française and trained in computational chemistry at the Collège de France. Despite their identical genes, these two adults would have tremendously distinct mentalities, and these differences would become irreversible at maturity. The twin metaphor is apt, because any of us can easily imagine a physical simulacrum of ourselves with an utterly different mind. At birth an infant can develop into an infinity of selves, and its brain is equipped to deal with that uncertainty. But once the floodgates of culture have been opened, our personal encounters with it will determine a great deal about how our minds unfold. Plasticity is not an option for this process. It is essential.

Why would such plasticity evolve? Presumably to support better learning and memory. But our superplasticity would not have much, if any, adaptive value outside the context of a highly unpredictable cultural world. In most species, extreme plasticity would probably prove to be a liability, leading to instability. Only a complex culture can generate the kind of expanding cognitive universe that rewards superplasticity. But it also raises the developmental stakes because each child must be crafted to fit the ever-changing mold of culture. To achieve this, the human brain has evolved a maximally flexible, self-constructive developmental strategy, which can track a moving cultural target. Every human child is equipped to blend invisibly into its surrounding culture, however weird or wonderful that culture may be. Whether it is born to a cave-dwelling bear cult, a nest of Caribbean pirates, a community of sun-damaged Web surfers, or the slave ships of Ali Pasha, the budding young mind must prepare for a lifelong odyssey of fitting in. The epicenter of this adventure is the conscious mind.

THE THIRD MAN: DEEP ENCULTURATION

The incorporation of a human child into a symbolic culture is an enormously complex affair. Shortly after birth, the infant is wedded to a specific culture that takes control of its cognitive development through a series of transactions. This may sound improbable because cultural linkages are invisible to the child. They hide behind many surrogates, such as parents, family, tribal customs, institutions, and so on. These are the carriers of the culture, the front lines of the infant's encounter with vast collective forces that it never sees and whose existence even the surrogates may not suspect. This process is radically different from anything that occurs in other species. The enculturation of the mind amounts to a third factor in development, unique to our species. This third factor (the first two being genes and environment) might be labeled "deep," or "cognitive," enculturation, to distinguish it from the shallower forms of enculturation that are common to many species and have no structuring effect on the mind.

Deep enculturation is something like the Third Man in Graham Greene's story, a mysterious presence whose whereabouts are unknown and who turns out to have been there, undetected, all the time. It is so close to us that we are normally unaware of it. We don't tend to think of our friends and families as "carriers" of cultural messages and customs, but that is exactly what they are, and they affect us early and deeply. We might be able to experience the magnitude of culture's impact on our minds if we were to undergo a radical change in culture during our adult lives, suddenly traveling from the twenty-first century to the Middle Ages, for example. However, people rarely travel long enough cultural distances to experience such deep metamorphoses. Some children are forced to make such journeys, and they are changed profoundly as a consequence. But most people who have experienced truly deep cultural changes as children have trouble remembering their former identities with any clarity. Still, a few individuals have experienced dramatic and culturally engineered metamorphoses after infancy, and they have something to tell us. One such person is Helen Keller, whose life is discussed later in this chapter.

Deep enculturation is quite distinct from the effect of the social environment as it is normally defined. The social environment includes many factors that impinge on development, from bonding and competitive stress to the social facilitation of learning. These can affect brain functioning in many ways, but usually they have no direct influence on functional brain architecture. However, symbolizing cultures own a direct path into our brains and affect the way major parts of the executive brain become wired up during development. This is the key idea behind the notion of deep enculturation. As I pointed out in an earlier book on cognitive evolution, cultural influences can lead to the installation of totally new cognitive architectures, such as the neural wiring diagram that supports mathematical or musical literacy. This process entails setting up the very complex hierarchies of cognitive demons that ultimately establish the possibility of new forms of thought. Culture effectively wires up functional subsystems in the brain that would not otherwise exist.

When broken down into their components, the skills we acquire from deep enculturation can be reduced to chains of algorithms that control attention and emotional valences. Attention determines the sequential flow of memory fixations and perceptual comparisons, and these determine the precise quality and sequencing of subjective experience, producing unique juxtapositions in the mind's eye and influencing what habits we form and what interpretations we place on events. The emotional valences attached to various objects, events, and people are an important part of the same process of conditioning the conscious mind. Such algorithms establish the continuity of experience. There is a coherence, an interconnectedness, about conscious experiences that makes them very different from unconscious ones, where ideas and images can coexist in a pell-mell, disorganized manner and no drive for continuity tries to impose order. This can be seen in the strange, irrational chaos of dreams, where there are seemingly no rules. Gravity may be denied, personal identities may change in midstream, and the fusion of images and ideas may seem to have no more shape than a school of fish in a feeding frenzy. The flow of the unconscious can be very revealing at times. But it is easy to explain the unstructured stream of the unconscious with a few rules of association. It is much more difficult to explain how the flow of ideas in consciousness has such cogency. One possible explanation lies in the attentional algorithms we osmose from exposure to a culture. They tell us what to look at, even internally and in what order, what comparisons to make, and what conclusions to draw. They give shape to everything, from the most basic events, such as greeting one's relatives, to the most esoteric, such as reading a paper in electrophysiology.

Those deep attentional habits can push the mind inside itself, into memory, or outside, into the public cultural arena. They can also specify the mental operations that will be carried out on such cognitive voyages. These habitual sequences are automatized executive skills, woven into scripts and scenarios of considerable length. They are the vital formative algorithms of the mind. They are written, stored, and edited entirely by culture and accumulate rapidly over generations. The most significant influences of deep enculturation occur early, but they continue building, at a slower rate, throughout a life-span. It is a startling thought that the effects of culture can reach so deeply into the heart of human nature. The extreme forms of cultural relativism that have dominated so many academic disciplines in recent years have completely missed this point. Cultural relativism makes no sense in the cognitive domain, in view of the fact that cultural differences can sometimes become manifest as very deep cognitive differences. They don't always show in this way, of course. Most cultural differences are cognitively trivial. But not all of them.

Since the decoding of cultural signals places a heavy demand on conscious capacity, all the major restructuring effects of culture are channeled through the window of consciousness. The width, quickness, and depth of our conscious capacity affect the process of deep enculturation in the individual case because they establish limits, determining which algorithms might be accessible to any given child. This leaves a great deal of room for individual differences in basic capacity. But it also makes the child's consciousness the hot spot of the human cognitive universe. This is the contact point at which we all encounter and spar with the cultural colossus. By means of it, we can successfully penetrate its knowledge bases or decipher its baffling cues. We might even enter into a dialogue with it. Or we might not.

THE MUCH-MISUNDERSTOOD STATUE AND THE BIRTH OF CONSTRUCTIVISM

It is hard to imagine how our mental universe might look at its very beginning, before this process started in earnest, and how we might have emerged from nothingness. Two centuries ago a French philosopher was the first to speculate about how early experience might appear from the inside. His ideas were joltingly original then, and they still have an exciting feel of discovery about them.

Étienne Bonnot de Condillac is best known as the interpreter of John Locke's psychological ideas to the French-speaking world. Like Locke, he was an Empiricist, and believed that knowledge and ideas originate entirely in experience, rather than exist in the mind in some form prior to experience, as Descartes and Leibniz had earlier suggested. He was the first to understand that experience was so complex that it could not be deciphered without a powerful set of executive powers to start. In 1754, Condillac wrote his *Treatise on the Sensations*, in which he proposed one of the most interesting thought experiments in the history of psychology, known simply as the Statue. It is an intriguing introduction to the buzzing, blooming confusion of an immature mind, long before it can think or speak. This simple exercise ultimately extended the theoretical reach of Empiricism because it changed the way we think about thought.

Before we begin our revisitation of the Statue, there is an important distinction to be made between weaker forms of Empiricism and a school of thought known as Constructivism. Condillac did not simply propose the boringly obvious notion that all our specific habits and memories are learned. The fact that I speak English and not Inuktitut, or play tennis and not jai alai, or tend to stare impassively and fidget with a pipe instead of shrugging my shoulders and chain-smoking Gitanes is a pure historical accident, the product of my particular life history. This is not news now and was not news even in 1754. This is weak Empiricism, so obvious as to be uncontroversial. Even the hardest of Hardliners now concede that knowledge originates in experience (although many will not admit that the framework of knowledge might also originate there). Condillac proposed a much stronger, Constructivist version of Empiricism, one that is far from boring and remains highly controversial. It asserts that the very structure of the mind is set up in experience and that many of its highest capacities are not innate, in the sense of being implanted at birth, but instead are generated by the appropriate sequencing of early experience. In this view, the developing mind emerges in adulthood with a specific structure, not necessarily because that structure was innate but because it flowed naturally and inevitably from particular sequences of experience. This idea applies especially to higher symbolic capacities. These will self-assemble in a growing mind if it is provided with the right epigenetic conditions.

Condillac foreshadowed the modern Constructivist view, elegantly articulated in this century by Jean Piaget, that the mind of the child constructs itself in ontogeny. He proposed that the adult mind emerges from its relatively unstructured initial state in the infant through an active assimilation of the structuring effects of the developmental environment. Condillac anticipated Piaget in many ways. He held that many of our fundamental mental faculties are ultimately constructed from experience or, more accurately, assembled in an ontogenetic hierarchy, whose primary source lies in sensation and whose more abstract aspects are assembled from simpler ones. Condillac's model of mind was thus Constructivist long before there was a Constructivist movement.

His key assumption was an unwritten one, also adopted by Piaget, that the active constructional process is a conscious one. In his attitude toward consciousness, Condillac differed from his Idealist enemies of the seventeenth and eighteenth centuries. Leibniz, for instance, was a Nativist, who believed that the mind was born with a very specific innate structure. He placed consciousness on a completely different plane from sensation and disconnected it from the more mundane functions of the mind. In contrast, Condillac placed it on the same plane as the rest of the mind, in the very eye of the cognitive tornado. He also recognized that its two distinct aspects, process and content, should be kept separate. He saw that the conscious process must be largely built in, while its content is acquired entirely in experience. However, for Condillac, even the conscious process itself is not entirely innate. In his view, it extends and elaborates itself with experience and does not acquire a mature structure until after it has attained a fair amount of experience. Eventually it develops a very complex structure, capable of formal thought. However, very little (only the building blocks) of that adult complexity is innate. This leads to a conclusion that was too revolutionary for Condillac's time and even, perhaps, for ours: To a remarkable extent, we are our own conscious creations.

In placing such emphasis on consciousness, Condillac and his contemporaries were very different from their twentieth-century successors, the radical Behaviorists. The latter gave learning a central role but held to a largely unconscious, or automaton, model of the learner. They even managed to ban the word "consciousness" from American experimental psychology for more than fifty years. Behaviorism's experimental agenda was otherwise quite derivative of Condillac and reflected the faith of all Empiricists in the explanatory power of a general-purpose learning mechanism. Not all Behaviorists rejected conscious cognition, but the more radical ones hobbled themselves by insisting that for scientific purposes, we must behave as if consciousness did not exist. This resulted in a drastic oversimplification of the processes of learning and memory. As a result, Strong Empiricism was left only half explored. Cognitive science has since picked up the slack on the mechanisms of consciousness, and we are trying once again to understand the executive brain. But we still haven't fully grasped the relationships between the latter and our vestigial modular mind, let alone how its various subdivisions are modified by deep enculturation, in the development of the modern adult mind.

Like all generations before them, modern developmental theorists have tried to define themselves by rejecting their master, Jean Piaget. But they have retained much more than they have dropped, despite the fact that many of Piaget's proposals have not been confirmed. There are still many recent adherents to Constructivism. Above all, Jerome Bruner has been the bridge between Piaget and the current era, braving the slings and arrows of Hardliners and Minimalists alike throughout his colorful career. New York City developmentalist Katherine Nelson has built a broad synthesis of the theoretical issues involved in the self-construction of human minds. Prominent Constructivists, such as Annette Karmiloff-Smith, Elizabeth Bates, Jeffrey Elman, John Flavell, Terrence Sejnowski, Steven Quartz, and Michael Tomasello, are working to create a merger of neuroscience, evolutionary theory, and cognitive science within a Constructivist framework.

Condillac's thought experiment might justifiably be regarded as the original inspiration of this important line of thought. His Statue was conceived as a virtual mind whose sensory organs could be added or subtracted at the whim of the experimenter and whose ability to self-construct under various conditions could thus be explored. The Statue had two built-in components: sensory pathways, which provided its sole source of experience, and innate capacity, which endowed it with the ability to carry out certain operations on its sensations. Its innate capacities were very much like the conscious functions we have been discussing (the possibility of unconscious modular cognition had not yet occurred to anyone). The Statue's marble exterior prevented it from sensing the outside world, but it could be "opened" selectively and given sight, hearing, touch, smell, or taste, as the thought experimenter wished. Sensory modalities could just as easily be subtracted by closing the gaps in its marble cladding. The *Treatise* was a way of thinking through the question of whether basic concepts, such as the ideas of space and time, and fundamental skills, such as language and rational thought, are innate or acquired through experience. It forced us to explore our underlying assumptions about what tools a mind must originally have, so that it can profit from experience. Condillac had to define precisely what the growing mind's interface with the environment, on a moment-to-moment basis, actually entails. In the process of this exploration, he produced a model of the unfolding, or ontogenesis, of the individual mind. This was a purely conceptual exercise, but it examined with great clarity the possibility that a human mind might actually build itself out of a succession of experiences.

The Statue has sometimes been misunderstood as a paradigm for a tabula rasa mind, an unstructured blank slate waiting passively for experience to write something in its empty memory banks. But it was not conceived as either passive or empty. It had a built-in tendency to take an active stance toward understanding the world. Equipped with a conscious capacity much like our own, the Statue could construct a fairly complete mental world from very little sensory evidence but only if it took an active, interventionist stance. To translate this idea into slightly more technical jargon, we might think of the Statue as a powerful, but initially unstructured, nervous system, equipped with a powerful domain-general toolbox that allowed it to reconfigure its own functions continuously, so as to become able to perform increasingly more complex and abstract cognitive operations. The Statue started with a few basic inborn operations but eventually achieved a capacity for formal abstraction, building most of its higher capacities from experience. This is essentially the kind of theory that Piaget tested empirically two centuries later.

Condillac was subtle enough to realize that his Statue could not start its life cycle in a state of complete neutrality or indifference toward sensation. He gave it an innate bias to notice things and to care about what it noticed. Curiosity must be a basic property of any cognizing system, and this presents us with a strong scientific challenge. It is not obvious how a mere brain could have such a property, but clearly many brains have it. There seems to be a self-reinforcing joy in discovery, and curiosity has a subjective, sensual aspect. Its pleasures can be as intense as any others. Without this self-reinforcing feature and the intolerable pain of boredom, there would presumably be no motive force for a mind to exert any effort to understand the world.

This was recently recognized by one of the major theoreticians of connectionist computing, Jeffrey Hinton, who has tried to design a self-reinforcement feature into his simulations. In fact, we have no idea how real nervous systems acquire their self-reinforcing drive for knowledge, but there is no doubt that even simple nervous systems must be able to register the positive or negative consequences of experience and attach emotional valences to each episode.

Condillac also realized that his Statue needed an additional associative capacity, in order to remember its feelings of pleasure or pain in connection with the other sensations that accompanied them. These were part of its design specifications from the start, because without them the Statue's experience would have had no coherence. Condillac saw that once we were equipped with this associative feature, our need for coherent mental content would effectively constitute a drive system. He acknowledged that the semantic engine that drives our cognitive adventures is not an inert bank of memory modules; instead it is a strange kind of pleasure-seeking animal, an informavore that sniffs out the world's patterns and shapes and finds the experience of simply knowing things intensely pleasurable. It tolerates neither chaos nor boredom for long.

In conceding these innate propensities, Condillac avoided the tabula rasa assumption and gave his Statue a very specific engineering design. One key element in his design was the domain generality of conscious experience, which could include virtually any combination of sensations. The Statue's capacity to learn from any source or combination of sources had to reside in a set of elementary operations with general access to many channels of experience, so that it could draw comparisons between them. This concept, of a domain-general central processing capacity, amounted to a rudimentary version of the more sophisticated models of consciousness reviewed in Chapter 5. Its most basic operations included a built-in drive to seek out experience, store selected data, and carry out simple operations on them. This central architecture, uncomplicated as it appears, was a pretty good guess. It corresponds to much of what we know about conscious capacity. Moreover, it had many key elements found in modern computers, including a central processor (roughly corresponding to the whole domain of consciousness), permanent memory storage (in some unspecified medium outside consciousness), an input-output interface (the sensory and motor systems), and certain elementary executive operations (such as search, store, switch, and compare).

The Statue started as we all do, with no content—that is, with no specific experiences or memories—and in the absence of these, Condillac surmised that it could have no awareness. Conscious capacity, to be realized, must become the consciousness of something. Condillac tried to show how given its initial raw capacity for consciousness and a stream of sensations, his Statue might become truly conscious of something. In his Preface, Condillac instructed the reader on how the thought experiment should be conducted: "I wish the reader to notice particularly that it is most important for him to put himself in the imagination exactly in the place of the statue we are going to observe. He must enter into its life, begin where it begins, have but one single sense when it has only one, acquire only the ideas which it contracts: in a word he must fancy himself to become just what the statue is. The statue will judge things as we do only when it has all our senses and all our experience, and we can only judge as it judges when we suppose ourselves deprived of all that is wanting in it."

With these rules, the *Treatise* begins, with a section titled "The First Cognitions of a Man Limited to the Sense of Smell." In the range of things it can know, smell is the least of the senses. Olfaction has the smallest cortical representation and the most diffuse internal organization of all the human sensory systems, save taste. Along with taste, it is also the most archaic and deals only with the proximal environment. Thus anything that can be built from experience on the basis of smell alone could presumably be exceeded by the other senses. The Statue's smell mind could only smell, remember past smells, and feel pleasure or pain in response to each new or remembered smell. Such a consciousness, wrote Condillac, "can no more have ideas of extension, shape or of anything outside of itself, than it can have ideas of color, sound or taste. Only relatively to itself are odors smelled. In regard to itself smells are its modifications or modes. It has no idea of matter. It could not be more limited in its cognition."

Condillac's strategy here is to simplify the problem of development by making it as uncomplicated and unidimensional as possible. Instead of a mind of many senses, embedded in a world too complex to describe in words, he reduced the Statue's experience almost to the point of extinction. For those of us with a normal array of senses, it is hard to imagine a "consciousness" that consists only of smell, a sort of cognitive black hole without sounds, companions, places, events, or actions. The Statue even lacked a sense of its own embodiment. It had no experience of touch and no internal body senses by which to track the activity of its muscles, joints, tendons, or viscera. It was also paralyzed. It could not even generate simple voluntary movements, lacking the quadriplegic's capacity for eye, head, or facial movements. Thus it faced the world without any possibility of reacting to it, exploring it, or expressing itself overtly. It was nothing but a passive, nearly empty container into which nothing but odors could be dumped. It was like the bandaged, immobilized Soldier in White in Joseph Heller's *Catch-22*, except that unlike that unfortunate soldier, the Statue had never known anything else.

Astonishingly, Condillac found that despite being limited to smell, such a mind buzzed with growth, which flowed from the stream of experience itself. Condillac had zeroed in on the assumption that defines Constructivism, that the conscious mind is the amodal governor of its own domain; it weighs the evidence it receives, however limited, and constructs a world from it. This particular domain is initially unspecialized, at least with regard to the kinds of conceptual material with which it works, and can take many shapes, depending on the particularities of its own experience. Since it receives input from any incoming channel that happens to be opened, it might experience a variety of possible mental worlds. During the two centuries since the *Treatise* was published, this amodal domain has become known as the central processor or central executive, depending on whether you prefer engineering or boardroom metaphors. It was the grandfather of the Global Workspace, mentioned in Chapter 5, and anticipated its relationship with a complex executive brain apparatus. Condillac realized that this active central process was the control center for mindbuilding, or cognitive self-assembly.

However, the Statue began with more elementary steps, which connectivist programmers will recognize as important and dauntingly difficult to achieve. It had to start by learning to distinguish memories from sensations. The significance of this step was made clear in a little section titled "How It Would Be Limited without Memory": "If there remained no recollection of former modifications, then on occasion of each sensation it would believe itself to be feeling for the first time. Whole years might be swallowed up in each present moment. Were its attention always limited to one mode of being it would never be able to take account of two together, and never be the judge of their relations."

But how would this stripped-down mind, this purely olfactory consciousness, remember anything at all if it could only receive a series of smells? At this rudimentary stage in its development, memory would be confounded with sensations. It was, after all, only one of the two ways the Statue could feel, one being a state of specific smells related to the past, the other of smells related to the present. But how could it know whether a given smell was present or past in its origin? Condillac surmised that when the Statue receives a new odor, "[it] has still present that which it had been the moment before. Its capacity of feeling is divided between the memory and the smell. The first of these faculties is attentive to the past sensation, while the second is attentive to the present sensation. . . . [It] usually distinguishes the memory of a sensation from a present sensation only insofar as it feels the one feebly, the other vividly. I say usually because memory is not always a feeble feeling, nor is sensation always a vivid feeling."

Condillac thus held that memory, in its earliest stages of development, should be regarded as *active* feeling because the cause of the memory is internal, whereas sensation is *passive* feeling since it is externally triggered. The Statue could not initially differentiate between the two states because "it never perceives anything which is not or has been its own state of being." But with repetitive use, memory—that is, the active recall of past experience—became a habit, and the Statue could routinely recall its past states. The lesson is an interesting one. In consciousness, memory must be gradually differentiated in experience from other types of feeling, primarily by the fact that it is less vivid, but this realization comes only as the result of experience. Memories feel different from sensations. This is a learned discrimination that would have to occur very early, and although such initial steps in mental development would not be recalled later, they would nevertheless provide a foundation for a great deal of subsequent experience.

Another capacity emerges out of the differentiation between memory and sensation. This is the ability to make comparisons between experiences (again, Condillac obviously refers to conscious comparisons). With several past states to recall, the developing Statue reached a point where it could compare past states with one another, for "[c]omparing is nothing else but *giving attention to two ideas at the same time*. When there is comparison there is judgment, the perception of a relation between the two ideas which are compared. Thus it acquires the habit of comparing and judging. . . . Consequently it is sufficient to make it smell other smells in order to make new comparisons, form new judgments and contract new habits. [The italics are mine.]"

Condillac realized that a diffuse cognitive system could not compare two past memories unless it was innately able to divide its attention between two mental objects. "It experiences no surprise at its first sensation because as yet it is unaccustomed to any kind of judgment. . . . But it cannot fail to notice the change when it passes suddenly from a state to which it is accustomed to a quite different state, of which it has as yet no idea. Surprise makes it feel more strongly the difference of its modes. Its attention, determined by the pleasures and pains it feels, is now applied with the greatest liveliness to the succeeding sensations."

Some readers will recognize in these words the prototype for the modern idea of automatic attention. This is a built-in feature of the mind, a "system interrupt," which cuts into the stream of consciousness whenever inputs violate its predictions. "Where odors have attracted attention equally, they will be retained in memory according to the order in which they occurred and that order will be the order of succession. If the succession includes a great number, the last impression, because the newest, will be the strongest;

the first impressions will be weakened by insensible degrees, and at last extinguished altogether. Should there be some to which only small attention has been given, they will leave no impression, and will be forgotten as soon as perceived."

Experiences are thus held in a temporary holding tank, their freshness reflected in the vividness of each trace. The very short-term trend of experience is projected from this sample, and deviations from the expected pattern automatically attract attention. The interconnections between short-term storage, automatic attention, and mental modeling all are recognized in this short passage. Condillac's message is that it would be impossible to construct a complex mental model of an amodal perceptual world without these basic executive capacities.

Recognizing the connection among drive, motive, and memory, Condillac placed an additional design requirement on his Statue. Drive makes memory active, and without it, its activity ceases. Memory itself takes two forms, one weak, the other vivid. The latter is called imagination, and if it is vivid enough, it can capture attention completely and distract it from sensations. Thus there are three basic states of attention in the Statue, one directed toward sensation, another toward memory, and a third toward imagination. These are learned states. The Statue cannot initially distinguish between them. The distinction comes only with experience, through serial experiences and comparisons. These serial experiences might be called Condillac sequences, as shown in Table 6.1.

TABLE 6.1

Condillac sequences: the field-differentiation of awareness into sensation, memory, and imagination. This occurs as the direct result of specific challenges to the conscious process (left column), which stimulate its components to set up various operations and working memory spaces. (t=time)

Content	**Process**
(Experiences, in approximate order)	(Learned executive operations)
t. 1: Sensation X	Focus attention externally
t. 2: Memory of prior sensation X	Focus attention internally
t. 3: Compare sensation with its memory	Split attention between two fields
t. 4: Fading memory of sensation X	Distinguish levels of vividness of trace
t. 5: Weak sensory trace	De-focus external attention
t. 6: Attempted imagining of sensation X	Exercise active memory (imagination)
t. 7: Imagine sensation X	Focus attention internally
t. 8: Compare active and passive memory of X	Distinguish imagination from perception

Such sequences serve to establish the internal cognitive structure of the executive brain. Condillac achieved something important here. He showed that the conscious apparatus starts as a very simple device, which reflexively attends to input, holds it in memory for an indeterminate period, and makes serial comparisons of its experiences. This feature allows the nascent mind to differentiate the three major fields of awareness. Even processes like attention and voluntary imagery are not entirely inborn in the Statue. They must be exercised, configured, and structured in early experience. The splitting of attention must be practiced, although it eventually becomes automatic, its early learning promptly forgotten. The basic machinery of comparison and recollection is built in, and its own development takes place so early that its unfolding is lost to memory. This period leaves no retrievable record. This early differentiation of the child's inner world guarantees that memory and attention can never be neutral. From birth they acquire bias.

Condillac assumed that the olfactory Statue's imagination must necessarily be more active than a normal one because it could not be distracted. He assumed that a much richer sensory experience would produce a feebler imagination because the mind would be constantly under the spell of the external world. Again, an interesting assumption is revealed here. Condillac evidently thought that trade-offs are endlessly involved in the process of mindbuilding and that no mental ability is absolutely fixed in its dominance, relative to other abilities. Anything may be changed in epigenesis. Mental capacities compete for space in the head, and some inevitably crowd others out of the picture during development. Prolonged domination by one basic attitude of attention would lead to the dwindling of others. Too much sensation might prevent us from building a habit of active imagination and make deeper memories feebler on recall. Mental faculties compete with one another early in development. This idea has a modern parallel in Gerald Edelman's theory that neural networks compete with one another as the brain develops.

Condillac then tackles the problem of how the Statue's imagination could have developed its autonomy from the outside world: "Our senses, always on guard, keep us constantly warned against the object we are trying to imagine. The ease with which we can avoid objects hurtful to us, and secure those we want, brings about a still further weakening of our imagination. But since our statue can only escape a disagreeable feeling by vividly imagining a more pleasurable state, it will exercise its imagination to procure effects which its actions cannot procure."

One might think of dreaming as an example of a state in which distractions are normally reduced, and the imagination is relatively unconstrained by external reality. The smell-bound Statue, with little information about external reality, would have no distractions, even when fully awake, since even the most unpredictable event would, from its viewpoint, consist only of new smells. Assailed by a strong sensation, however, the Statue would become completely and perfectly passive and would not recover its imagination until the sensation lost its vividness. Note the emphasis on the power of the imagination to review and alter experience in memory. The Statue could change the order in which sensations were registered, for instance, or change the associative patterns between them: "[O]ften the Statue will remember less of the order in which it has experienced its sensations than of the order in which it has imagined them." Only by "the exercise of the faculties of imagining and remembering" can we learn to recognize sensations that we have already had. In this, Condillac anticipated Frederick Bartlett's idea that recall is an imaginative activity, which can reconstruct reality, rather than a simple matter of pulling veridical memories off the shelf.

The smell Statue's mind developed into a surprisingly more elaborate adult than Condillac had expected. Starting with a diffuse capacity and restricted to smell sensations, the Statue had acquired six operational capacities through the early differentiation of its experience: attending, recollecting, comparing, judging, imagining, and recognizing. All these had to be rehearsed and reused constantly to be maintained. They represented a simplified picture of the normal developmental sequence of a child with a full range of sensory experience, but the principle stands. Although Condillac couldn't have known about genes, we now realize that a plastic self-assembly process such as this would require relatively little specification at the genetic level. An engineering design with a small number of innate features might nevertheless result in a complex adult operational structure, if it is provided with the necessary attentional and memory resources.

His Statue might not have experienced very much by our standards, but Condillac was not really interested in how rich its perceptions were. He was less concerned with content than with process. He wanted to know how a mind might naturally, and consciously, grasp the distinctions between the various sources of experience—internal versus external, present versus past—that constitute a framework for all complex mental operations. He never suggested or implied that the Statue had any representational awareness of what it was doing, only that it was aware and drawing on its conscious reserves to build its own mental universe. Still, even with this amount of progress, the Statue was locked in a very tight box. Would such a sensory-deprived Statue linger in a diffuse psychotic state, unable to distinguish fantasy from reality or past from present? Not so, thought Condillac. Even a smell-bound Statue could have abstract ideas, although they would be very different from the kinds of ideas we might expect in a more typical mind in possession of all its senses.

Of course, a one-dimensional Statue could know nothing of the event structure of the outside world. It had no way even to judge that there were objects out there, and people, and entire societies, that were separate from it. Moreover, Condillac concluded that in the absence of a distinction between self and nonself, the Statue could never acquire language. But this might have been entirely due to the fact that smell is a very limited sensory modality. Would it have done better if it were given the more powerful distal senses, seeing and hearing? To answer this question, Condillac extended his thought experiment, granting the Statue access to taste, sight, sound, and various combinations of these. His conclusion was surprising. None of these senses, not even vision, could give the Statue a sense of the external world as long as it remained completely passive. Vision and hearing alone would follow the same developmental route as smell and could not provide any basis for discerning objects or distinguishing the difference between self and the outside world. A mind limited to passive sensation cannot even establish that there are external objects. It can know experiences only as states of itself. It can experience its own mentality only as a series of variations on sensory quality.

This situation changed once Condillac added action and touch to the Statue's sensory capacities. His definition of touch included all the somatic senses of heat, cold, light pressure, deep pressure, and vibration, plus what we now call proprioception and kinesthesis, the internal body sensations from receptor organs in muscles, joints, tendons, and internal states. It also included active, or haptic, touch. In the chapter titled "How We Pass from Sensations to the Knowledge of External Objects" Condillac gave his Statue this capacity and stripped away all its other senses. The Statue became deaf, blind, and without taste or smell, but it was suddenly capable of movement and active touch. Under these circumstances, he concluded that the Statue would quickly develop a sense of its own embodiment:

> It is evident, then, that we pass to the knowledge of bodies only when our sensations produce the phenomenon of extension, and since a body is a continuum, formed by the contiguity of other extended bodies, the sensation which represents it must be a continuum formed by the contiguity of other extended sensations. We have not found this character in any of the sensation which we have observed. . . . Although the statue ought to have sensations which it perceives naturally as modifications of its organs, yet it has no knowledge of its own body, at the time it experiences such sensations. In order to discover itself it must be able to analyze, that is to say it must successively observe its self.

Once it could differentiate the self, the Statue could distinguish among space, objects, and a range of abstract ideas as well. The active-touch, or haptic, Statue proved to be a far more knowledgeable being than any version of the passive Statue. Condillac went on at length to establish that once the Statue had active touch, the introduction of other senses would have far-reaching consequences. In other words, action can greatly amplify the informative value of sensory data. Once vision is complemented by a capacity for active exploration, the entire visual world can be differentiated into events, actions, and objects. The same applies to hearing and the other senses. This conveys one central tenet of Constructivism: Action begets self-knowledge. Structures develop in an amodal manner, building and accumulating into the domain-general functional apparatus of adult thought, through active exploration.

But Condillac's bottom line was really about the absolutely essential role of consciousness. The assembly process relies heavily on the resources provided by conscious capacity. Those resources are not in themselves conscious content. Rather they are the process that generates the content that gives awareness its specific shape. These ideas are still alive in the current theoretical literature. Assuming, in our modern context, that the Statue's virtual perceptions must be the product of some form of neural binding, we should not expect it to bind together complex percepts without the necessary attentional tools and sufficient working memory capacity to cut to the chase. It needs working memory and split attention to select, evaluate, and compare the raw sensory data that must be bound together to form very large, unified event representations. Following Condillac's developmental logic, the enlarged capacities I outlined in the section on level-3 awareness would have enabled the human brain to bind unusually large event percepts, relying on our intermediate-term working memory capacity. In other words, the development of awareness is all about scaffolding the binding process into successively higher structures and forming a hierarchy of executive skills.

Condillac also realized that the major fields of experience are not given but derived from experience. By this he meant imagination, memory, and sensation, which are three vastly different arenas of consciousness in adults but do not begin that way. In fact, he probably underestimated the degree to which these fields are later cultivated and the degree to which they remain vulnerable to cross talk, even in adults. The confusion of imagination and reality is a common experience and usually a disturbing one. The stream of consciousness is sufficiently complex that a child could not be expected to ratchet itself up through successive stages of development, define the fields of awareness clearly, and differentiate them endlessly into finer and finer territories without formidable executive resources. The underlying unity of our own mental world is something that we must invigilate furiously, even as adults. We do not tolerate lack of unity for long unless we are suffering from certain forms of psychosis.

Condillac may come out of this with the appearance of an almost frightening prescience. But he did miss some important things. With two centuries of hindsight we can say that he did not realize some of the formidable obstacles that faced his Statue's advance to fully symbolic cognition. Oddly, he never thought to question the myth of the isolated mind and left his Statue disconnected from society. As a result, the Statue was abandoned in a cultural cellar for a long time and the beating of its Telltale Heart fell on deaf ears until well into the twentieth century, when Vygotsky suggested that immersion in culture might be the key to its intellectual liberation. Condillac, and most of his successors, failed to realize the possible significance of culture as a means of raising the Statue's awareness to a new plateau or connecting it with a wider cognitive universe. But he took the first important step toward such a possibility by recognizing that consciousness is the arena in which all major cognitive battles must be fought.

MANDLER'S DICTUM

Condillac's notion of conscious self-assembly converges with an idea that has a long, distinguished, and independent history in psychology, the principle that learning and skill acquisition occur only, or largely, in consciousness. It is telling that laboratory studies of infant development rely heavily on tracking attention to assess what children know. In adults, under most circumstances that demand new learning, attention coincides with awareness. So it is, undoubtedly, with infants. Infants shift their attention, and as they do so, they move their eyes and heads and pause in whatever they are doing, just like adults (and they do this without any of the duplicity they will acquire later on). This allows us to judge what they know about the world by finding out what surprises or bores them. One common technique is to give them pacifiers and measure changes in their sucking rate. When infants become interested in something new, their attention goes into high gear and they start sucking furiously. Once they have become used to their new state of mind, they slow their rate of sucking. This method can reveal how sensitive they are to subtle changes in their environment. An even more powerful index of their attention is tracking eye movements. This method has taught us a great deal because it not only reveals when infants shift their attention but also shows where it is being directed.

The assumption implicit in this methodology might be called Mandler's dictum, after one of the founding fathers of the Cognitive Revolution, George Mandler, who argued long ago that consciousness is essential for learning and memory. His proposal might be summarized thus: As conscious attention goes, so go learning and knowledge. A corollary of this is that virtually all learning involves conscious processing. Another is that infants must learn where to attend, and when, very early in life. As they acquire new attentional tricks, their connection to the social environment is given shape, interlocking the child's attention with its internal and external worlds. This follows logically from what Condillac wrote about his Statue. Because of the close synergy between attention and learning, the direction of attention invariably reflects the models we already have in our minds. Attention is almost always guided toward the solution of a problem or challenge according to a learned strategy. The mind might resolve a novel perceptual challenge by creating new perceptual and conceptual categories, but in the process, it will also generate new markers for the future direction of its attention. Consciousness helps us carve pathways through the apparently incomprehensible mazes of culture, by leaving behind a chain of very specific directional markers. The child has to find those markers and learn how to follow them, or the cultural world will remain opaque. As the mind gets better at this, it will find more difficult paths that escaped its earlier notice. Attention seeks out more exotic details in this way. As we grow older, our cultural world expands because our attentional paths lead us there.

Attentional sequences can thus become automatized. This assumption has been verified many times. The fine structure of an activity, such as reading, is dictated by a maze of attentional algorithms that have become completely automatic and guide the internal and external scanning of what is being read. They are buried beneath the surface, but without them we become like children again, baffled by the complexity of the printed page. These automatized routines are never absolutely rigid. Eye movement patterns are never exactly the same on two occasions, nor are our internal scanning operations. They have to make accommodations for the specific conditions of each new context, but once these are established, adjustments are made without much, if any, conscious monitoring. In this sense, selective attention can become independent of conscious regulation. But this does not discount the role of conscious supervision in attention. Rather it enlarges it. Only consciously supervised rehearsal can fix those attentional demons firmly in our memory banks, so that our consciousness gains higher-level control over its own attentional routines, governing the whole hierarchy that it set up in the first place.

Instant boredom is one of the hallmarks of the restless, relentlessly curious, novelty-driven infant mind. Its attentional responses are being programmed and automatized at an incredibly rapid rate, as it churns its way through the field of its own experience, like some fantastic threshing machine, sorting, baling, and discarding impressions and ideas. Its awareness sits juxtaposed between old knowledge and new. When its interest in something dies down or habituates, it looks for something else, preferably recognizable but somewhat unpredictable, from which it might learn something new. Children allocate attention in essentially the same way as adults, deliberately and voluntarily. When infants are newly born, for a few weeks their attention is largely under automatic control, dictated by innate reflexes. During that time they respond to the brightest, strangest, or loudest stimulus and to certain special stimuli, such as facelike shapes. But very soon after birth they begin to inspect and examine the world more deliberately. Their attention wanders. They try this and then that, and they don't like being bored. They guess at the structure of their world and develop and test expectations. They habituate to routine events very quickly, and their preference for novelty speeds their attention toward those situations where learning opportunities abound. This leads to new expectations and circular interactions with the world, in a developing chain.

The point is that the child's consciousness of the world doesn't appear suddenly, in an infantile epiphany. Nor does it appear by philosophical fiat, for instance, by having the child pass an exam in language at some specified level, to be granted permission by the Academy to become aware. Nor is it a question of attaining a certain kind of selfhood, along the lines of our recent Western notions in that regard. Nor must the child's concepts of time and space meet our cultural standards. All these are important factors in expanding personal awareness, but the elements of conscious capacity are there from the start, directing the self-assembly process and ordering the mind's growing world. As the mental world grows, so does awareness, and vice versa.

There is no evidence that infants learn anything of significance outside of awareness. For the most part, the vector of attention follows the vector of learning. The mind attends, registers, and changes its bias, and the next encounter with the same situation fixes the chain of attentional habits that directs future learning in that context for a lifetime. Hardliners may argue that children acquire language without conscious effort, but they can't really mean what they seem to be saying. Surely they imply only that a child has no way of representing what it is learning. I hope they do not mean to suggest that children learn language without investing any conscious capacity. Aside from the absence of any empirical proof of this curious notion, modern research has contradicted it flatly and conclusively.

Peter Jusczyk has shown that infants as young as eight months have expectations, even in the realm of language. This is revealed by their pattern of curiosity, which can be tracked by their eye and head movements in a forced choice situation. They generally spend more time looking at novel stimuli, and Jusczyk has exploited this tendency very effectively to show that they build a mental model of language in incremental steps at a very young age. Early on infants pay particular attention to the sound envelopes that surround the use of their native tongues, such as the identities and emotionalities of the voices they hear. They learn to recognize the average spacing of pauses in the speech stream and its texture. They notice violations of what they expect to hear and direct their attention toward those violations. They know about the "prosodic," or emotional, features of speech a very long time before they can understand its meanings. Later they use this knowledge as an attentional hook to help them home in on the intended meaning of what they hear. The speaker's tone of voice tells the infant where and when to look and marks the significance of the utterance. The situational context usually provides the meaning. This reduces the difficulty of tracking the content hidden in words and sentences.

The infant brain is thus modeling the world and constantly building new versions of it, which in turn affect its future biases, in an unbroken iterative loop. This allows infants to scaffold their learning in a developmental hierarchy. Children predict, monitor the environment for exceptions to their predictions, and update them accordingly, all under the guidance of their conscious capacity. Far from being passive absorbers of linguistic knowledge, children are active, vigilant, curious, and, above all, deliberate in their approach to acquiring skills, including language. As their fund of knowledge grows, they are increasingly selective in the way they allocate their attentional resources, and their competence grows in a scaffolded, hierarchical manner, at different rates in different domains.

This does not deny that some reactions are inherently automatic. In fact, we have many innate reactions, especially emotional reactions. But most of the higher cognitive operations of the mind are acquired, painstakingly and deliberately. This does not imply that children can articulate what is happening while acquiring them; of course, they can't. That kind of sophisticated representation of what is known is a specialized linguistic behavior and is culturally quite recent. Sometimes it cannot even be found in adults with full language mastery.

After learning has taken place, consciousness does not necessarily continue to play a role in a particular behavior. Complex behaviors are assembled in stages. Initially they draw heavily on conscious capacity and dominate awareness. But after extensive practice, they become so automatized that they no longer demand much conscious monitoring. A common example is learning to drive. This skill makes tremendous demands on consciousness while it is being learned, and its component routines must be mastered separately. Thus we may learn to start, steer, brake, park, back up the car, and look out the rearview mirror in separate sessions, then gradually assemble these into an integrated skill. Eventually we learn how to drive in city streets in traffic. This requires painstaking self-assembly. It takes time and a great deal of focused concentration. But time and effort pay off, and after several years the skill becomes so automatic that we can drive through heavy traffic while carrying on a conversation, listening to the radio, and searching for something in the streetscape, without much conscious supervision. The dozens of little subroutines that were consciously wired together into "driving skill" have become automatic, and our awareness can be directed elsewhere. The whole skill hierarchy, including some high-level evaluative routines, has been sent downstairs to the unconscious. This automatized package is built on top of another, older hierarchy that includes myriads of attentional algorithms learned earlier in childhood, such as those that told us where to direct the eyes, how to sit, how to judge distances, how to scan the visual field, what to look for in it, and so on. How do we pack such complexity into our brains? That is a question for future generations to pursue. But we know it cannot be done outside of consciousness.

The ultimate example of a complex skill hierarchy is language. It is also assembled in stages and makes heavy demands on conscious capacity. When children are first learning to talk, they direct inordinate amounts of attention toward the imitation of the spoken sounds they hear. Over a period of months they progress through various stages of babbling, gradually narrowing down the range of sounds they produce, until their repertoire begins to resemble the selection of speech sounds that are normal in the home environment. Videotapes taken during this period show how carefully children attend to their communicative environment. They hover, watch, and listen in fascination to the mysterious games being played around them. Moreover, during the so-called critical period for language acquisition, their attention is more easily captured by the complex language-related behaviors they observe. They certainly do not look as though they were acquiring language unconsciously! Infants study the speech of people around them just as intensely as, say, medical students watching their first surgical operations. They need multifocal attention to find out what words mean. Yet complex as this process is, once the speech hierarchy is acquired, it also goes downstairs, and it becomes just another automatized skill, albeit the most complex one in the universe of learning.

Reading is an even more important case in point because theoreticians do not normally suggest that there could be a built-in genetic program for reading (reading is so new that there has been no time to evolve one). The acquisition of reading makes great demands on conscious capacity and involves a great deal of off-loading to the unconscious. The total conscious load imposed during the learning of advanced literacy skills is enormous and absorbs our attentional capacity for many years. With practice, reading should become automatic, and the process itself should become virtually invisible. But if the learner cannot achieve this automaticity, the result is a crippling dyslexia. Many so-called dyslexics suffer from an attentional disorder, rather than a specific deficit in reading per se. An expert reader must be able to focus on the meaning conveyed by the text, not on the nuts-and-bolts attentional algorithms that deliver the final perceptual product. As in the case of learning to speak, its whole painstaking acquisition history is soon forgotten. But it is easily recalled when we are faced with having to master reading a truly novel form of writing. To remind you of this, I would simply ask you to learn a new script. Learn to read, say, the text of *Macbeth* in Chinese ideograms. I guarantee this much: You will never be more conscious, and more deliberate, than while you are learning such a demanding set of operations. But once they are learned, the demons will go downstairs, the ideograms will become invisible, processed down below, in the automatized basement of the mind, and your attention will focus only on the meaning of the text.

If we combine these two notions, conscious self-assembly and conscious learning, we have a solid rationale for assigning a Constructivist role to consciousness. Conscious capacity is effectively a filter through which all significant developmental events, at least cognitive ones, must pass. This gives awareness a frankly functional and highly adaptive role in development. It also grants the mind eventual control over what it can become since in the long run, contextual knowledge gradually dominates the uses of the executive system. Moreover, at a stroke this provides a strong alternative to the Hardliners' passive version of consciousness. Instead of a helpless spin-off, the conscious mind dominates. It fashions most of the unconscious demons of mind, from inception to demise. In fact, one could say that the unconscious exists largely at the whim of consciousness. It consists, for the most part, of flotsam and jetsam, off-loaded attentional chains, scripts, scenarios, and so on. They are kept in the basement, dragged out and used when needed, and then stored away until the next time they are summoned into awareness. The relationship is two-way, but asymmetrical. The unconscious can intrude into conscious life, but the active, constructive agent is consciousness, not the unconscious.

THE EXTRAORDINARY MIND OF HELEN KELLER

The Statue of course was pure fiction. It stood facing the world with infinite potential, as we might expect in a thought experiment that didn't have to deal with the inconveniences of empirical proof. As fiction it could exist on a shoestring, so to speak, and Condillac imagined that it could construct a fairly complete mental world from much less than the usual evidence, climbing the cognitive heights even without the two great sensory systems, vision and hearing. But clever as this tale may have been, it was mere fantasy, blue-chip fantasy perhaps, but no more. Where was the empirical proof that a mind, trapped inside a body without sight or sound, could achieve any significant capacity for thought and awareness? How do minds in the real world cope with the problem of inventing a personal symbolic universe from minimal experience? What does their experience tell us about consciousness?

The most deprived a person can be, without sinking into nothingness, is to be without sight and sound. The blind can fall back on hearing, and the deaf on vision, and this allows them to maintain contact with the cultural environment. However, people who lack both vision and hearing are isolated. They cannot see the gestures and signs of the visual world and lack any access to speech, so that they rarely develop language of any kind. They often have no significant contact with their culture and lack a capacity for symbolic thought and awareness. There have been a few exceptions to this, and understandably, they have attracted a great deal of attention. They have pushed the limits of the human mind and challenged our most basic assumptions about its natural form. Among these exceptional people, the most extraordinary was Helen Keller. She has been popularized, idealized, and romanticized, almost to excess, but has not received the serious attention she deserves in this generation.

In a sense, the initial predicament of Helen Keller, who became suddenly deaf and blind at the age of eighteen months, was similar to that of Condillac's imaginary Statue, except that she had experienced normal development as an infant, and this undoubtedly worked to her advantage in the long run. Nevertheless, disaster struck while she was still an infant, and her two major channels of experience, vision and hearing, were permanently closed off. As a result, her speech development not only came to a halt but reversed. She soon lost all but one of the few words she had learned. She became withdrawn and difficult to manage. Frustrated and angry, she retreated into a shell, and despite the efforts of a loving family to help her, she remained there for more than five years.

But Helen was very different from most cases of early deafness and blindness, where the prognosis is very poor. She achieved some remarkable, some would say incredible, intellectual feats. She acquired a peculiar kind of sign language, learned to read and write and to type. She learned several languages, attended high school and university, and was graduated with honors at the head of her class from an elite Ivy League college. She became an author, activist, and celebrity and traveled the world to raise money for the cause of educating the deaf and blind. She gave countless speeches through an interpreter and promoted many causes, including suffrage for women, workers' rights, socialism, and Maria Montessori's ideas on education.

This alone would have qualified her as one of the wonders of the modern world, but she surpassed even these achievements in what she apparently knew. As an adult she had strong opinions on many issues that she could not possibly have understood in the ordinary way. For instance, she became an ardent believer in Swedenborgian philosophy, wrote eloquently about her religious beliefs, and formulated strong views on child rearing and practical psychology. But what can such ideas mean if you have never seen the world, observed society directly, or heard any of its sounds since infancy? What is external reality if your only contact with the world is by touching, tasting, and smelling it? What is thinking if your only way of thinking is to move hands and fingers that you cannot see? Helen defied common sense and threw away the rule book. She used expressions such as "I see" and "I have heard" appropriately, while being unable to see or hear. She perceived the subtle social nuances of human relationships, while remaining unable to enter society directly. How could this have come about? What is the nature of human abstraction that we can understand and even formulate such ideas in the absence of all the usual sources of imagination and memory?

Helen did not come by her knowledge easily. Everything she did was so difficult that most people would have given up early in the learning process. But she worked furiously at mastering all she encountered. This was evident even when she was very young. She seemed possessed by an inordinate ambition, and one can only speculate what she might have been like had she retained her vision and hearing. As it was, her achievements were the product of enormous discipline and work. She maintained a huge correspondence with many of the major personages of her era and wrote several books. Now, writing a book—even a conceptually simple book, such as a children's story or an autobiography—is a grueling enough task for anyone. It takes a great deal of organization and stretches memory to the limit. As someone who has always depended heavily upon sight and hearing I find it hard to imagine how a person without them could even imagine the possibility of creating a book, let alone actually produce one. Moreover, she wrote not one but fourteen books, typing out her ideas and laboriously proofreading them in various tactile media. This took her years of incessant day-in, day-out concentration.

In fact, the discrepancy between what Helen should have been able to know, given her direct experience, and what she obviously knew greatly troubled many of her detractors. Could she have cheated? Although she received more than her share of medical and scientific study, there is an element of mythmaking in her case that tends to raise the hackles of scientists. Many of her contemporaries were skeptical of her achievements and looked for some way of devaluing what she had accomplished, especially during her first decade of fame. While her performances were accepted by the public, they were quietly disbelieved by many experts because they knew what a gauntlet she had thrown to the establishment. The scale of Helen's achievement is like finding an exception to the law of gravity. It has enormous implications for theories of the mind. For this reason, we need to examine her case carefully, with a healthy dose of skepticism.

We should confront our most obvious suspicions immediately. She was very close to her teacher, Annie Sullivan, who served as her eyes and ears for several decades. Could Annie have cleverly deceived the public? Could the two of them have collaborated in perpetrating a fraud? How do we know that Helen really said what she apparently said or that her books or opinions were really her own? I confess that these questions were prominent in my mind when I first became interested in her case. But after spending considerable time in the archives of several institutions that hold her papers and records, I am convinced that her achievements were indeed her own and that her story is, if anything, more believable than most, precisely because her life was played out in a glass jar, under intense scrutiny.

An anecdote might convey the rigors and pressures of the strange, strict atmosphere in which she lived. Helen decided that she wanted to attend Radcliffe, one of the most academically demanding women's colleges in the United States (she later wrote that she would have preferred Harvard had it been possible for a woman to attend). The president of Radcliffe was hesitant. Worried about the reputation of the college, he was concerned that Helen should have no "advantages" over the other girls attending his prestigious institution and that she should meet precisely the same entrance requirements as all the other applicants. He didn't want to be vulnerable to charges of favoritism. He might also have suspected fakery. This was par for the course in Helen's life. To guarantee that there could be absolutely no possibility of cheating, he stipulated that Helen's tutor, Annie Sullivan, could not be present in the room during the entrance examinations. So young Helen was led, stumbling, by the hand into a large room, where she was seated at a large table. The only other persons in the room were the proctor and an interpreter fluent in braille, who copied the questions into braille notation, but was not allowed to communicate with Helen. Helen had to complete the examinations in the usual time, with the additional stress of knowing that her every movement was under observation. She wrote the tests and, as everyone knows, did very well, despite the draconian precautions taken.

Well, then, the diehard skeptic might suggest that she couldn't have been totally blind and deaf. Maybe she had a small degree of residual function in one or both senses and took advantage of that fact. This possibility can be raised legitimately because she did not have a full neurological examination in her early life. However, it is extremely unlikely, given her family history. Helen was the child of a devoted mother and father, who had financial resources and political connections. She was described by relatives as an unusually bright, affectionate infant, who walked at one. She had been completely normal before her illness. Indeed, she was described as a very active, spirited, and occasionally mischievous child, who already understood a fair amount of language. At the age of eighteen and a half months she contracted a fever and, within two days, became unresponsive to both light and sound. Her mother discovered this while bathing her, after her fever had abated somewhat. She noticed that Helen's eyes did not follow her hand, and her head did not turn to her voice. This fact is hard to doubt. It was backed up by all members of the family and by their doctor. The disease, although diagnosed only as fever and congestion, was later thought to have been meningitis, which would indeed have produced similar symptoms. No specific diagnosis exists because in those days medical knowledge was limited, record keeping in rural Alabama was poor, and the family refused an autopsy on Helen's death.

These substantive facts were never seriously questioned by any of the doctors who examined her during the following years or by any of her caretakers, teachers, or biographers. The definitive neurological description of Helen Keller's sensory loss was written by the well-known neurologist Frederick Tilney, who was president of the American Neurological Association at the time. He published an exhaustive examination of Helen when she was in her forties and stated that she was totally blind. In fact, during the examination of her eyes, Tilney found no retina. He also judged her to be completely deaf, with no residual hearing by either bone or air conduction. With the cooperation of several agencies that hold her records, I attempted to track down her earlier medical history, especially anything held by the physician who saw her in Baltimore, or by Alexander Graham Bell, but without success. Thus I don't know whether she retained some residual sensation in her earlier years, but in view of her history, this seems very unlikely. Helen later remembered having occasional diffuse visual sensations for a period following her fever, but these apparently took the form of an oversensitivity to bright light, which produced severe pain, and would not have been useful.

Could her other senses have become overdeveloped to compensate for her loss? Tilney had expected to find that she had extraordinary sensitivity to smell and touch. But contrary to his expectations, her thresholds were generally within the normal range. However, he might have missed something important, especially when measuring her touch sensitivity. His examination was restricted to very basic measurements of touch thresholds, consisting of only the lightest, faintest things that she could feel. A more comprehensive testing procedure, including tests of active, or haptic, touch and of her capacity for touch imagery, might have yielded very different results. In her own words, which would have warmed Condillac's heart, active touch was her main road to knowledge. "My hand is to me what your hearing and sight together are to you. . . . All my comings and goings turn on the hand as on a pivot. . . . It is not for me to say whether we see best with the hand or the eye. I only know that the world I see with my fingers is alive, ruddy, and satisfying. . . . In my classification of the senses, smell is a little the ear's inferior, and touch is a great deal the eye's superior."

To judge from these words, she defined touch in the same way as Condillac: as an active and extensive modality that included the body senses and internal feelings. In light of this, one might expect that her haptic senses would have been more refined than the statistical norm. However, we shall never know since Tilney did not test them.

Tilney did find one exceptional area of sensitivity in Helen. She was much more sensitive than most people to vibrations in the frequency range of zero to five thousand cycles per second. In this she resembled another famous deaf-blind person, Laura Bridgman. Laura, who lived in the late nineteenth century, was the first documented case of deaf-blind achievement. In her time she had been as famous as Helen Keller. There was a personal link between the two women. Annie Sullivan was trained by the man who had successfully taught Laura the system of hand signs, and Helen once met her famous counterpart, whom she found to be a "cold" person. This might have been due to Laura's more limited communicative talents. However, in her sensory profile Laura was remarkably similar to Helen. Laura's senses were thoroughly tested by a psychologist in 1890, and the published results indicated that she had normal sensory thresholds in all modalities except for her remarkable sensitivity to vibration. Both Laura and Helen apparently relied upon their vibration sensitivity to feel what people were saying by placing their fingers on the speaker's throat. Helen also claimed the ability to feel the deeper proprioceptive rhythms generated by trains, storms, galloping horses, and so on.

The age at which she became sick is an important factor in Helen's later successes. At eighteen and a half months most normal children have acquired some limited understanding of spoken language, such as the names for key persons, objects, and events in their lives. In fact, their understanding of the emotional and gestural aspects of language is well established at that age and they often have some limited speech. Thus, although they usually do not understand the meaning of long sentences, they understand many stock phrases, and understand whether the utterance is a question, a command, or a declaration, whether the speaker is friendly or menacing, and so on. Eighteen-month-old infants can produce a number of words, usually in single-word utterances made in relation to familiar scenarios and situations. For instance, there is a going to bed routine and an eating routine that might include a few words expressed in context. Most children of that age also have very good nonverbal expressive skills, including pointing, the use of eye contact for gaining and controlling the attention of others, other forms of attention getting, imitation, playacting and event reenaction, some reciprocal games (common examples are pattycake and making faces), and some role-playing games. This pattern seems to fit Helen's record of development before her illness, as reported by her mother and father and later relayed by Annie Sullivan and herself.

However, despite what her skills might have been at eighteen months, they were largely lost once she became deaf and blind. Annie Sullivan documented this as follows:

> . . . the disease checked her progress in the acquisition of oral language and, when her physical strength returned, it was found that she had ceased to speak intelligibly because she could longer hear a sound. . . . She always attached a meaning to the word "water," which was one of the first sounds her baby lips learned to form, and it was the only word that she continued to articulate after she lost her hearing. Her pronunciation of this gradually became indistinct, and when I first knew her it was nothing more than a peculiar noise. Nevertheless, it was the only sign she ever made for water. . . . The word water and the gesture which corresponds to the word good-by, seem to have been all that the child remembered of the natural and acquired signs with which she had been familiar before her illness.

Despite this disastrous turn of events, her mother persisted in trying to communicate with her. But Helen was largely isolated now; there was no possibility of speaking to her or of fixing her gaze, showing her pictures, or playing the many visually mediated games mothers normally play with their children: no peekaboo, no tag or catch. She became unresponsive to her surroundings. Her world became silent and dark. Her movements became awkward and uncertain. She was very poor at navigating her way around, even in familiar places, such as her own home. Initially she was withdrawn, and later she became manipulative and unruly, throwing tantrums whenever she failed to get what she wanted. Recognizing Helen's social isolation and emotional difficulties, her mother insisted on lots of touching and hugging and maintained as much of her routine—bathing, feeding, and so on—as possible.

Helen also developed forty or fifty "home signs" by which she could communicate with her parents and playmates. These signs and gestures consisted mostly of reenactments, such as imitating the act of washing the hands, and iconic signs, such as "father," which she indicated by miming her father's eyeglasses. She had few playmates but learned a few games and even managed a few pranks. She developed what might euphemistically be described as very bad table manners. At mealtimes she would throw food, grab things from other people's plates, and stuff whatever she could find into her mouth. Otherwise, her physical development was normal, and her mother insisted that she should walk, play, and participate in family life to whatever extent she could. She played outdoors and seemed to love nature, especially animals and flowers.

There is nothing in this history to suggest the possibility, let alone the likelihood, of malingering. In fact, some of her family members were not supportive. Her older brother found her embarrassing and wanted her placed in an institution. One would have to come up with a massive conspiracy theory, involving her whole family and all the doctors and teachers who tested and instructed her. Toward what end could such a conspiracy have existed? Not for profit since no one could have predicted her later success. Nor did the family need the money. In fact, her prospects at age six probably did not look any better than those of other deaf-blind children until a half-blind Irish orphan named Annie Sullivan arrived at the Keller home.

CONTACT

The deaf have their own criterion of what language is. Helen Keller certainly did not meet that criterion before Annie Sullivan entered her life. In the profoundly deaf, language is usually defined by the possession of a formal sign language, such as the American Sign Language (ASL) for the deaf, which confers the ability to engage in conversational exchanges. Once you have ASL or its equivalent, you are judged by the deaf community to have language. Helen did not have this.

However, she had many of the gestural precursors of language. In her first eighteen months she had acquired many basic nonverbal communication skills, such as pointing and gesticulating. This gave her a head start in the communicative basics that are essential steps in language acquisition, such as joint attention and the idea of taking turns, so that both speaker and listener do not try to speak at the same time. Following a definition I developed elsewhere, I call Helen's nonverbal communicative skills mimetic. Mimesis is an analogue style of communication that employs the whole body as an expressive device. Mimesis is really about acting. It manifests itself in pantomime, imitation, gesturing, sharing attention, ritualized behaviors, and many games. It is also the basis of skilled rehearsal, in which a previous act is mimed, over and over, to improve it. To judge from the earliest reports in the Keller archives at the Perkins Institution, her autobiography, the writings of Annie Sullivan, and several biographies, notably the one written by Joseph Lack, Helen was a consummate mime long before she met Annie Sullivan. Examples of her skill include her gestural labels for various people and objects and her capacity to learn and reenact long scenarios, such as making a bed, assembling a toy, and playing a game. She could concentrate her attention, for fairly long periods of time, on activities such as folding clothes, feeding the turkeys on her parents' farm, and sewing eyes on her dolls.

Before she had language, Helen's world was quintessentially mimetic. She enjoyed role-playing games with her dolls. She explored other people's faces and bodies, to feel and share emotion, long before she was able to use that same tactile skill for language. She displayed considerable acting ability, using her emotions deliberately in her interactions with others. She had ways of mocking people, provoking them, and manipulating them. She indulged in her share of angry protests and deceptions, showing herself to be normal in this regard. In sum, she had mastered most of the mimetic games of childhood, soaking up the body language, custom, and ritual of the household, to the extent that she could take its measure and occasionally even rebel against and challenge it. She did all this without words, sentences, or explicit symbols of any kind.

Language was completely absent from her life. There was no storytelling, no questioning, no gossip, no wordplay, no stream of word thoughts. However, this did not prevent her as an adult from remembering a fair amount from this period. Her later recall of events that occurred during her early life was sometimes quite vivid, and she recounted many memories that must have been formed without any mediation by language. For example, she recollected in detail a shattering incident at the age of four, when she broke her favorite doll. This was one of many episodes she described in her book *The Word I Live In*. Oddly, in that book she wrote that her early life had been a period of meaningless chaos. However, her rich episodic memories contradicted her own theorizing about her early life. She remembered many early events in some detail, despite her lack of language. This is not surprising, in view of the fact that recovered aphasics can often construct and recall memories of complex events that they experienced while lacking language. Helen's memories of her early life were not very well articulated because they were not integrated into an ongoing autobiographical narrative, as her later ones were. Her recollections of that time were strictly concrete, episode-bound flashes of imagery and emotion. But they stayed in her memory and could be recalled, even though her mind at that time had been stalled in the mimetic mode. Of course, without the powerful retrieval cues provided by language, her memories from that time were much harder to find than her later ones.

Helen's parents identified as many as sixty of her early mimetic signs. Within weeks of arriving at the Keller home, Annie Sullivan had listed thirty signs, including the personal identities of people Helen knew (she mimed her mother's distinctive hair bun at the back of her neck), and similar iconic gestures for food, locomotion, water, eating, dressing, sewing, and places. She understood and produced signs to go upstairs, to her room, or to her trunk downstairs. However, despite the presence of these considerable mimetic skills and a limited but truly interactive social environment, her gestures never blossomed into full-fledged language. She had no words, grammar, sentences, or inner speech.

Her manual and constructional skills were considerable. She once built a three-dimensional model of her house, which she used as a guide, to very good effect. She also maintained favorite hiding places all over the house, where she stored her dolls and other possessions, revealing a remarkable spatial memory. Yet her knowledge of the world and her facility with mimetic expression were apparently not sufficient in themselves to permit her to cross the Rubicon to language. She needed a cultural linkup to achieve this. On her own, without a lifeline to the culture, she was trapped.

To their credit, Helen's parents never gave up on her. When she was six, she was brought to meet Alexander Graham Bell, who recommended that they contact the Perkins Institute in Boston. This proved to be a lucky choice. The director of the institute, Michael Anagnos, assigned Annie Sullivan as Helen's teacher. Although Annie was almost blind herself and poor and inexperienced, she had received excellent training with Anagnos's predecessor, Samuel Gridley Howe, and was about to become a brilliant teacher in her own right. She traveled to Alabama, moved into the Keller home, and remained with Helen as her constant companion and communicative channel for most of her life.

When Annie arrived, the family's main concern was not to teach Helen language. Their immediate goal was a much more modest one, simply to gain some control over her difficult behavior. Annie's first attempts to establish some discipline resulted in a tantrum and a very upsetting scene, and Annie requested that she and Helen should be sequestered outside the main house in a guest cabin on the Keller property. For the first few weeks Annie worked hard at establishing a firm routine. She imposed strict discipline and forced Helen to carry out basic daily tasks, such as eating, dressing, making her bed, and keeping her things in order. Helen's upbringing at home had lacked discipline. Understandably, her parents were too emotionally involved to enforce rigorous rules. Annie's new regimen took weeks to establish, but it eventually worked well. Within a month Helen had calmed down and accepted the strict rules that Annie had prescribed.

At the same time, Annie started to teach Helen the simple hand alphabet that Howe had used with Laura Bridgman. This rather primitive system had been originally invented by a contemplative order of monks in Spain to enable brief communications while living in silence. The hand alphabet was read by placing the reader's hand lightly over the speaker's hand. The speaker spelled out the words to be communicated, letter by letter, while the reader's hand read the speaker's message. In its original form the hand alphabet assumed that both speaker and reader could already speak and spell the words. But Helen had never learned to speak and had absolutely no concept of writing or spelling. Nevertheless, she learned it as her major language modality. It is difficult to conceptualize what form a word or, especially, a sentence could take, using a hand alphabet as the primary communication channel into the tertiary cortex. Still, it worked. It seems that the superficial aspects of the signing system, the specific forms of the codes and grammars it employed, were irrelevant.

Helen's later recollections of hand signing were revelatory in this regard. She claimed that experienced practitioners of hand signing could communicate automatically with this technique and that they thought *only of the message*, not the communicative technique itself. In other words, hand signing became for her just as fully automatic a means of communication, as spoken language is to most people. This is how she described her use of hand signing in the classroom: "The lectures are spelled into my hand as rapidly as possible, and much of the individuality of the lecturer is lost to me in the effort to keep in the race. . . . But in this respect I do not think that I am much worse off than the girls who take notes."

This was obviously a highly automatized skill, but understandably, it was hard for her to acquire this level of proficiency, and it took years of deliberate rehearsal. At her first exposure, hand signing did not work at all. Annie repeatedly produced the appropriate patterns, day after day, week after week, but Helen did not even realize that the shapes that Annie traced into her palm were meant as symbols for something else. She simply mimicked Annie's movements and learned to repeat a number of lengthy signing sequences without realizing that the actions were meant to be words. This was like the babbling of babies, who babble for the sake of babbling, like good mimetic creatures, just to join in, without knowing that the sound sequences are meaningful. In her later recollections Helen said that she initially thought Annie's hand signs were just a new game. She found it interesting enough to play it repeatedly, with her dog as well as with Annie.

As these first tiring weeks went by, Helen passed an important attentional milestone. She acquired self-discipline and developed a trusting relationship with Annie. Annie had cleared out a lot of bad habits of behavior that might have prevented Helen forever from understanding the communicative attempts that her teacher was making. A structured pedagogical environment was the crucial linkage that Helen needed to make contact with the symbolic dimensions of culture. But it could not be loose and informal, like the normal social environment, for a very good reason. Like the Statue, Helen had only a few openings through which knowledge could enter, and thus her experience came serially, in very small chunks. To decode this relatively barren sequence of touches, smells, and feelings, she needed a fixed routine and a habit of compliance, which provided an invisible attentional structure within which remote abstractions had some possibility of being perceived clearly. The complex algorithms of communication, which are more salient in a normally rich perceptual world, would have been much less so to Helen, and without a rigid framework within which to perceive it, the world seemed full of noise. Her routine narrowed the range of possibilities, focused her attention, and reduced the load on her short-term memory.

Why was fixed routine so important? Helen's dilemma might be compared with that of parity in communications engineering, which amounts to a matching problem. Before senders can transmit messages successfully, they must match the signs of the codes of both receiver and sender. For example, a computer linked to a network must have the same parity settings as the rest of the network. Early in the design of computers, there were many problems in this regard because codes had not yet been standardized, and programming was not smart enough to seek out and automatically match codes. Helen's problem was a great deal more complex, but in principle, it was a matter of getting on the same wavelength as everyone else. Most people do this with nonverbal signs and mimetic behavior. This allows them to verify that they have the attention of their listeners and to check whether the message is getting through (if you just told a great joke and they all look puzzled, it seems fair to conclude that it isn't—or that they didn't get it). This need to verify contact can lead to some irritating behavior, such as having to check after every sentence, using glances, nudges, eye contact, and so on. People like Helen can't do this, and as a result, they can easily lose track of the communication game and end up talking earnestly to walls. Other than actually holding you by the throat, they have no way to verify whether you are paying attention or getting the message.

But Annie's structured world solved this problem, which is primarily an attentional one. An anecdote might clarify what I mean. A few years ago Harlan Lane, an expert on the psychology of deafness, brought me to a meeting of the Cambridge Deaf-Blind Contact Center. This is an organization of people who, like Helen, have lost both vision and hearing. Its executive committee was meeting to organize its fifteenth anniversary celebration, and I felt privileged to attend. It was then that I began to understand some of the communication obstacles facing people who are deaf and blind. Many of the people in this remarkable group suffer from Usher's syndrome. They were born congenitally deaf but began to lose vision as young adults. Thus they had all learned sign language as children and could still sign. However, most of them could no longer see the signing of other people. Nevertheless, their primary means of communication was still sign language. They could read one another's signs by placing their hands over the hands of the person talking to them. For those relying entirely on signing, no communication was possible unless two people were within touching distance of each other.

Their meeting was a remarkable feat of organization. Seven members were present, and each needed an interpreter, fluent in sign, and in some cases also in English, to mediate communication. The chair's interpreter could see and hear and was fluent in both sign and English. In total, fourteen people were involved, so that the seven executive members of the Contact Center could talk to one another. Whenever one of them wanted to say something, an interpreter would read the intended message and catch the attention of the chair's interpreter, who would transmit the message to the chair. The chair, through her interpreter, would then attract the attention of the other interpreters, who would inform the other members of the executive that a message was coming. Everyone would then pause and wait for the incoming message, which was sent from the speaker to the interpreter, to the other interpreters, and finally to the targeted receivers. The same routine had to be repeated for each reply. Since there was usually a backup of incoming messages and comments to be sent, all exchanges had to be coordinated through the chair's interpreter, with the chair's approval.

This conveys something of the dilemma of single-channel communicators who lack vision and hearing. Everything must come into the touch system in tandem. For people who have a normal range of experience, language is embedded in a multichannel, massively parallel and amodal process that can simultaneously maintain parity, direct attention, and verify intent, mostly through vision and sound, but also through other modalities. Because of this, our communication networks operate fast and smoothly even when they involve many simultaneous senders and recipients. On the floor of a stock exchange dozens of people can communicate simultaneously because they share the same nonverbal framework. Deaf-blind communicators lack this, and their messages must be lined up, in single file, for sending and receiving. In the absence of the rich contextual cues that allow most of us to anticipate and focus on what is relevant, they need a formal system of regulation that can structure the incoming flow of information. Otherwise, it is unintelligible. There is no way to distinguish signal from noise.

This had been exactly Helen's dilemma. She lacked a contextual framework, so she needed a rigid structure and clear attentional cues to be able to sort out the flood of stimuli she received from her family and now from Annie. She might intuit what people were trying to tell her, but only if the world's underlying communicative structure could be laid bare for her. Annie knew that and dictated that all Helen's messages had to be sent through her, within a rigid system of rules that excluded noisy signals as much as possible. This focused the attention and memory of this little girl for the first time. Helen had needed a systems manager to sort out her communications with the rest of the world, and Annie served that role. Given the serial, single-channel nature of her world, Helen had to keep everything in her prodigious working memory (which, unfortunately, no one ever tried to test) for much longer than most hearing or sighted people do. Annie's structured environment allowed her the luxury of focusing on the message and not having to choose between a faint signal and all sorts of attractive and distracting noise.

At this point a critical development took place, an epiphany of sorts that has become excessively mythologized by Helen's biographers. Both she and Annie described this experience in their writings, and the incident was seized by William Gibson and used in his play *The Miracle Worker*. One day, after many unsuccessful attempts to teach Helen words, Annie combined the spelling sequence for the word "water" (a degraded word that Helen had somehow retained from her infancy) with the act of immediately plunging Helen's hand into water. Although they had rehearsed this word often, without her realizing what Annie intended, this time Helen stopped for a moment and suddenly understood the connection between the hand signs and the meaning. As Annie Sullivan later wrote, "As the cold water gushed forth, filling the mug, I spelled 'w-a-t-e-r' in Helen's free hand. The word coming so close upon the sensation of cold water rushing over her hand seemed to startle her. She dropped the mug and stood as one transfixed. A new light came into her face. She spelled 'w-a-t-e-r' several times. Then she dropped to the ground and asked for its name and pointed to the pump and trellis, and suddenly turning around she asked for my name. I spelled 'Teacher.' All the way back to the house she was highly excited, and learned the name of every object she touched, so that in a few hours she had added thirty new words to her vocabulary."

Helen had grasped the general principle of connecting a sign with a meaning. Moreover, she had grasped this principle *consciously*. The awareness of someone else's intention to communicate had kicked in, and the result was a major step toward acquiring language. Her sudden awareness was accompanied by astonishment (both Helen and Annie vividly recalled this long afterward), followed by frenzied attempts at communication. It is crucial not to overinterpret this. Helen had shown some progress on previous days, and this incident was not unique. Moreover, her signaling was still not language. But it was an important step toward protolanguage, a preliminary form of language that has words but lacks grammar.

This was a brief moment of glory in Helen's otherwise slow-moving daily grind. Her temporary explosion of new words lasted only a week or so, but the principle of naming things had been established in her mind. Annie wrote that a few of the signs learned in that initial rush were undoubtedly residual gestures and that most of the things that Helen wanted to sign were ideas for which she had previously developed mimetic gestures. But a few of her requests seemed to be for completely new, hitherto unlabeled objects and actions. She was apparently able to invent new expressions very soon after she grasped the principle that hand signals could be used to communicate ideas. These efforts to build an expanded repertoire of word signs started her on a longer, much harder journey toward full language.

Helen was unusual because she learned how to write shortly after learning to sign, and writing soon became her main channel of communication. Her handwritten messages form a remarkable record of her progress. She printed simple letters to her family and friends within a year and quite sophisticated ones within two. She also learned to read braille and several kinds of raised print (at that time the supremacy of braille was not yet established). In addition, she could type in several languages and could lip-read with her fingers by tracking a speaker's vibratory signals with her hand on the speaker's neck, lips, and nasal area. If we were to draw a graph of her language skills, it would look radically different from the normal profile of language in every way. Her development followed a different time course and used completely different inputs, outputs, rules, and morphology. She is our best evidence for the enormous plasticity of the human brain and the virtually limitless ways it can assemble language skills.

Helen could formulate her thoughts using simple hand signals, by typing, by punching braille, by printing her thoughts on paper, and finally, to the dismay of many, by speaking, or at least trying to speak, in English, German, and French. She desperately wanted to learn to speak and trained intensively with a speech teacher for several years during her adolescence. However, her attempts were unsuccessful. Although she seemed to make herself understood, in a very limited way, to a few intimates familiar with her slurred style of vocalizing, her public speeches were not intelligible to most observers.

Helen learned all her symbolic skills after the normal critical period and leapfrogged over many phases considered by psycholinguists to be essential to acquiring language. She went straight from initial word learning to grammar and sentence structure, simultaneously, in several different expressive media, none of which was normal. The rules and habits needed to print and read braille, for instance, are completely different from those needed to use hand signs or raised print media. Helen eventually learned to express herself in English so well that her work was favorably reviewed by Mark Twain, who became a friend. He was particularly impressed by her understanding of nuances and her mastery of the deeper meanings that lay beneath the surface of literary works. She achieved this partly because of raw talent, undoubtedly, but also because Annie was a tough teacher and disciplinarian who never let Helen take shortcuts. From the start Annie insisted that Helen always communicate in full sentences, never in abbreviated signs. Attention to the correct grammar of expression was part of her linguistic routine from the start. Of course, none of this would have worked if it were not for the fact that Helen had the capacity to meet Annie's demands. The likelihood of ever again pairing two such people, each with such heroic dedication, is not very high.

What was the nature of that capacity? We can be sure that it was not a dedicated language module, since Helen did not meet the standard conditions for acquiring language. Her symbolic capacity must have been guided by a domain-general mechanism because her language skills eluded the normal auditory and visual channels and took an unusual sensory route. Her brain employed an anatomical path that was far removed from the normal patterns of linguistic communication. This suggests that her language learning was managed by a truly amodal system; that is, one that could capitalize on any available input or output channel. Moreover, the capacity demands of learning to sign and read entirely by touch are very considerable. The short-term memory requirements of tactile sensation are qualitatively different from those of sound, and the absence of vision meant that Helen had to pack all her spatial knowledge into the same sensory system she used for communication. This is a radically different attentional task from that facing a hearing person. Somehow, in Helen's conceptual world, "inner speech" meant "inner touch," and the imagery that drove rational thought in her mind must have been largely haptic. Nevertheless, this did not limit her ability to think about very abstract concepts such as morality, pedagogical principles, and human rights.

It is significant that Helen lived at home for more than five years without developing any of the representational elements of language, despite having an extraordinary talent for it. Again, this militates strongly against the idea of an innate module for language. A built-in language device, fully equipped with metagrammars, would surely have kicked in with a few simple words or expressions. After all, her experiential base, in Condillac's terms, was sufficiently rich, and she had plenty to narrate. But this did not happen, not even after she had several dozen mimetic gestures and seemed capable of some two-way nonverbal communication. Apparently, neither grammatical nor lexical invention can spring fully armed from elementary mimesis. Additional principles are involved, and above all, a link to a symbolic cultural web is needed. The symbolic framework must be established from outside the nervous system. Otherwise, a mind will stay in a holding pattern until the conditions for internalizing symbols are met. Helen might never have achieved grammar and might have stalled exactly where most similar cases have done, at the level of very simple sentence comprehension, if it had not been for enforced enculturation. She was required to participate, and reciprocate, by Annie.

Her mimetic skills were her salvation, and this vindicates Condillac's insight. She had retained her sense of embodiment and her physical selfhood, and this gave her a foundation for her sense of self. Her remarkable capacity for inventing gestures and expressive tricks undoubtedly came in handy later, when she needed the tools to acquire a large vocabulary. But they were not enough, without a means of tracking the cultural codes that had become invisible to her. Despite the fact that she had so much rich cognitive material and all the apparent ingredients needed for language, including the raw objects of narrative description, event perceptions, agents, actions, situations, and models of three-dimensional space, she never used any of these to construct descriptions. Her isolated mind could not invent language as an internal means of thinking and representation, even under conditions where it would have been extremely important to do so. There are good reasons to consider Helen's case as a paradigmatic experiment of nature. She epitomized the dilemma of the isolated mind in a way that is unlikely to be matched.

Helen Keller's success is the argument par excellence for a domain-general, constructive, conscious mindbuilding process. Conscious processing is effortful, uses capacity, requires high levels of self-monitoring, perspective taking, multitasking, and so on. Unlike most of us, Helen had very few moments when she could afford to run in "neutral gear." Every waking moment was a challenge. The main lesson to be learned from her life is perhaps the simple fact that even a mind as talented and remarkable as hers was incapable of generating the basis of symbolic thought on its own. The notion of labeling events and objects *simply doesn't occur* to an isolated mind. In order for a mind to become able to profit from the tremendous intellectual gains that come with symbolic competence, it must be programmed from the outside; that is, by an existing culture. Once spontaneous symbolic invention actually begins, it is particularly striking that the principle of naming and labeling does not automatically trigger grammatical competence. Grammar is hard won, even with intensive enculturation. Before its rules can be learned, they must be found. To find them, the mind must be able to split its attention, use its imagination, and test its conclusions, over and over, spinning off memories at a dizzying rate and storing away what it learns in a form that can be easily recalled.

One striking aspect of Helen's life is that once she had language, she experienced a revolution in the subjective quality of her conscious experience, especially her memories of it. She reported that she had never been so conscious before. In fact, in her book *Teacher* she referred to herself before she met Annie as "Phantom" and after, as Helen. She claimed that her early experience had lacked the sharpness and definition that language later brought to her life. Yet she clearly remembered many of her experiences during that so-called Phantom time and even recalled her feelings and reactions to very specific events. Her descriptions are a giveaway because they are so full of detail. Without conscious attention, our minds simply do not remember such details. But it is significant that even without the self-reflective form of intelligence that depends on language, she could still remember some early events, beautifully parsed and in proper context. She was still only Phantom when they occurred, and Helen did not yet exist, but the memories endured.

Helen had a genius for vanquishing intellectual adversity. But first and foremost, she was a supremely vigilant, proactive person. Her great capacity for conscious control was the ultimate constructive force in her long-term development. It was her interface with the liberating treasures of culture and the key to the greatest transformations in her development. It is only through the interlocking of consciousness and culture in a continuing cycle that she could build the hierarchies of consolidated habits and skills that she assembled into a vastly complex self. Embedded in the unusual maze of attentional, social, athletic, rhetorical, literate, and intellectual skills that ruled her actions, there was an incredibly strong sense of self. Helen gradually acquired her adult personality, after endless difficult metacognitive bouts of agonized self-evaluation, in loop after loop, as she internalized the norms of her culture. She developed nothing of what we would call inner speech, or reflective intelligence, until she was guided through the long, convoluted labyrinths of deep culture to its symbolic core. This brings us to the central principle that this process entails.

OUTSIDE-INSIDE

From her own testimony, deep enculturation freed Helen Keller's mind, let her out of prison, and allowed her to think. All her depth and richness as an adult came from the emancipating effect of symbolic culture. This presents us with a dilemma. Such cultures cannot function without languages, and brains cannot generate languages without preexisting symbolic cultures. How could such a symbiosis have started? There is only one possible answer. Short of invoking an evolutionary miracle, *expressive culture must have taken the first step*. Some archaic cultural leap, deep in hominid prehistory, must have set the stage for our transition toward a symbolizing mind. But at first glance this doesn't seem possible. Doesn't cognition always lead, and culture follow?

One way out of this dilemma is to apply a principle first proposed by Vygotsky in his pioneering studies of children. This is the so-called Outside-Inside principle. Vygotsky observed that children always copy the externals of language first and do not initially have inner speech, or silent forms of symbolic thought. The developmental rule is that symbolic thought first represents external action, and only later reconstructs it so that it will occur internally. Every function in a child's development thus appears twice: first interpersonally, then intrapersonally. Thus the silent thinking skills of adults might be very misleading when we want to specify their origins. Given the self-centered nature of symbolic cognition in adults, we might easily gain the impression that they used their skills in a solipsistic manner from the start. But this is wrong. Vygotsky's studies have suggested that when they first appear, children's own symbolic performances are completely public, *even to themselves*. Only later do these operations become internalized, and independent of a specific social-mimetic role. The direction of the flow is clear: from culture to individual; from outside to inside.

The evolution of human symbolic skills must have emerged in a similar way. This must be so because even now our modern brains cannot gain symbolizing skills without going through an externalizing phase. Our dependence on enculturation would surely have been greater two million years ago, when our cerebral volume was only 70 percent of its present size. The first symbolizing algorithms must have been impressed on our brains from outside the nervous system. Presumably these consisted of communication patterns that developed out of group behavior and were gradually internalized. One can legitimately ask where those patterns came from, of course, and I shall elaborate on this topic in the next two chapters.

7

The First Hybrid Minds on Earth

Language is like a cracked kettle on which we beat out tunes for bears to dance to, while all the while we long to move the stars to pity.

—GUSTAVE FLAUBERT

Metaphor is not just a matter of language, that is, of mere words. On the contrary, human thought processes are largely metaphorical.

—GEORGE LAKOFF AND MARK JOHNSON

We acquire our symbolic skills from the outside in. Therefore, we had to evolve them in the same way. Symbolic thought and language are inherently network phenomena. Thus their existence cannot be explained in a solipsistic manner. The problem calls for a paradigm shift, away from our mainstream theories of human evolution, which tend to assume that language evolved inside the brain box; that is, from the inside out.

ABANDONING SOLIPSISM

The theory outlined in this chapter might be called biocultural. It places the origin of language in cognitive communities, in the interconnected and distributed activity of many brains. Such communities have their own independent dynamics. A brain that evolved specifically for this type of environment would not necessarily resemble one that evolved to produce language on its own. This is partly because human culture includes a great deal more than language and partly because the minute you embed a brain into a cognitive community, you change what it must do in order to remember, think, and represent reality.

The evolution of human cognitive communities can be taken as an excellent example of emergent evolution, in which a series of small incremental changes at one level can result in a radical change on a different level. The formation of cognitive communities was undoubtedly one of the most extraordinary events in the history of the biosphere, yet it seems to have been caused by a relatively simple expansion of the executive brain, with a corresponding change in developmental plasticity. The specific form of human consciousness was fixed by the demands of this adaptation. We are culture-mongers, driven by the very nature of our awareness to seek refuge and solace in community. We connect with and learn from others to a unique degree. Symbolic thought is a by-product of this fact, and so is language. Both result from the collision of conscious minds in culture.

The evolutionary origins of language are tied to the early emergence of knowledge networks, feeling networks, and memory networks, all of which form the cognitive heart of culture. Language was undoubtedly produced by Darwinian selection, but it evolved indirectly, under conditions that favored those hominids who could make their shared cognitive networks more and more precise. To judge from its impact on the development of infants and apes, the emergence of language could not have initially been an end in itself and would not have initially changed the nature of cognition. It would merely have made the operations of the same system more exact and less ambiguous. But since language obviously had great survival value, natural selection would have increasingly nudged hominids in the direction that humans finally took, with the emergence of a vast, high-speed speech capacity. The strangeness of this final direction (primates are poor vocalizers) should not distract us from the need for smooth continuity in the early transitions from primate to human. The first priority was not to speak, use words, or develop grammars. It was to bond as a group, to learn to share attention and set up the social patterns that would sustain such sharing and bonding in the species.

It may seem odd, but the almost inconceivable complexity and beauty of language are better explained this way than by the reverse scenario, which demands the evolution of a complex language module first, and then the use of language to create cultures. The machinery of language is far too exquisite to have been encoded entirely inside something as unpredictable as the genome of a developing brain. It was more probably shaped by the demands of a communicative universe that was much larger than one contained inside a single brain and was instead provided by a community of brains. Under the right circumstances, such a community could generate a virtual infinity of possibilities. Just as the physical environment drove the evolution of perceptual capacities, so cultural energy drove the evolution of sophisticated communicative capacities. It also generated forward pressure on the developing mind. Recent developmental research has provided some important clues to the origin of language. The mental abilities most central to enculturation, those involved in joining cognitive communities, develop very early and are logically and empirically prior to language, both in development and in evolution.

The biocultural approach does not challenge the obvious fact that humans have a unique capacity for language. Language may still be regarded as an "instinct," for those who care about such things. But it is not an isolated module; it is embedded in a wider set of instincts for culture and blended into the cognitive system as a whole. Oral language, or speech, may be seen as a late specialization within the context of a much wider and earlier cultural adaptation, and this is why it has remained heavily dependent on cultural networks for its evolution and development. The great divide in human evolution was not language but the formation of cognitive communities in the first place. Symbolic cognition could not spontaneously self-generate until those communities were a reality. This reverses the standard order of succession, placing cultural evolution first and language second. It also suggests that human ancestors could not have evolved an ability to generate language unless they had already connected with one another somehow in simple communities of mind.

This makes the mechanism of language highly dependent on culture. The same may be said of all forms of symbolic processing. This is the ultimate irony: We could not even begin to think our most private thoughts without satisfying our brain's cultural habits first.

CONSCIOUSNESS AND COMMUNITY OF MIND

The relationship between consciousness and culture is a reciprocal one. While culture emerges from the attempts of an expanded awareness to connect with others, it is immersion in culture, rather than any feature of the brain, that defines our truly human modes of consciousness. Enculturation dominates human cognitive development. Moreover, this principle must have been in play from the start of our cultural adventure. We must reject Condillac's notion that active sensation alone could have provided a sufficient basis for developing a fully human mind. In this aspect of his theory he was dead wrong. The Statue needed much more than a capacity for actively experiencing the world before it could achieve a fully human mode of thinking. It needed to steal algorithms and symbol systems from culture, as Prometheus had to steal fire from heaven.

But stealing fire from heaven is not as easy as it sounds. It is a matter of parsing the cultural landscape, discovering its hidden secrets, and overcoming many obstacles to understanding. This cannot be done before someone reveals, directly or indirectly, where the codes are located. These codes can be found only if early experience prepares the brain for the task. Enculturation necessitates the early differentiation of working memory into multiple fields of awareness, as well as a linking of its powers of recall to the appropriate cultural subroutines. These are crucial in preparing a brain to find the attentional flags and markers that will eventually lead it to the heart of culture. Although we still do not know enough about how these sequences unfold in detail, we know, from observing the limitations of wild-reared apes and feral children, that if these are not programmed into the brain early enough, the system will later become impossible to train.

A key step in this process is the interlinking of the infant's attentional system with those of other people. Reciprocal eye contact, or gaze, is one of the channels through which this develops. Voice and touch are also common channels. These are all involved in familiar circular routines, such as greeting, hugging, and playing. Moreover, they must become reciprocal, so that the infant is an active participant in these routines. These interactions are learned by the infant primarily through intimate personal encounters, starting with early mutual imitations with the mother and broadening to include more elaborate exchanges of facial expression, voice, and gesture, which lead to playacting and various social games. Such encounters serve to build a complex repertoire that will regulate shared attention for the rest of the child's life. The cultural repertoire constitutes a scaffolded system, with each new level adding to a vast cultural edifice of control. The result is to interlock the infant's growing mind with those of its caretakers and ultimately the broader society. A hierarchy of habits for shared noticing, caring, feeling, and remembering are cultivated in this way and prepare the child's growing mind for more ambitious forays into culture at a later age, when the learning of new skills will depend on more intricate interlinkages of attention, such as those that allow us to notice, and sometimes to share, group intentions. They are also vital for verifying that one's interpretations are indeed aligned with those of others.

These patterns of behavior are transmitted across cultural generations by repeating the same basic sequences in the context of pedagogy. The distributed texture of this shared pedagogical process constitutes a kind of cultural memory, which records the huge unwritten fabric of shared feeling, group bonding, and common behavior that underwrites the deep enculturation of every infant. Despite its apparent position at the bottom of the cognitive hierarchy, this system of cultural bonding operates at a very high level of abstraction and is well beyond the capacities of other primates. It demands an extraordinary degree of conscious control, especially during the learning and teaching process, but it does not need to be executed in a self-aware, or deliberately guided, manner. We need to keep in mind the distinction between controlled processing and explicit self-awareness, which is entrenched in very old cultural habits that are implicit in their influence.

These basic tools can lock a child into a largely invisible framework of attentional control that shapes every subsequent encounter with culture. Every nuance of gaze, gesture, tone of voice, or facial expression acquires a cultural meaning and, if understood, might serve as a cue to what is important. If the child can learn how to read these signals, evaluate them, and respond to them, it can learn to navigate the cultural labyrinth. Once acquired, the same skills will also enable a person to become a pedagogue later in life and assume the role of cultural guide for someone else. This will happen without reflection since it will simply involve the reapplication of basic habits learned in infancy.

Under normal circumstances this process constitutes a distributed social learning device that goes far beyond the old-fashioned concept of a general-purpose learning ability. It is very different from the learning capacities of other mammals, in that it is design-built for living in culture and incorporates group cognition into its routines. This early connection with culture enables human infants to track and understand many of the complexities of cultural survival before they learn to speak. These basic routines are the basis on which more difficult self-supervisory habits can be acquired later in childhood. During their first eighteen months, infants are already beginning to learn how to review their own actions over the longer run, in the light of what others think and what the consequences of these actions will be. This is a generic metacognitive skill whose importance cannot be exaggerated.

Infants must also develop an early familiarity with other cognitive domains that matter in a social context. For instance, to prepare for language, they must first learn about such things as the significance of different tones of voice, of taking turns, and of pauses in conversation. They must learn the overall sound contours of their native tongue and the links between these prosodic aspects of speech and the events they perceive in the social world. In the process they must also learn something about society's attitudes toward emotional expression and various kinds of role playing. Infants learn to accept these things as integral parts of the theater of human life. In turn, this will enable them to find the significances that lie behind the bewildering behavior of adults. In the familiar scenarios of childhood, children repeat and reenact the basic rules and expectations of social life until they have been written deeply into memory.

Note that this is all domain-general training. Virtually any cognitive module might prove relevant to cultural life, and thus the habits of social interaction tend to engage the generalist aspects of brain function. Without this training, children will not acquire language normally. In cases of extreme deprivation, they will remain deficient in symbolic thinking and autobiographical memory, for life. Helen Keller and Laura Bridgman were able to learn language long after the critical period precisely because they enjoyed a normal stream of experience during their first eighteen months and were able to acquire these basic executive skills. This gave them the broad foundation they needed for their later reconnection with culture. Without it they would have failed. There are no recorded cases of successful late language development in anyone who has been deprived of this vital early stage. The early programming of shared attentional habits and working memory skills is an absolute precondition, not only for the development of language but for achieving fully human consciousness later in life.

THE CULTURAL RELEVANCE OF A MULTIFOCAL, MULTILAYERED CONSCIOUSNESS

In order to acquire our deepest cultural linkages, humanity needed to evolve a major change in the nature of working memory. Human cultural evolution could not have progressed very far if we had not been able to subdivide working memory into several subdomains. It would have been impossible to develop a collective mentality of any kind without some means of subdividing working memory into at least two fields: self and other. Even in a simple two-person interaction, it is important to control and monitor the attention of the other person. This applies even to nonverbal communications, including gesture, pantomime, imitation, and skilled rehearsal. To make even a simple gesture, such as turning the hand while reaching toward a doorknob to indicate a desire to open the door, infants must be able to keep the gesture itself distinct from the reactions of others to it in their working memories, while switching back and forth between the two perspectives. There is considerable attentional dexterity at the heart of the self-other distinction.

Moreover, pedagogy, so vital in human culture, relies on our ability to subdivide our working memory system into even more compartments, juxtaposing in awareness not only self and other but also past and present, all within the same time frame. Thus the teacher tracks the student, remembers the previous state of learning, and adjusts the lesson accordingly. Meanwhile the student tracks the teacher's intentions and intuits where the lessons are leading. And so on. Every move on the part of either player must be noted and remembered, as interactive signals bounce back and forth between them. They could not engage in this complex sequence of mutual cognitive control without a multiregional working memory system with different streams dedicated to the teacher's actions, the student's responses, the previous and current state of the learned task, and the relationships between these, with each stream updated regularly. Without this, the pedagogical sequence for any complex skill would break down because both teacher and student would lose track.

Thus it is clear that fairly early in our cognitive evolution, hominid working memory must have differentiated into a system that was more flexible than that of apes, so that it could split into several open-ended subregions, each with an extended time span. This adaptation would have helped sustain the pedagogical interactions needed to teach the difficult skills practiced by early hominids, such are spear throwing and the manufacture of stone implements. Early hominids must also have evolved a capacity to switch between several of these subregions at once. This would have allowed them to parallel process and thus compartmentalize their conscious mental activities, so that they could run several trains of activity concurrently, while reviewing and comparing experiences. This would have been necessary even for relatively simple achievements, such as group coordination, organized big-game hunting, and long migrations.

The existence of this adaptation is confirmed in the nature of modern human consciousness. Although our experience is subjectively unified, our awareness stream is rarely one-dimensional in a structural sense. Under the right circumstances, we can maintain several parallel lines of thought, each in a different mode. You and I can read the newspaper, eat breakfast, respect the social niceties, make conversation, take a few phone calls, plan our day, and ruminate about a possible trip to Argentina all more or less simultaneously and within the same intermediate time frame. Running frames within frames concurrently is routine for our species. Moreover, it is a flexible process. Our conscious activities seem to be stacked, or arranged in cascade, whereby our working memory system can be subdivided into several simultaneously active zones, or narrowed to one, as the situation demands. This resembles the pull-down menus and windows on computer screens. However, the metaphor should be reversed: Software interfaces have been designed to match our existing mentalities, not the other way around. The idea of maintaining multiple ongoing agendas came from observing how people normally work and think, not as a result of inventing computers.

Our human cognitive style is linked to this multifocal consciousness, and language, in particular, is highly dependent on this feature. Many other adaptive changes were important for the evolution of language, including a modified vocal apparatus and an expansion of memory capacity. But the executive system was the key. It was always at the center of the cognitive storm because with the growth of culture, our conscious universe started to move at a new speed, generating many more facts and encompassing many more players and a more complex social system. Above all, the speed at which knowledge could accumulate depended on better working memory management.

Modern symbolic technology capitalized on this multifocal capacity and radically transformed our knowledge networks in the process. It also changed the rules of the cognitive game by plugging us into more powerful distributed systems. Thanks to those changes, we can now play faster and more complex games in awareness. But the increasing number of potential foci, the higher turnover rates of information, and the speed with which we can shift time perspectives or change our locus in the memory stream have driven our conscious capacity to the wall.

THE STAGES OF HUMAN CULTURAL AND COGNITIVE EVOLUTION

In a previous book I proposed that the present form of the human mind evolved over the past two and a half million years, in three major cultural stages, or transitions. My central hypothesis was that during that time, hominid cognitive evolution was increasingly tethered to culture. Our remarkable evolutionary drive was presumably sustained by the many advantages of having a collective mentality, and our brains went through a series of modifications that gave them this strong cultural orientation. That orientation was something new in evolution and has made us very successful as a species. It has profoundly affected everything we do for a living, whether hunting, gathering, toolmaking, fighting enemies, organizing migrations, adapting to climate change, or protecting ourselves from perils. The close linkage between brain and culture has accelerated the rate of human evolution. As communities became better able to store and disseminate knowledge, they evolved a Baldwinian strategy that hijacked the normally slow-moving process of natural selection and caused it to speed up. Cultures are more efficient than individuals at exploiting the fitness value of genetic variations, which might otherwise have a negligible impact. This is the basis of Baldwin's effect. In this manner the speed of evolution of the human brain was doubly increased not merely by its extraordinary ability to learn but by its capacity to share knowledge, in distributed networks that can accumulate knowledge at blinding speeds.

There are three major stages, or transitions, in my version of our cognitive emergence. These are shown in Table 7.1. Each of these transitions changed the nature of human consciousness in a major way. The scenario of human evolution seems to be one of tension between culture and conscious capacity, with culture steadily pushing that capacity to the edge, so that it continuously expanded. Culture was a radically new presence, and the mind kept adjusting itself to the new reality of distributed cognition. The result of that tension, in the long run, was the emergence of a symbolizing mentality.

The first transition started a little more than two million years ago,

TABLE 7.1

Successive layers in the evolution of human cognition and culture. Each stage continues to occupy its cultural niche today, so that fully modern societies have all four stages simultaneously present.

Stage	Species/Period	Novel Forms	Manifest Change	Governance
EPISODIC	Primate	Episodic event perceptions	Self-awareness and event sensitivity	Episodic and reactive
MIMETIC (first transition)	Early hominids, peaking in *H. erectus* 2M–0.4 Mya	Action metaphor	Skill, gesture, mime, and imitation	Mimetic styles and archetypes
MYTHIC (second transition)	Sapient humans, peaking in *H. sapiens sapiens* 0.5 Mya–present	Language, symbolic representation	Oral traditions, mimetic ritual, narrative thought	Mythic framework of governance
THEORETIC (third transition)	Modern culture	External symbolic universe	Formalisms, large-scale theoretic artifacts, massive external storage	Institutionalized paradigmatic thought and invention

when the species *Homo* first appeared on Earth. It is not absolutely certain which of several hominid species made the first great intellectual breakthroughs, but there was a general increase in hominid brain size during that transition period, and all these early hominids were more humanlike than the australopithecines. The first stone tools appeared simultaneously with the species *Homo*, along with evidence of a drastically changed diet. *Homo* was an omnivorous species from the start. There is evidence that this species ate much more meat than any of its predecessors. Some of this meat was obtained by hunting big game, a remarkable achievement for a nearly naked, relatively small creature. Early hominids could not have achieved this feat, or obtained so much meat, without improved tools. To this end they invented stone tools. The archaeological record also shows some fossilized wooden spears from this early period, but it is likely that they also had tools made from other perishable materials, which vanished in the intervening two million years. They had a communal style of living and used home bases as a fixed point from which to venture forth, to hunt and gather foodstuffs. They left behind toolmaking sites, seasonal hunting camps, and continuously occupied fire sites, all of which indicate a group-oriented way of life, in which both material and cognitive resources were shared. Thus, very early in our evolution, hominids had become highly social and evidently used a cultural strategy for remembering and problem solving. In other words, although we have no evidence of language at this time, there is very good evidence of culture.

The achievements of early hominids revolved around a new kind of cognitive capacity, mimetic skill, which was an extension of conscious control into the domain of action. It enabled playacting, body language, precise imitation, and gesture. It also acted as a mode of cultural expression and solidified a group mentality, creating a cultural style that we can still recognize as typically human. Mimesis enabled early hominids to refine many skills, including cutting, throwing, manufacturing tools, and making intentional vocal sounds. Although not yet language, these sounds were nevertheless expressive. We call such vocal modulations prosody. They include deliberately raising and lowering the voice, and producing imitations of emotional sounds.

The second transition started with the arrival of archaic *Homo sapiens,* about half a million years ago. It culminated in the evolution of our particular subspecies, *Homo sapiens sapiens,* about 125,000 years ago. During this time the brain and vocal tract underwent a great change. Sapient humans started with the rather primitive material culture they inherited from their predecessors but then began to innovate at a much higher rate. They invented a wider range of sophisticated tools and produced beautifully crafted objects, improved shelters and hearths, and elaborate graves. Within another 10,000 years, they had started to use several forms of self-adornment and were manufacturing a very large variety of multipart objects, including weapons, hafted tools, boats, complex dwellings, ritual quasi-symbolic artifacts, and simple musical instruments. They had also migrated over much of the world, using various technologies to adapt to a variety of climates and ecologies. They came to dominate the Earth, and spoken language was undoubtedly the special power that favored them over their rivals and predecessors. Spoken language produced oral culture, which was the universal form of human culture until very recently.

The third transition started about forty thousand years ago, and revolved around a revolution in the technology of symbols. Cognition continued to evolve, but this time it was mostly driven by technology and culture itself. The main cognitive driving force underlying this transition was the externalization of memory. Whereas earlier humans had to depend entirely on their biology—that is, on their brains—to remember, modern humans can employ a huge number of powerful external symbolic devices to store and retrieve cultural knowledge. This revolutionized the way humans think and the kinds of distributed cognitive systems we could construct. Thus modern culture contains within it a trace of each of our previous stages of cognitive evolution. It still rests on the same old primate brain capacity for episodic or event knowledge. But it has three additional, uniquely human layers: a mimetic layer, an oral-linguistic layer, and an external-symbolic layer. The minds of individuals reflect these three ways of representing reality.

In effect, these three transitions are major checkpoints on a long evolutionary road, which may have had many stops and starts. They caused three shifts in the nature of consciousness during our evolution: (1) more precise and self-conscious control of action, in mimesis; (2) richer and faster accumulation of cultural knowledge, in speech; and (3) much more powerful and abstract reflective cultures, driven by symbolic technology.

THE FIRST TRANSITION: ESTABLISHING THE MIMETIC FRAMEWORK OF HUMAN CULTURE

Unfortunately, we have no way to time-travel and actually observe the first members of our species, but we know a lot about their nervous systems, lifestyles, and major cultural achievements, and when we place them on a curve, extending from primate to human, we can narrow down the number of feasible theories about their mental capacities and social organization. By drawing on our knowledge of cognitive science and neuroscience, we can infer what their cognitive options were. On the basis of these considerations, the first cognitive transition seems to have revolved around one central issue, the invention of culture as a collective means of accumulating experience and custom.

The first humanlike culture was associated with a new species, *Homo.* This was a culture of public action, without languages or symbols but equipped with mimetic expressive skills. This was the birth of the actor, the tribe, and the gesture. Given its evolutionary proximity to primate social styles, it must have been dominated by direct expressions of emotion, and indeed, the control of emotion by public action and gesture would have been one of its first priorities. Thus contrived public displays, public competition for control, and a great deal of deception would have been the rule. It would also have been possible to share attention and knowledge, albeit indirectly, by means of gesture, body language, and mime, any of which can communicate an intention quite effectively, without words or grammars.

Mimesis must have come early in hominid prehistory because it was a necessary preadaptation for the later evolution of language. It provided the underpinnings of social connectivity and conventionality. It took the primate mind one step farther in the direction of improved social coordination and collective cognition. The group was primary, and thus having an accurate sensitivity to group feelings was a survival-related skill. Mimesis is still the elemental expressive force that binds us together into closely knit tribal groups. Of all our human domains, mimesis is closest to our cultural zero point. It is also closest to emotion. Mimetic capacity has huge emotional ramifications because it involves both the conscious elaboration and the suppression of emotion.

Mimesis is the result of evolving better conscious control over action. In its purest form, it is epitomized by four uniquely human abilities: mime, imitation, skill, and gesture. These are direct offshoots of the expansion of the human executive brain system and the Executive Suite. The most basic form of mimetic action is mime, the imaginative reenactment of an event. It is exemplified by the pretend play of children, in which they might, for example, feed a doll in exactly the same way they were fed themselves, a few minutes earlier, by their parents. Pretend play is not a fully symbolic mental representation, but it has implicit reference in that it refers to something other than itself. In the above example the child's actions with the doll represent a previous event, the scenario "eating," which is being reenacted. Adults engage in many institutionalized versions of pretend play in theater and film, and imaginative role playing is integral to adult social life. Indeed, what is a "career" or a "vocation" except a role-playing game extended over an adult lifetime? Even though adult life is dominated by a heavy overlay of speech and literacy, our basic drives and ambitions are shaped by these kinds of mimetic scenarios.

We can also reenact emotional events, and here we can see how emotional mimesis differs from the innate emotional communications of other species. Innate expressions, such as crying and laughing, are stereotyped, universal, and part of the actual emotional experience itself. This gives such expressions a signal, or cue, value, but it does not make them into intentional acts. Mimesis is different. A child miming the act of crying is deliberately acting out a role; that is, pretending to be sad. This is a case of cognitive, not emotional, control over action. It is also a quasi-intentional act and a very abstract representation, independent of the emotional state to which it refers (even though it may restart the emotion). Actors may vary widely in how they do this, depending on what aspects of the emotion they choose to emphasize. This unpredictability of mimetic style applies, whether we are playing a simple game of charades or performing in professional theater. There is always an element of arbitrariness in how a mime is performed.

The second product of mimetic capacity, precise imitation, is more challenging than simple mime, because it is usually seen as an attempt to replicate an event that has some instrumental purpose. To copy a purposeful action, we need to understand the other person's objective. For example, to imitate the sharpening of a spear accurately, we have to understand what the spear is used for and why the point needs to reach a certain degree of sharpness. For certain kinds of cultural transmission, it is not enough simply to mime the movements themselves. A child can mime spear sharpening long before it can actually copy the act of sharpening a spear, and this is not always explainable by the child's lack of strength and coordination. It takes time to develop an understanding of purpose, and some purposes are easier to grasp than others. These distinctions are a matter of quantity, not quality.

The third product, skill, is closely related to mime and imitation. Skill results from rehearsal, systematic improvement, and the chaining of mimetic acts into hierarchies. Thus, to learn to play tennis, we must master a number of elementary action chains and then piece them together into a very complex contingent arrangement, so that the right action will result on very short notice in every possible future situation. The same rule applies to any skilled craft. Whether we are learning to weave, manufacture tools, or cook food, we must learn a set of basic action sequences, generalize them, and rehearse them until they become second nature. To do this, we must mime our previous actions and modify those productions until we approximate an ideal. We must usually go well beyond rote mime or imitation to do this well. We may actually have to modify the action sequence for ourselves and generate new ways of executing it. We take for granted our ability to do such things because we are so good at it. Children commonly practice actions for no apparent reason other than the pleasure of refining them. They might practice hopping on one foot, skipping stones, or kicking a ball for an entire afternoon just to become better at doing these things. By its very nature, our capacity for skilled rehearsal is potentially creative because it can generate variation; that is, contribute to the overall range of actions in our social group's repertoire.

The fourth product, gesture, is a natural derivative of the first three. Gesture and mime are closely related, except that we usually define gesture as an explicitly communicative, or intentional, act. The easiest gestures for infants to learn are simple mimes that refer to specific things or people by miming events. Developmental psychologist Linda Acredolo has shown that a child as young as nine months can easily learn, for example, to represent a fish by miming its puckered mouth. Gesturing becomes much more subtle and complex as children grow older, and they eventually become able to express complicated things in sequences of iconic gestures, which might trace the path of objects in an imagined space or mime the opening of a door or a box. Susan Goldin-Meadow, who has studied gesturing in depth, has shown that it can also contain implicit grammars, which resemble the iconic grammars of sign languages. That particular kind of gesturing is very abstract and was probably a late development in mimetic culture, possibly the main lead-in to the evolution of language. It is an advanced form of mimesis and must have been preceded by simpler reenactive mime, imitation, and skilled practice.

Mimetic capacity produces a layer of cultural interaction that is based entirely on a collective web of conventional, expressive nonverbal actions. Mimetic culture is the murky realm of eye contact, facial expressions, poses, attitude, body language, self-decoration, gesticulation, and tones of voice. It is still the primary public dimension that defines our personal identity, and in it, style and tradition matter, to the degree that these things establish who we are, who our friends are, and where we stand in society. Mimetic styles vary tremendously from one cultural group to another. The mimetic elements of culture are so distinctive that they are often the most difficult features for foreigners to master.

Mimesis is also the cultural layer that children encounter first. The distinguished developmental psychologist Katherine Nelson has assembled an extensive body of evidence that shows how mimetic event representation is the universal means of interacting with very young children. In fact, she has called the world of the two-year-old an almost purely mimetic culture. The exaggerated facial expressions, tones of voice, and role-playing games that children enjoy are mimetic in character. A child's life is really a series of mimetic set pieces, in which various cultural scenarios are being learned, acted out, rehearsed, and varied at a very fast pace. As a child goes through a typical day, it is expected to act out various roles, such as "eating at table," "washing," "dressing," and "going out." Other societies might have very different scenarios, but family life provides the first stage on which our actor minds become locally famous, interlocking with other minds, engaging in roles, joining attention with others in common causes, sharing emotions, customs, and action memories. By means of mimesis, we can share tacit knowledge to a remarkable degree. Mimesis is the level of cultural interaction on which we first assume a basic tribal identity and become conscious of ourselves with reference to our primary social group.

Mimetic capacity was primarily the result of merging the executive brain with the action brain, when the hominid executive brain system extended its anatomical territory into the frontal and subcortical regions that control voluntary action. This anatomical change had major social ramifications. Active social networking, even of the exclusively mimetic variety, makes heavy demands on attention and memory, and we can corroborate this with evidence drawn from studies of child development. Oxford psychologist Paul Harris has examined the close interplay between children's early mimetic involvements and their development as social beings. Early in development, the child connects with a mimetic social network ruled by custom, convention, and role taking. The family is a small theater-in-the-round, featuring a series of miniplays, in which each member must assume various roles. Children understand these theatrical productions so well and so early that they can act out any role, within the limits of infantile acting. This is shown in their fantasy games, where they might choose to play the father, the mother, themselves, or even the dog or the family car. Children become excellent mime artists and actors, long before they can verbally describe or reflect on what they are doing.

Even in adults, mimetic role playing and group identification remain in place as powerful forces for regulating social coordination. As long as the members of a culture understand the roles they are playing in a situation, the basic frame of group activity will work well. Without that frame, groups become unstable and quirky, to say the least. A high school class will run smoothly if all can agree on what the roles "student" and "professor" entail, in terms of nonverbal behavior. It will soon fall apart if they cannot agree on who is playing which role or what those roles imply. Social structures depend heavily on mimetic consensus for their smooth operation.

We can see this at work, even today, at the institutional level, especially in large corporations and the military, two institutions that have discovered how important it is to communicate the nonverbal dimensions of their corporate culture to their new employees or recruits. This is why rituals, such as initiation rites, and marks of hierarchy, such as saluting and bowing, are so enduring in such settings. Institutions will simply break down if there is no mimetic scenario to define their styles. This is especially true when we wish to communicate the basic social customs of a society to a new generation. Mimetic transmission of custom is crucial to the long-term durability of any culture. Without it, the social order can break down, and it is doubtful whether a complex social order could function, or successfully re-create itself in the next generation, without a set of mimetic conventions.

Mimesis thus stretches far back into prehistory, and we can confirm this by the nature of the cultural skills our archaic forebears evidently had. For over two million years hominids shared food, disseminated skills by imitation, created mini-industries for activities such as tool manufacture, and held group behavior to very stable long-term patterns, such as the continuous habitation of a site and the maintenance of fire. These things are not easy to achieve, and there is no good evolutionary rationale for attributing them to language. On the contrary, language is thought to be the main reason why our particular species, *Homo sapiens*, broke away from this early pattern. But to achieve what they did, archaic hominids needed an expressive skill that enabled them to bond, coordinate group activity, transfer and refine skills, and establish a network of custom and convention. Such things take considerable management, and even though they did not have the modern cognitive tools we enjoy, they had mimesis, which is a surprisingly effective management device. It requires, or perhaps it is more accurate to say it evokes and enforces, a pattern of consensual action, and this is controlled largely by mimetic action, such as pointing, vocalizing, and eye movements. This still works. In fact, we cannot do without it. Without a mimetic consensus, even NASA could not operate efficiently. Identification, role playing, and social organization function by mimetic actions, as part of the group theatrical production that we call social life.

The emergence of mimesis was our first step toward evolving an effective distributed knowledge network, which could coordinate the actions of groups of people. However, it is unlikely to have emerged in a single leap, or saltation. There is a very good reason for assuming that it passed through several smaller stages, or degrees of approximation. Natural selection cannot produce a trait in any species except through incremental change, that is, only if the preceding species is already close to having it. We cannot seriously hope to evolve, say, Cicero from Kanzi in a single step because the cognitive distances are just too great. We have to reduce those distances. We might conceivably have evolved Cicero from late *Homo erectus*, or Neanderthal, in a single step. But we would still have to explain how those species evolved from their predecessors and those from the gracile australopithecines. Evolutionary reconstruction is a process of finding the most feasible series of interpolations and testing them out on available data. What we are trying to do is establish a set of logical intermediate stages, or preadaptations, that could have bridged the huge evolutionary distances between the minds of modern humans and those of our very distant ancestors. Mimesis is my best guess at the nature of the first step. Someday, if our genetic methodology allows a much finer-grained classification of various subspecies of *Homo*, we may identify others.

Mimetic expression can be shown to function independently of language in children and in certain people with language disabilities. People who lack language because they were born deaf and never learned to sign can nevertheless manage the purely mimetic aspects of human culture with ease. They can hold jobs, as long as these don't demand the kinds of knowledge mediated by our oral-verbal cultural sphere, and they can master the elements of living in society and maintain stable long-term emotional relationships.

Mimesis is logically prior to language because without it, we cannot rehearse or refine any skill, let alone one as complex as speech or language. Group mimesis is basic to all education and training. Without a very efficient capacity for imitation, we could not hope to disseminate complex skills or cultural conventions. Mimetic gesture, at its most advanced, is a direct precursor to grammar. Sign language expert David Armstrong has argued that on the basis of its close relationship with gesture, sign might even have been the first kind of language to evolve. Gesture is inherently grammatical and has a complex structure that can express abstract relationships by using action metaphors. Moreover, without gestural capacity there could be no spoken language since the phonology of speech is assembled from tiny gestural components, the so-called "articulatory" gestures. With no mimetic capacity in place there would be no possibility of evolving phonology because phonology consists of a hierarchy of demons, or gesturelike actions. Also, unless hominids could assemble such hierarchies of demons, languages could not have come into existence. Languages are built following a "particulate" principle, not unlike that of genes. They are made up of lexicons of words, and words are assembled from a finite number of syllables, which are assembled from a finite number of phonemes. This requires self-supervised rehearsal and the rapid remembering of contingent relationships between sound and meaning. Without mimetic capacity such a hierarchy could not be assembled. Mimesis had to come first to create a lexical morphology—literally, the forms out of which words are made.

Emotional regulation was another crucial achievement of mimetic culture from the start. We can be tremendously moved by mimetic display and find it impossible to resist massive public waves of feeling. The funerals of kings and popes, the triumphal parades of athletes and generals, the sheer overwhelming power of military marching, priestly ceremony, or grand athletic spectacles attract us and force us into roles and attitudes, whether we like it or not. A subculture without a highly visible mimetic dimension would be difficult to recognize as truly human. And it probably would not function smoothly because it would have no effective mechanism for dealing with group feeling. Unable to harness the power of human emotion or to channel our deepest tendencies to form tribal alliances and hierarchies, such a culture would founder.

Mimesis was a self-sufficient cultural adaptation in itself. It could not look ahead to language, and it stood very well on its own, securing the foundation of the pyramid of human culture. Its vestiges persist today, as the unspoken foundation of all cultures.

THE GERM OF SELF-CONSCIOUSNESS

What is the underlying cognitive basis of mimesis? How might it be related to consciousness? One way to approach these questions is to consider the four major cognitive elements I classified as mimetic: skill, mime, gesture, and imitation. How do they cohere in terms of their implicit cognitive structure? The bottom line is that they all require us to map actions onto event perceptions. When I want to reenact a scenario, say, dropping the ball at the goal line, I have to be able to perceive from a different perspective my own imagined action, in this case dropping an imaginary ball after marking an imaginary goal line. I have to see it from the vantage point of my audience. Moreover, I have to perceive the implicit metaphorical relationship between my performance and the scene it represents. It is not easy to conceive of a neural system that is so flexible and so able to control an infinity of possible forms of action involving virtually any muscle group.

Because of the complexity of the motor brain and its many semi-independent subsystems for actions such as walking, running, gripping, looking, and vocalizing, the physiological challenge of evolving such an integrative motor capacity in the brain is not trivial. But that is precisely what hominids did. They evolved a mimetic controller, a whole-body mapping capacity that is truly domain-general in its reach and able to bring a variety of action systems under unified command. Unified voluntary action schemes of this sort would not be possible without a centralized brain map, a virtual space where the actor can deliberately review and modify every action in imagination. Think of this map as a "model of models," an image of the self-in-world that is connected to the motor systems. Since mimetic performances are often very fast-moving and complex, involving the cross-modal coordination of many different muscle groups, this map must have the capability to override our lower-level reflexes.

The existence of such a controller can be verified by analyzing the actual performances of highly skilled dancers and athletes. For example, classical Indian dancers are able to combine precisely modulated facial expressions with very fine manual movements and simultaneous limb and trunk movements, each of which has an independent meaning. This kind of performance cuts across many motor systems, and the entire voluntary musculature must be brought under unified control. Such skilled performers are able to integrate many different action systems under a unified plan. To construct these kinds of skill hierarchies, the dancer needs a substantial metacognitive, or self-evaluative, capacity. This is a highly conscious process of review and rehearsal. Dancers, actors, and athletes routinely review the success or failure of their efforts, over and over, and refine their performances accordingly. It is this extraordinary ability to become directly conscious of the body, in detail, that allows them to rehearse a scenario, again and again, until it meets a criterion of success. It also leads to the eventual automatization of the performance. With enough practice, those skills can be automatized, and at that point their execution requires less direct conscious supervision.

These considerations suggest a possible scenario for the evolution of mimesis in archaic hominids. Before their motor productions could become more skilled, they had to evolve direct executive governance over action. *Attention had to be redirected inward, away from the external world, and toward their own actions.* This was a significant break with the primate past. Whereas primate consciousness had occupied a largely perceptual domain and was directed mostly at the outside world, hominid conscious capacity invaded the domain of action. The executive brain system improved its ability to monitor the state of the physical self and gained access to much more detail, so that precise attention could be paid to the body's own movement patterns. Thus the first rung in our distinctively human ladder of awareness is the physical self, a supraordinate form of body awareness, born of a need to refine action. When I first suggested this idea about a decade ago, there was limited neurological evidence to sustain it. Now, however, there is much better support for it, stemming from recent research on the motor systems of the brain.

We are getting very close to developing a primitive neural model of imitation, and this will be a significant step forward in solving mimesis. It is highly likely that the key to human mimetic skill will be found in the extensive anatomical interconnectivities among the premotor cortex, the supplementary motor area, and the cerebellum. Above all, it will be found in their overall regulation by the prefrontal cortex. The human prefrontal cortex has evolved in such a way that it radiates greater control over the many brain areas that regulate action than it does in apes or monkeys. Neuroscientist Terry Deacon has called this phenomenon a "leveraged takeover." He has shown that through a developmental or epigenetic strategy, the human prefrontal cortex invaded many brain regions that formerly dominated the control of action in primates. It even invaded the subcortical nuclei governing the vocal apparatus, which receive very little direct cortical input in the ape brain. Since the human frontal cortex is generally regarded as the governor of conscious self-regulation, these anatomical differences amount to a major evolutionary change in hominid physical self-consciousness. Another crucial executive structure in mimetic capacity is the cerebellum, the performance optimizer we reviewed in Chapter 4. The other components of the executive brain—the thalamus, the basal ganglia, and the parietal cortex—are also involved in imitation. As the executive brain expanded its domain and applied its powers of control and review to a wide range of self-initiated actions, mimetic skill became a reality. This change did not require any qualitative innovation in primate brain anatomy, only a further differentiation of structures that are known to exist in the ape brain, but it made a large difference in functional terms. By means of these changes, archaic hominids became able to construct a much wider variety of action patterns consciously, and this trait persists. We are incredibly inventive with our bodies, including our faces, hands, voices, and postures. We delight in creating actions that have no practical purpose and spontaneously generate infinite numbers of expressions of all kinds.

KINEMATIC IMAGINATION

The cognitive core of mimesis is kinematic imagination, the ability to envision our bodies in motion. This is the basis of conscious review and rehearsal, which are central to mimetic action. To vary or refine any action, an actor must carry out a sequence of basic executive operations. That sequence is fairly standard: (1) generate the intended action, (2) observe its consequences, (3) remember them, and (4) review the original action pattern in imagination, in order to (5) generate the action again. The new patterns are supposed to avoid the negative consequences of the previous actions, and they may or may not succeed, so the loop continues, driven by an idealized image of the movement we wish to generate.

Figure 7.1

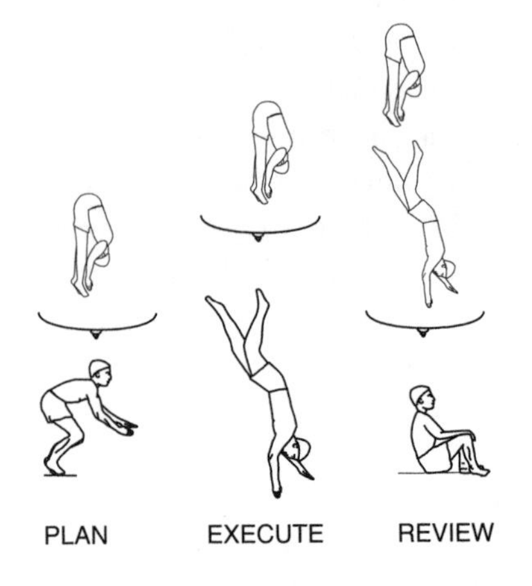

Does this sound familiar? Of course it does. It is essentially the same mechanism described in Chapter 3, when we discussed Olive Chancellor's flexible attention span. This is the most basic form of intermediate-term governance. In humans, it extended its reach from perception into the domain of action, but it is the same control system. Working memory, attention, and explicit recall are combined into a review routine that can evaluate the success of a self-initiated action in context, and modify it accordingly. In this extended sequence the focus of attention is not the reward or punishment that follows an act, or its social consequences, but the *form* of the act itself.

The most compelling example of this ability is our unique sensitivity to rhythm. Rhythms are a real giveaway, revealing the domain-general nature of mimesis like no other demonstration. Rhythms can be played out on any muscle system, in any combination, which shows how abstract mimetic performances can be. A rhythm is a perceptual template that expresses temporal relations; it can originate in a sound, feeling, or something seen, and it can be played out vocally, manually, or with the whole body (Figure 7.2).

Rhythmic dance can extend over long time periods and incorporate context. Think of a pair of flamenco dancers; their routines are usually full of mime and gesture, and they sometimes play out entire mimetic scenarios, adhering throughout to a rhythmic framework. Such performances are usually holistic in style and can be transformed into many variations, which cannot be reduced to discrete or digital elements. This is a fuzzy skill, where the Gestalt, or overall pattern, dictates the shape of the action, and a metaphoric principle rules.

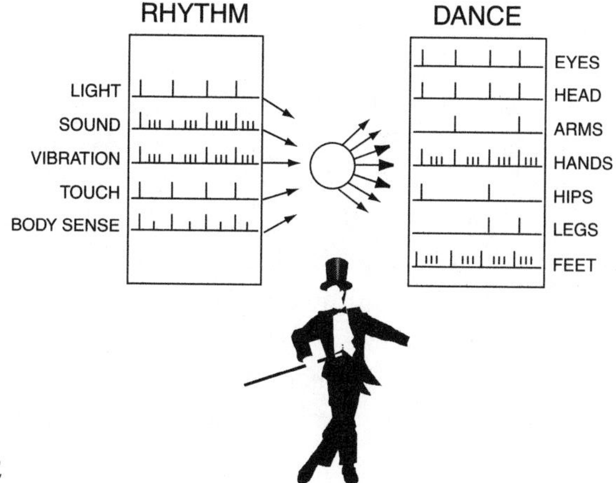

Figure 7.2

Kinematic imagination is a very peculiar capacity. It is our real specialty as a species, our true Cartesian Theater. It is this, our imagined kinematic image of self, that anchors our experience. It is also the bedrock of our orientation toward culture. In the seesaw struggle between brain and culture, the key surely lies in the first transition, without which the whole scenario could never have started. This was the initial *externalizing* phase from which our symbolizing mentality later emerged. It was the basis of our elementary cultural matrix, our characteristically human Ur-culture of the expressive actor, including simple ritual, custom, and nonsymbolic conventions. It was the direct result of a fairly straightforward anatomical change to the brain, but it produced a revolution in the conscious control of action. Conscious processing had previously been concerned mostly with perception and short-term memory. Now it switched its focus to the voluntary control of action. That shift, from perception to action, had an enormous impact on group behavior because it created patterns of public activity that were increasingly complex and unpredictable. From the relative anarchy of ape social groups came toolmaking industries, fixed campsites, complex group routines regulating fire use, more efficient hunting and gathering techniques, and a variety of customary expressions that served to maintain an enduring collective memory for what worked.

Paradoxically, the hominid mind achieved this stage of primal culture by turning further in on itself and evolving better self-representations. The result, our capacity to imagine the self-in-its-environment, was fundamental to human cultural change at that stage. This improved the hominid's ability to imitate, gesticulate, and acquire complex skills and generated waves of novel activity, which reflected back and forth, creating an increasingly complex and unpredictable public theater of convention. This relatively simple event triggered the long give-and-take between brain and culture and established a universal mimetic framework for human life.

Moreover, it achieved this cultural change without languages or symbols, which were still far beyond the zone of proximal hominid evolution. Mimesis fell very comfortably within that zone, so much so that it may have started to evolve long before the appearance of *Homo,* perhaps in australopithecines, without leaving a trail at that early stage. The revolutionary impact of mimesis should not be underestimated. All previous brains had been designed to fend for themselves, not to form cognitive networks with other brains. Mimesis reversed that tendency, started a distributed cognitive process, and triggered a series of cultural revolutions.

THE SECOND TRANSITION: THE SPIRALING COEVOLUTION OF THOUGHT AND SYMBOL

Language is different from mimesis, but it has mimetic roots. It is a collective product and must have evolved as a group adaptation, in the context of mimetic expressive culture. Given the conventional, collective nature of language, it could not have emerged in any other way. Language is above all a cultural product. Just as our bodies conform to the mimetic conventions of culture, so our minds conform to its linguistic ones. By internalizing these conventions, we gain new powers and can trade, amplify, and crystallize thoughts, remember with greater clarity, share memory, and enjoy membership in an enduring cognitive entity that transcends the individual. Like mimesis, language proceeds from outside to inside, and children must first master its public forms before internalizing its use. Like mimesis, it imposes massive conformity and welds a group of people into a cohesive entity. But here the resemblance stops. Language changed the rules of the cognitive game. It introduced an infinitude of new possibilities. It borrowed much from mimesis but started a whole new level of cognitive-cultural interaction.

Modern culture runs on languages and symbols, the way our economies run on money or computers on Boolean algebra. In the vernacular, language is often equated with full consciousness. When I am made aware of something, I usually mean that I am now able to speak about it. This is an idea with some merit. Certainly, there is something about capturing events in words, talking about them or, better, writing about them, that gives us the subjective impression that we are sharpening our awareness of those events and bringing them into relief. In so doing, we experience a clarity of mind that does not come in any other way. Fully human consciousness is inconceivable without language. Indeed, inconceivability is itself inconceivable without language.

This remains true as long as we float on the cognitive surface in the domain of words. But below the surface this idea runs into trouble. The surface of language might be defined as its form, and its form is open to many interpretations. For example, if I read Shakespeare's line "The quality of mercy is not strained," I can put many different spins on those specific words, depending on context. I might be baffled and find no meaning in it (I wouldn't be the first, in this case). I might completely miss Shakespeare's point. Or I might overinterpret it, reading in subtleties that Shakespeare did not intend. The problem is that language regularly betrays us, misleads us, or proves completely inadequate to the task of capturing what it is supposed to capture.

Our best writers have frequently written that language is hopelessly inadequate to their expressive task. Flaubert, quoted at the top of this chapter, typifies this sentiment. He despairs, convinced that he always falls short of his highest aspirations. No one understands what he intends. He is fated to "beat out tunes for bears to dance to" when what he really wants is to "move the stars to pity," a sentiment that seems to be shared by virtually all writers. Jackals pissed on his works, said Flaubert. The bourgeois climbed his literary pyramids, he complained, without knowing and, worse, without caring whether or not they understood what his works were really about. Presumably they climbed up for the view.

But what kind of view could such hapless readers hope for if they could not see what they were supposed to see and even the author himself was unhappy with the final product? The answer may be that rather than get us to our destination, intelligent reading only tells us where to look and roughly what to expect. It hints at possibilities, at novel states of being that we might, with luck and hard work, attain, if only for a moment. This outcome is never guaranteed, and we accept this. Reading seldom achieves what either author or reader originally intended, with any precision. Instead it elicits an intuition of *possible* knowledge and provides a motive force to move the imagination in a certain direction. A good reader can hope to discover not so much the exact content of a work as the possibility of a new domain of understanding. In this, there is no longer any boundary between author and audience. They become one and the same in their desire to realize raw, unexpressed possibility or, more accurately, in their deliberate pursuit of enhanced awareness of something. The same motive force that drives writers to write in the first place drives their future acolytes to read what they have written.

Reading is only one of the myriad of symbolic games we play—games that inspire our mutual exploration of the possible and our collaborative extension of the boundaries of subjective awareness. Such games are played by all of us, whether or not we can read. We employ all manner of symbolic expressions, from song, art, dance, ritual, and myth to speech, writing, graphs, tables, data banks, printouts, and equations, all to the same end: to ratchet up awareness in the service of various personal and collective agendas. Those agendas are often not realized because we are generally terrible at formulating most of our ideas and even worse at understanding what we hear and read. But this does not matter. It is the game itself that drives us, as our cognitive depths, both individual and cultural, strive for wider and wider horizons. Collisions of ideas are messy, chaotic, and full of grasping desperation, like the crowded raft in Delacroix's monumental painting of survivors at sea. Some ideas survive the ordeal, and despite our poor mastery of language, we occasionally experience a victory or two. Now and then thought triumphs, despite its irrational roots and our linguistic ineptitude.

This complaint, that language fails us when we need it most, may seem trivial from a Darwinian perspective. After all, language cannot have evolved for making complex propositions because our most advanced thought games are historically very recent and rare. When language stays closer to its original adaptive purpose, it is more successful. Language is spectacularly good at performing its bread-and-butter functions, such as communicating gossip and issuing simple imperatives (Caesar loves Cleo, No Parking, etc.). If it seems inadequate when we venture into more esoteric domains, we should not be surprised.

But its inadequacies are a tip-off. They allow us to conclude that the surface form of language is subsidiary to a deeper process, which cannot be, in its origins, linguistic or even symbolic. We could never detect our common failures in capturing thoughts in words if these were nothing but strings of symbols in the first place. If thoughts were simply words or their functional equivalent, our detection of expressive failures would itself have to depend upon properly formulated propositions, expressed in symbols, and so on ad infinitum. Thought would be locked into a self-enclosed circle of symbols. However, as we have seen, this cannot be. How could a reader anticipate a hypothetical state of understanding never before experienced, or validate that experience when it arrived, if it were only words that resulted from the reading? In their creative origins, symbols are a product of thought, not vice versa, and in their interpretation, symbols get their meaning from thought, not vice versa.

This dependency is especially evident in the case of original or creative work. The intuitions that create a truly novel symbolic expression, say, the

Pythagorean theorem, come from somewhere else in the mind. The fuzzy insight precedes the formal equations, and the latter can be validated only by matching them to another fuzzy insight, which then might become less fuzzy, through the influence of the symbol. And so on, in circles, until the symbol teases and twists the thought into shape. But the engine of the thought process must evaluate its own final state. There is a reserve capacity in the mind that can stand outside any symbolic product of the cultural system and evaluate its success. This is what allows the writer, or the reader, to judge the adequacy of the result. To achieve this, the symbolic circle must be broken within the virtual space of a single individual's awareness.

Language cannot "explain" thought at any level. Primates can obviously think, but they have no natural language. They can solve some fairly difficult problems without symbols. Humans can also think without language. This can be seen in very young children. The same is true of some aphasic patients who lose language altogether yet continue to be able to think in an artistic, musical, mechanical, social, or spatial sense. More important, language cannot even explain the existence of verbal thought. As I have just pointed out, it cannot explain our own innovative uses of words and other symbols or our ability to judge their success. In every case thought is the arbiter, and language is the child of thought, invented in the service of thought, employed forever as the amplifier and mediator of thought. The human mind is infinitely wider and more supple than all its languages and symbols.

The quality of thoughts can be improved with language. But thoughts do not start there or end there, nor are they judged there. Words and sentences define and clarify knowledge that resides elsewhere, in foundational semantic processes that we share with other primates and where the motive force for the evolution of language must have originated. It is easy to swim in our symbolic-cultural sea without noticing this, but our ability to perceive the inadequacies of spoken and written language reveals its deep cognitive roots. Our conscious mind strives after clarity to placate its own inner semantic processes. Somehow, it can judge when clarity has been achieved.

But what about very abstract thoughts, such as "Fiduciary instruments seldom control the money market" and "Deconstrained quarks can only roam free for microseconds because matter reverts quickly to one of its more common states" and "The quality of mercy is not strained"? Are these not "pure" symbolic sentiments? How can they be understood in any other part of mind? But they are. Fiduciary instruments can be reduced to a series of operations, and however abstract its formal definition, the money market is actually a very concrete idea, nested in some very practical facts and consequences. A money market is no more or less abstract, say, than a social relationship between two parties. And deconstrained quarks can be visualized in relation to our visual model of the atom; the subatomic scale can be imagined by a simple transformation. Concepts such as roam, freely, microseconds, matter, states, and mercy all have metaphoric anchors in experience, and the roots of meaning lie in countless such metaphoric comparisons.

This deep semantic loop applies to other symbolic domains that are very different from language—for instance, the visual arts. When artists judge that they are failing or succeeding in their pursuit of a previously nonexistent mental image, they must be able to evaluate their expressive failures and intuit their unexpressed intentions in much the same way that most of us evaluate spoken expressions. This also applies to music and mathematics. The nascent melody must be recognized when it reaches "perfection," an intuition that composers make with a part of their creative minds that lies far outside their musical notations or the formal rules that guide their composition. When mathematicians formulate a completely new set of equations to solve a problem, they cannot evaluate what they have done until they break the circle of symbols and relate them to other mental structures, usually visual images, that reside outside the equations themselves. Driven by a deeper intuition and a need for closure or differentiation, they evaluate the new symbolic expression and modify it out of necessity. The success of a truly new expression can therefore be judged only by a part of the mind that intuits the successful clarification of its own inner state.

Thus comes an important generalization: We evaluate all symbolic expressions from outside the symbol system, from a region of mind that, in its principles of operation, is different from, and much more powerful than, the reach of any consensual expressive system. Consensual symbol systems exist for the purpose of satisfying our deeper semantic (metaphoric?) intuitions. We can even judge the failure of an expression without having words or symbols for the as-yet-unexpressed ideas that we know we have failed to express! We can recognize the value of ideas whose mere possibility we suspect, and whose reality still eludes us completely, to the point of pursuing them passionately for a lifetime. This is like the Klondike gold rush in a more esoteric domain. Hundreds of physicists seek a Grand Unified Theory, and they expect to be able to judge its success when it comes. But how could they arrive at such a judgment? Scientific fundamentalists seek a savior, a conceptual messiah, whose form they cannot yet describe but that they will recognize when it appears. And so it goes with all our original uses of language. The ordinariness of this dilemma was the basis of Ludwig Wittgenstein's devastating attack on theories that tried to make language the sole foundation of thought. Language is harnessed to a thought skill. It amplifies the primary drivers of thought and serves as the mediator, but never as the inspiration, interpreter, or final judge. Words and languages are mysterious things if we insist on closing the linguistic circle and keeping it closed. But once we open the circle, we are left with a system that is at least solvable in principle.

The creative tension between a symbol and its cognitive domain is the source of the cognitive work the symbol does. This tension works in both directions. Boston linguist Ray Jackendoff has explored the complex standing relationships between languages and the various domains they represent. He and his colleagues have uncovered the implicit grammars that apply to symbolized musical and pictorial vision, and these are presumably representative of the kinds of rules and conventions that inevitably follow, once a group starts to construct a symbolic consensus in some domain. Symbols are produced by the tension between the two sides of a symbolized cognitive system, the form of the symbol itself, and the meaning systems to which it is attached and on which it acts.

Our neural models of the process whereby forms are mapped onto meanings are at best superficial, and we have almost no detailed knowledge of what is going on inside the vast neural networks that produce these complex states of tension. Nevertheless, it is evident that states of tension between large neural networks seem to parallel the state of our conscious semantic awareness. Awareness watches over and initiates all major symbolic performances (there are also small, repetitive, automatic performances, but these do not usually draw on conscious capacity). Consciousness is always on watch, and what is expressed is almost always less witty, less elegant, and less precise than what was intended. To our relief, our awareness knows this and compensates for it. Language has no autonomy from this conscious, evaluative process, none. Our games with symbols may tease and twist our deep semantic spaces into novel shapes, creating new spaces between existing meaning states. Truly novel meanings have no other way to come into existence. But our symbolic universe serves only one ultimate master, our drive for clearer states of awareness, in both sender and receiver.

PIGGYBACKING LANGUAGE ON CULTURE

There are several good reasons for developing a culture-first approach to language evolution. First, it allows us to remain consistent with what we know about the evolving hominid brain. Second, it provides both continuity and a radically new evolutionary factor, enculturation, that might account for the qualitative shift in human cognition. Third, it accounts for more than language and explains the robust mimetic framework within which language exists. Fourth, it explains why language does not self-install in the minds of isolated human beings. Fifth, it gives us a semantic base, a means of referring language outside itself. This is important because hominids could not have evolved a language capacity as an isolated adaptation. They had to maintain a self-sufficient lifestyle at all times. Hominid symbolic capacity cannot be disconnected from the cultures of which it has always been part. Language is only a part of our larger cultural adaptation.

Finally, a culture-first approach provides us with an efficient mechanism, call it a piggybacking strategy, for the self-replication of the entire human cognitive system. Early computer programmers used the term "piggybacking" to refer to a strategy whereby they stored as much information as possible on digital tape and imported it into memory only when necessary. This allowed them to overcome the limitations of the very small memory systems available thirty years ago. Over the infinitely longer stretches of evolutionary time, hominids have off-loaded as much storage as possible to the cultural attic and brought it onshore (into the brain) only when necessary. With that strategy in place, the hominid brain had to evolve the equipment needed to read the current state of the cultural storage system. It could thereby replicate the current version of the hominid mind, in both its individual and distributed aspects.

Is this a return to Constructivism? Of course, but with a difference. Individuals construct themselves with the pedagogical guidance of culture. This is cultural Constructivism, and I don't mean it in the narrow, literal sense that critics like Steven Pinker have used. Constructivists do not deny the need for automaticity and deep structure, as he has suggested. Pinker has skewered educational Constructivism by equating it with poor pedagogy, especially a failure to attend to the basics. But sloppy educational policies were the product of the egalitarian ideologies of the sixties, not of Constructivism. As even the briefest review of Condillac will reveal, Constructivism has always paid close attention to exactly what is being learned and in what sequence. Modern Constructivism is much more sophisticated than the straw man Pinker has attacked. This is evident in books such as *Rethinking Innateness*, and other recent Constructivist manifestos.

The idea of a culturally linked Constructivism mediated by the executive brain is compatible with multilevel theories of evolution because it off-loads replicative information to culture on a sliding scale. Early hominids stored very little in culture. But with the passage of evolutionary time they started to do so more and more. As evolutionary psychologist Henry Plotkin has pointed out, this sort of storage can be fitted nicely into a conventional

Darwinian model of evolution. Cultural storage feeds back onto other levels of storage, such as genes and epigenetic processes. Our most powerful cultural storage device is language itself (this includes more of its implicit codes than of its explicit meanings). However, language is far from the whole story. We store cultural information in many other ways, in customs, institutions, and material objects, among other things. All this is part of our replicative machinery as a cognizing species.

It is crucial to remember that language is only a secondary level of storage in this system. We cannot even replicate the language mechanism itself without using some of our other culturally stored information. A humanity stripped completely of its languages and deprived of any memory of its existence in the past, like those unfortunate human survivors in *Planet of the Apes*, might be able to re-create a full-fledged language, after an unknown number of generations, by revisiting the stages of cultural evolution. But there is obviously a great efficiency to be gained by copying an existing culture. To regenerate all the communicative habits that support full-fledged language, all the relevant aspects of the enculturation process would first have to be restored. This might take a while. In his studies of creole speakers on an isolated island, Derek Bickerton found that they had taken at least two generations to invent a true language (the first generation spoke only a "pidgin" language that lacked grammar), even though the founders all had normal mother tongues to start with, and normal early enculturation. We do not know how many generations it would have taken if they had all been deprived of normal early development.

Language can be compared with visual perception, where the physical environment, not the brain, has generated the complexities we perceive. The visual brain has evolved the equipment it needs to read the environment, but it does not try to store it or anticipate exactly what it will receive. It has some biases and special sensitivities, such as stereo and color vision. But these are very basic. The human brain has only what it needs to become optimally sensitive to survival-related aspects of light. This is an efficient way to evolve a complex system, and it undoubtedly works just as well for language. We don't need to fill the brain with prepackaged Kantian categories, Platonic ideals, or Universal Grammars to enable it to think or speak with words. The communication environment itself is the storehouse of what it must learn. The brain needs to have the innate capacity to find, filter, and remember the essential features of that environment.

Universal Grammar is somewhat akin to the universal laws of learning so beloved of Hullian Behaviorists: an elegant phantom. Such a grammar may exist, in the sense that the rules of language follow certain general principles, but tell us nothing about the mechanism of language or its relationship to cognition. They do, however, tell us something about the real-world constraints that shaped language evolution. A list of the basic elements of Universal Grammar includes the parts of speech—namely, words, which are invariably sorted into verbs, nouns, and other categories. It also includes the phrase structure of speech and the fact that phrases are arranged into hierarchies, whereby some of them can be made subordinate to others, as in "The cat that ate the rat slept under the porch," in which the phrase structure enables us to understand that it was the cat, not the rat, that slept under the porch. Then there are the grammatical rules themselves, which always include some way of indicating negation, agreement, combination, and sometimes tense. Other complex rules govern the use of inflections or modifications of words that change their use. Anyone who has studied Latin, as I did for four years, will appreciate inflections. They also apply in English; for instance, I can make the word "child" into a plural by adding the suffix "ren."

Given the nature of what languages describe, this may seem like a rather obvious list of features. How else could we label the agents and actions that characterize an event like, say, throwing a strike in the World Series than by having words for pitcher, ball, catcher, throw, and so on? How else could we specify who threw the ball than by having some way of distinguishing subject from object? How else can we represent space than by somehow specifying up, down, beside, and above? The parts of speech and the rules by which they are governed seem to emerge naturally from the progressive differentiation, or parsing, of event perceptions. In this case, we can say that language begins by simply putting labels on specific aspects of an episodic perception. In fact, it is the latter, episodic cognition, our vestigial mammalian inheritance, that has imposed this universal frame on language. Presumably, if we could free ourselves from that constraint and from our cultural history, we could devise languages without these universals. Indeed, we have done precisely that: In the context of engineering, we have invented artificial languages, showing that humans are not constrained by these universals in all their linguistic inventions. But in early evolution languages had to emerge from the mists of episodic and mimetic cognition. They were tied to the representation of concrete events and episodes. Their potential for abstraction had not been realized yet.

We have good evidence for this deeper cognitive foundation in the well-documented relationship between language and metaphor. Language floats on a sea of metaphor, and this dependency suggests that language is usually not the captain but the passenger in this vessel. Princeton psychologist Julian Jaynes was probably the first to point out the metaphoric nature of thought and coined the term "metaphorizing" to describe the most elementary operations of thinking. More recently the functional linguists Mark Johnson and George Lakoff have shown convincingly that language is driven by metaphoric principles. Metaphor is usually defined as the use of analogy, or similarity, while forming substitutions in speech, as when we say, "We're in a real dogfight now," meaning, of course, that we will be faced with strong resistance, not that we will have to fight dogs. Another example would be to say, "He is a very big man in his field, head and shoulders above everyone else," in which we are not referring to stature but mean that "he" is important. Lakoff and Johnson have suggested that metaphoric expression taps a cognitive vein that is much more fundamental than language itself. In effect, metaphor is a dead giveaway (to use a metaphor) of the episodic roots of language.

Lakoff and Johnson hold that this is the normal, or canonical, form of human thought. Metaphor is also universal because linguistic meaning gains its natural structure from this metaphoric, or episodic, connection, which is determined by the very events that language tries to describe. This gives us a nice continuity from the ancient mammalian capacity for social event perception to language. Our experience base is still the same as it is in other social mammals. We can talk about it, and they cannot; but the existence of language does not change the fact of our common experience base. Our dependency on metaphor exposes the vestigial mammalian cognitive system that drives our use of language.

In the way they are constructed and assembled, artificial languages are free of the encumbrances of episodic perception and mimesis because they are not directly constrained by biological inheritance (although they have technological constraints). Early humans, on the other hand, had to scaffold language on the foundation of an episodic and mimetic inheritance. The so-called universals of language coincide beautifully with what would be imposed by evolving language within a mimetic envelope. I made that argument from a neurocognitive standpoint long ago. But the linguistic argument has been spelled out much more clearly and elegantly by David Armstrong and his colleagues and by Elizabeth Bates and her colleagues. It is not that linguistic universals don't exist. They do. But they do not demand the sort of explanation to which many linguists stubbornly adhere. Linguistic universals spring from the context in which real-world languages are learned and, more important, in which they evolved. Like any other set of conventions, linguistic conventions are shaped by the situations in which they originated. They have mimetic origins. Thus, once we change our paradigm, the features of Universal Grammar emerge smoothly from a close analysis of gesture, mime, and imitative behavior. The "language instinct" exists, but it is a domain-general instinct for mimesis and collectivity, impelled by a deep drive for conceptual clarification. This is not a clocklike device in the head that ticks out words and sentences; rather it is a perfect marriage of executive capacity and off-line storage.

Going back to our comparison with vision, we can say that human culture is very different in one crucial way. The brain does not generate our visual world (the Presocratic philosophers were wrong on that one). However, it does generate our cultural world. We do not have to explain the intricacy of the world of light as if the brain created it, whereas we do have to explain how culture came to be what it is, in terms of our brain's capacity to generate that complexity. Does that not imply a qualitative break and discontinuity in our evolution quite unlike what happened with vision? This is a serious objection, but it can be answered without conceding any ground to discontinuity theory. The answer is in the concept of distributed cognition and the kinds of social metaorganisms described by E. O. Wilson and others. Highly complex patterns can emerge on the level of mass behavior as the result of interactions between very simple nervous systems. Researchers in the field of Artificial Life have devised powerful computational models of evolution and have shown that complex and often unpredictable patterns can emerge on a metalevel from the interactions of very simple virtual organisms. The message from these simulations is powerful. Under the right circumstances, a small change in the individual nervous system can generate a huge change at the level of culture.

Languages and symbol systems are thus a mirror of our lower-level cognitive interactions. Complex patterns of use are generated at the group level, and there is no reason why they could not piggyback on much less complex brain adaptations (to risk a truly awful mixed metaphor, piggybacking is a two-way street). If we assume that the group matrix was driving the evolutionary process, a very reasonable assumption in this case, the adaptations that produced human language capacity would not necessarily require a simulacrum of language built into each individual brain. Language would only have to emerge at the group level, reflecting the complexity of a communicative environment. Brains need to adapt to such an environment only as parts of a distributed web. They do not need to generate, or internalize, the entire system as such.

Customs, languages, and semiotic codes are distributed across many individuals and across a multitude of external devices that constitute much of our material culture. The individual brain has only to track the emerging complexity of this entire matrix, not to copy it. The most efficient way to solve a distributed web is undoubtedly to become part of it and evolve with it, rather than try to simulate or duplicate the whole mechanism in each mind. Cultures and their brains have advanced by interlocking with each other. Language can be evolved and replicated only by means of this piggybacking strategy. This framework determined the way in which languages first emerged. Otherwise, we would have very different brains from the ones we have.

Our cultural storehouses far exceed what any brain could possibly copy or remember, and no single brain can totally comprehend every aspect of a symbolic cultural environment, let alone generate one. Cultural richness accumulates over time, from the interactions of many brains over many generations. In this, cultures are more like ecologies than they are like the millions of individual organisms that make up ecologies. Languages in particular constitute virtual ecologies that occupy the spaces between virtual organisms. Words and symbols are more like the oxygen or nitrogen cycles, which are "reproduced" by means of ecology-level events, than they are like palm trees or squirrels, which reproduce genetically, in single organisms.

OUR CEREBRAL BOXING MATCH WITH THE CULTURAL MATRIX

This places the creative source of language in a unit that might be called the individual-conscious-mind-in-culture. To shift back to the vantage point of the individual, symbols of all kinds are the playthings of a fantastically clever, irrational, manipulative, largely inarticulate beast that lives deep inside each of us, far below the polished cultural surface we have constructed. That passionate and devious intelligence drives all our collective enterprises, lobbing its semiotic hand grenades at various imaginary targets, while it lurks in the background, just out of sight, self-conscious, self-deprecating, self-doubting about its own value, and largely unaware of the detailed complexity of the cultural matrix. This intelligence is isomorphic with our conscious experience of the world, just as it is in other mammals. This is why the human brain cannot symbolize if it is isolated from a culture. This is also why the brain has always been embroiled in an ongoing boxing match with culture. This match has been filled with tension from the start, and there have been unexpected shifts in the balance of power between the brain and its cultures or rather between the culture and its brains. Now the mind leads with gestures and words to push one way, and now the culture pushes back in another direction, perhaps one that no one would have predicted. The sly semanticist in the head now pushes culture to provide the clarification it seeks, and now it is dazzled into acquiescence by the collectively generated result, sliding into temporary equilibrium.

The tension between cultural symbolic systems and the underlying intelligences that use them determines the quality of our uniquely human modes of consciousness. Awareness never resides entirely in what is said or in the ephemeral sensations that it captures and holds for ransom. It resides in the intermediate-term governor of mind that we endow with intellect and soul and call "me." But it is more than this since it also runs an elaborate cultural machine for the purpose of its own satisfaction. Culture shapes the vast undifferentiated semantic spaces of the individual brain. The brain takes on its self-identity in culture and is deeply affected in its actions by culturally formulated notions of selfhood. This is what theories of cognitive evolution must explain. Explain the knowing beast within, and we shall have a chance to build a scientific theory of symbolic communication. Ignore the beast, and we shall always fall short because without it, the rest of our theoretical edifice is left without a framework to support it.

The notion that consciousness is locked in an eternal struggle with huge cultural forces is acknowledged implicitly in Freud's concept of the unconscious, which is still important to those who study culture. Much of what Freud attributed to the unconscious is truly unconscious only in the cultural sense of the word—that is to say, things that were unexpressed or actively repressed at the level of culture, which is not the same thing as what we now mean by the term. In the terminology of modern cognitive science, the unconscious is a strictly solipsistic notion, not a cultural one. It refers to a region of the individual mind that lies permanently and by definition outside the reach of consciousness. The lines between consciousness and the mind's inaccessible unconscious modules are drawn very deep in the sand, and these lines are inviolable. By definition, a module, such as object vision, is permanently inaccessible to awareness. It serves up all the richness of the three-dimensional visual world to awareness, gratis and fully formed. But we can never gain access to the mysterious region of mind that delivers such images. It lies on the other side of cognition, permanently outside the purview of consciousness.

The cultural unconscious of psychoanalysis is another thing altogether. Its intuitions, drives, and emotional complexes are represented in art, writing, and theater. They are lived out on the cultural stage and are never, in principle, inaccessible to awareness. They can always surface in individual consciousness, even when poorly expressed at a cultural level. Repression is, more often than not, a failure of apperceptive or representational capture in the public arena. Some psychoanalysts, notably Jung, rooted the collective unconscious in cultural archetypes, but surely he meant only that the influence of such archetypes is usually implicit, rather than explicit. The deep cultural unconscious exists in a representational limbo that is temporarily uncaptured. The cognitive unconscious, on the other hand, is the golem, the automaton world of instincts and zombies. As we can see by observing the process of cultural invention in our own time, cultural symbols are always crafted products, refined by centuries of conscious effort. Such things as Earth Mother myths and ankhs never spring from the bowels of the permanently unconscious modules of mind. They are archaic cultural inventions that have hung around and may have sunk temporarily from view. However, they originated in consciousness and can return to it anytime. They are not permanently and by their very nature out of reach. This distinction is important.

This has implications for how we should conceive the relationship between culture and consciousness. I have placed the engine of culture in awareness and specifically in those aspects of metacognitive awareness that are summed up in the Executive Suite. But the patterns of culture, the mazes we must penetrate, are generated by the cultural matrix itself. Our cognitive interactions as a collectivity may not be as easily visualized as the physical interactions between migrating birds in the air or armies on the battlefield, but that doesn't make them any less real. The patterns that emerge at the level of culture are not only real but dominate the cognitive universe that defines what "reality" is.

THE MANAGEMENT OF IDEA-LAUNDERING SCHEMES

The most common way to bring something into cultural awareness is to find the right symbol to express it publicly. Language is often the principal source of such expressions. From early childhood it is the handmaiden of cultural awareness and mediates our thoughts. Language mediation is a good test case for describing the broader relationship between thought and symbol. To risk offending those who insist on elevating language to a semidivine status, I think that it really works more like an offshore bank, a tax shelter, or a money-laundering scheme than as a self-contained module of the mind. Language launders ideas rather than money. Even though it is a group-level ecological feature, it achieves this result through a process that can be reduced to a set of interconnected cognitive demons.

Linguistic idea laundering is essentially like any other hierarchy of interdependent cognitive operations. A child loops a thought into primitive symbolic form, produces an external utterance or gesture, evaluates the feedback obtained, modifies the expression, and continues until "clarity" ensues in the networks that generated the idea; that is, until they stop striving to improve their final states. This is another piggybacking operation, but an internal one; we learn words by hanging them on a mimetic fabric of action, based on experience.

The process of self-clarification and verification is not unique to humans. It is what most large waking brains do for a living. But the loop through a collective cognitive process is new to our species. Culture adds great power to our typically mammalian brain systems by filtering our cognitive activities through an elaborate public process that serves as a source of constant clarification and definition. The public loop is, metaphorically (and simplistically), a bit like the cleanup circuits used in neural net computing, which are employed to sharpen the state of another network. Like culture and language, they are parasitic on the networks that do the real semantic work. But they are also invaluable. A consciousness immersed in culture resides ultimately in the temporary state of a semantic network honed through a public process. The network itself stands in judgment of the success or failure of this whole operation. In effect, the semantic beast can stand back and evaluate its own expressive performances, just as it judges everything else about culture. The difference between us and other primates is only that our conscious-beast-within is much better equipped than theirs to understand and navigate the complex environment needed for public idea laundering. Our brains have been modified by millions of years of natural selection to do this well, and as a result, we can invent and employ languages to run our ambitious cognitive schemes.

At first glance, I may seem to be promoting a rather anarchic cognitive universe, a world in which words are empty abstractions that lack any absolute meaning and are entirely relative to an arbitrary cultural matrix and a still-unspecified concept of network clarity. This might seem to concede some legitimacy to the deep psychic drives that we have tried hard to lock in the attics and cellars of cognitive science. But if these deep drives exist, and they must, it is time to drag them out into the light of public debate because they are the principal forces that energize our entire cultural universe. Although our present understanding of cognitive drives is limited, there is no reason to think that they represent an unsolvable scientific problem.

I have defended the reality of intermediate-term governance in the human mind and its presence at the very core of all advanced mental skills in higher mammals. Now I shall up the ante. It is this capacity for intermediate-term governance that guides all our adventures with culture, including language itself. It is not the semantic network inside the brain, or the cultural matrix outside, but this level-3 metacognitive mechanism that enables such things to grow, both above and below it. Culture and its meanings, with all its languages and symbols, are the children of expanded consciousness.

Symbolic idea laundering takes a lot of management. A creative, interactive cultural process increases the sheer quantity of knowledge available to a person, and that knowledge has to be filtered, sorted, stored, and made retrievable. Moreover, as cultural knowledge networks become larger and more complex, they demand greater conscious capacity, especially with regard to voluntary recall from memory and selective attention. Our real-time, on-line mental lives become much busier because in addition to having to deal with the cognitive challenges that confront other mammals, we have to deal with a parallel cultural context every moment of our waking lives. We have to solve concrete problems, such as how to move about in space without injuring ourselves, find things, and fight battles, at the same time as we play out mental scenarios in imagination, talk to people, think about the social consequences of our actions, and plan ahead. This takes executive management. It also takes plasticity because cultures constantly reorganize the inner universes of their members and multiply their mental models. This loads us down with thousands of ideas and facts, and above all, it demands our close attention to the rules and codes that allow symbolic cultures to operate.

The multiplicity of our mental models seems to result from a finer and finer definition of reality. "Definition" is a superb word for this process of semantic differentiation, because it acknowledges the fact that symbolic idea laundering exists primarily to produce a state of equilibrium in the intermediate-term governor that drives the whole process. Right now I am sharpening my mental models using written language. But I am also talking to myself, mostly silently, as I do this. My innermost semantic layers are evaluating their state of inner clarity and generating counterideas because they know they must launder their hard-won knowledge through further public loops to clarify their own (my own) fuzzy thoughts, urges, and feelings. I do this over and over again, slowly titrating in on a satisfactory final state, a temporary equilibrium that will probably last for only a short time. I hope that by my doing so, my ideas will become sharper, clearer, more convincing, and more satisfying. Satisfying? How can mere ideas be satisfying? We have difficulty enough understanding how a meal or a sexual encounter can be satisfying, but ideas? The raw feel of intellectual satisfaction is yet another major puzzle for cognitive science to solve. Future generations of cognitive scientists need not fear any immediate danger of mass unemployment.

We rarely achieve perfection, and simple ideas typically succeed more often than complex ones because their apparent perfection is so intoxicating to the semantic beast within, which craves equilibrium. Making its powerful, fuzzy intuitions clearer is rewarding to the brain. There is joy in knowledge, and cultural contact is addicting. Culturally generated ideas, such as outside-inside, for instance, or superplasticity, or intermediate-term governance, have caught the attention of my own beast and had a huge impact on its (my) thoughts. These ideas have been productive for me, and seem to have slaked my drive for clarification, at least temporarily. But this habit of mine, of seeking clarity, is potentially dangerous because snappy ideological slogans, such as "Neural Darwinism," "Marxism," "Universal Grammar," or "Supply-side Economics," might pop up, throw my beast off its guard, and ambush its criticality, reducing it to the vanishing point (a cultish reaction) or raising its threshold of acceptance (a snobbish reaction). Had I been less lucky in the time and place of my birth, it might have swept me into barbarism!

I readily admit to a great deal of skepticism about my own place in all this, as I sit here in front of my computer, relentlessly pecking away at the keys, even though I am going to the immense trouble of writing this book. It is after all my own internal beast, with its own relentless drive for clarity, that I am trying to please. I feel like a lion tamer or, better, a dragon slayer whose dragon is within. I throw the beast a hunk of meat, say, a new fact, such as "there are mirror neurons in the premotor cortex," and it barely opens its eyes, snarling suspiciously, Not enough! More! More! And I toss it a libraryful of recent brain-imaging papers, or a packet of freshly baked Postmodern neologisms, or perhaps one of those labyrinthine novels by Umberto Eco. That usually throws it off balance for a while. Still, it sulks. It wants more. We are after all defining human nature here, and my beast is endlessly self-curious.

To paraphrase mad King George III: A curious fate, what-what? We must tread respectfully here, hold all our motives suspect, and recognize that the sources of our insights might not be what we think they are. Above all, our drive for publicly expressed cultural rationality and clarity is itself deeply irrational and unclear in its ultimate destination.

SYMBOLIC INVENTION AND THE GROWTH OF THE LEXICON

Where do words stand in all this? To some, words or, rather, word systems are the neurocognitive atoms of language. Each word system unifies all the different cognitive systems that operate in language. It serves as a focal point at which meaning, grammar, sound, and form come together. It is also an indispensable component of higher-level structures, such as phrases and sentences. Yet to others, words are fictions. Italian philosopher

Benedetto Croce held that sentences are the only meaningful units of language. Words, he said, are merely categories that we have invented after the fact. The disagreement between Croce's few followers and the hordes of grammarians that populate our universities might be resolved somewhat by appealing to our developmental and evolutionary roots.

Every word is a cultural invention, and individuals must learn the consensual maps that every culture uses to graft word forms onto meanings. The means of inventing such maps might cast light on their nature. Most words are invented in the theater of mimesis, within the framework provided by the typical scenarios of childhood. At first, words play a secondary role and merely complement mimetic exchanges. The driving force behind lexical invention is the need to disambiguate mimetic expressions. Mimetic expressions have a very low degree of specificity. Thus, when they fail, our cognitive beast looks for a way of disambiguating the message. If there was a fire in the building, I might normally be able to clear it with a few mimetic gestures and shouts, and my audience would have very few alternative interpretations of my expressions. But if the fire occurred after a terrorist alert and I ran through the building, flailing my arms and sounding very agitated, my message would be open to other interpretations. I would want to spell out my intent more precisely and find a sign that specified "fire." Or if I knew the source of the fire was downstairs, I might need to formulate a more precise expression, such as "Don't go downstairs, use the fire escape." The catch is, my expression would have to be understood by my audience. The map must be collective. Its success would depend entirely on a consensual process whereby the group agreed on a fixed set of symbols, mapped onto meanings.

This degree of disambiguation would not come merely by improving gesture. Gesture is not language. It resembles language in many of its features, but it cannot support a narrative. As mimetic capacity evolved further, it was transformed into language, at a cultural level, by roughly the same interactive sequence that children follow in acquiring language. They must hunt for clues to the intentions of others in order to discern what words mean. This effort is guided by pointing, mutual gaze, gesture, and vocalization. A vocalization channel would have been a very effective way to disambiguate a mimetic visual channel, without interfering with it. Initially, this would have been in the form of vocomimesis, with a specialized subsidiary role in mimetic communication, as a distinct disambiguating submodality of mimesis. In the run-up to full language it obviously assumed greater importance, for a variety of reasons that have been spelled out not only by me but by others, such as Michael Studdert-Kennedy.

The first disambiguating vocalizations need not have been specifically intended as parts of speech—that is, nouns, verbs, adjectives, or pronouns—

as we know them. Rather they might have been intended to specify entire chunks of an event representation, like the first utterances of infants, which often encompass full sentences in themselves, even though they may contain only a single word. Thus a child might say "ball" and really intend "Bring me the ball," or "Throw the ball," or "I dropped the ball," or "Daddy threw the ball," or even "That looks like a ball." The sentences are latent in the utterance, which needs only to be progressively disambiguated. We can often infer which of many meanings children might intend by monitoring their intonations, their gestures, and the context. Children eventually learn how to specify these meanings more precisely in full sentences, but they start with implicit sentences. From an evolutionary perspective, Croce appears to have been correct about the primacy of implicit sentential structures. In both individuals and groups, language moves from implicit sentences to explicit ones by progressive disambiguation. As implicit intentions are gradually transformed into explicit ones, vocabulary explodes, as do the rules for using it.

On a cultural level, language is not about inventing words. It is about telling stories in groups. *Languages are invented on the level of narrative, by collectivities of conscious intellects.* They went through many stages of evolution before they arrived at their present level of refinement, and categories such as nouns, verbs, and adjectives emerged only after episodic representations had been specified, first in mime and gesture and then in a less ambiguous form, in consensual symbolic systems. These took very long to evolve and emerged coextensively with the hominid brain. One might say that the ultimate complex structure of language was implicit even in pantomime and gesture but took time to evolve. Hominids needed considerably better attentional and working memory capacity and metacognitive skills to construct and maintain public systems of communication, with complex rules and thousands of words.

But at the start of this process, archaic hominids would never have heard anything like a word or a sentence and would have been driven only to clarify the existing mimetic scenarios with which they had dealt for many millennia. For example, when teaching toolmaking techniques, teachers might first have disambiguated the weak points in their pedagogical scenario by using visual emphasis and prosodic cues to direct their pupils' attention to what they were doing. This is a purely mimetic technique for controlling attention. But during the transition to language, those prosodic cues would have become more precise, and the teachers would then have employed specific gestures to label, for instance, which stage to rehearse or which striking method to use in edge sharpening. Such utterances would have initially constituted a shorthand for entire phrases, rather than words as we know them. It would have been difficult for a modern observer to classify such utterances as nouns, verbs, and adjectives, in a theory-neutral manner.

Simple grammars would have emerged naturally from this process, as the level of specification became more and more precise. Naming relationships such as "above," "with," and "below" is not a generically different activity from naming actions, persons, or objects. Thus we have no reason to expect that one class of word, say, nouns, evolved before another, say, verbs, or that the rules for word order should have evolved at a different time or independently from those that regulate inflections. Obviously, grammatical or function words, such as "with," "and," "because," and so on, would have appeared only when languages became sufficiently complex to require such specifiers. But specifying a plural, or the past tense, or a causal connection is not a generically different activity from specifying color, relative size, or form. Words are words, and once the cognitive habit of disambiguating expressive scenarios gathered momentum, there would have been tremendous pressure on the individual members of the founding group to adapt to the expanding language universe. Once we could specify and label relationships, we had an elementary grammar. Moreover, given the self-referential nature of mimesis, utterances would immediately have become potentially recursive; that is, able to encompass one another and split their semantic progeny into infinite numbers of self-referential parts. This was not because language is uniquely able to do this (mimetic imagination can also be recursive, as when we imagine boxes within boxes or envision unending embedded scenarios of action) but because when viewed from a certain perspective, events are themselves recursive. Recursion was not inherent in any specific process of representation; it was inherent in the structure of what was being represented.

What does all this have to do with hominid conscious capacity? Everything. Language came late. It coincided with the last great expansion of the hominid brain, which further increased the importance of the executive brain system, the same prefrontal-cerebellar system, with its many linkages, that expanded during the first transition. In addition, there was an expansion of other areas associated with language. These areas include the entire cortical region surrounding the base of the hippocampus and the peri-Sylvian cortex. We do not yet know exactly how languages are distributed within these enormous brain networks, but it is certain that the executive requirements of language were tremendous at every stage of its evolution. This demanded, at the very least: (1) a greatly extended and differentiated working memory; (2) a capacity for multifocal attention; (3) lifelong plasticity; (4) a huge expansion of long-term memory capacity, able to store thousands of neural word systems, in instantly retrievable form; and (5) a great increase in the amount of brain space devoted to semantic representation since semantic space would have become more and more finely differentiated under the growing pressure of an emerging language capacity. These are all domain-general requirements. The only modular element in language evolution appears to have been the vocal apparatus, but even this was not a truly modular change at the cognitive level since phonology, in its strictest definition, is not restricted to the vocal system (sign language has phonemes).

But it was the nature of conscious experience itself that changed most dramatically. Language differentiates experience for us. It refines the thought process and embeds it in increasingly precise culturally imposed algorithms. We know intuitively what words and sentences do for us. They define reality. They focus our attention. They elevate our awareness of whatever they specify. Moreover, when stories and ideas are juxtaposed, so that their meanings collide, they can shift our focus to new semantic spaces. Once the circular process of linguistic definition has been triggered, these spaces are constantly opening up. Once we have a sentence or a word for something, we seem to own it. We can roll it around in our minds, reflect on it, and conduct experiments with it. In a word, languages clarify the experienced world. They placate the internal beast of consciousness. The drive that energizes our use of language today, and invented it in the first place, is that same intermediate-term governor, constantly seeking greater clarity, conjointly with other conscious beings, in culture.

Ultimately, all this traces back to the event and its representation, on the one hand, and to cultures as idea-laundering devices, on the other. Just as the single episode was, and still is, the atom of mimetic experience, so it is also the atom of language. Episodes consist of smaller events, which consist of objects, actions, agents, and landscapes in a variety of arrangements. A major episode, such as "escaping from Farmer Brown's barn," might encompass dozens of microevents, including various visual images, sounds, and self-actions, such as falling down, getting up, finding the door, climbing, running, ducking, avoiding obstacles, and so on. The integrative challenge posed to the brain by such a complex set of simultaneous perceptions is huge, considering that the larger event structure supersedes the many possible versions of the event that might be perceived as equivalent. But the brain tends to integrate, not differentiate, the event. Language allows us to differentiate and decompose the experienced event.

This elaborate process, the linguistic definition of reality, is both the cause and the result of splitting not the atoms of the physical universe but the atoms of episodic experience into their component parts. Cultures that have languages can generate microfoci within the representational spaces of their members' minds. They abstract the components of event percepts from their concrete contexts and isolate common features of those percepts. Given our invisible habits of shared attention, and some cultural control over how experience is processed, a common language will allow us to share mind better, by defining a common representational framework. This gives us a new cultural domain, a stock exchange of the mind, where ideas and impressions can be traded, tested, and recombined at will.

THE VIRTUAL REALITIES OF ORAL-MYTHIC CULTURE

Once we have leaped into a narrative mind-set, our worlds become virtual ones. Language gives us a cognitive zooming facility, with which we can voluntarily alter the scale of an experience, plunging into microscopic detail or pulling back for the big picture. This is an awesome power, unequaled in the history of cognition. It allows us to bracket experience and survey our cognitive realm, with remarkable detachment. A cartoon I once saw, I cannot remember where, captured this perfectly. It showed a roomful of scientists watching a colleague fill a blackboard with equations. They were puzzled by what they saw, and little word balloons appeared over each member of the audience, saying things like "What is he doing?" "What on earth is he doing?" "What does he mean?" and so on. The lecturer, looking very worried and lost in his jungle of equations, was also thinking, "What in hell am I doing?" This is metacognitive evaluation. It may not look all that different in principle from mimetic review, but it is infinitely more precise in what it can specify. Linguistically mediated metacognition, with its powers of amplification and its capacity for recursion, is a far more powerful capacity, giving us a virtually open-ended ability to abstract. This has become the governing capacity of the human world.

If mimesis is our cultural glue, stories are the main by-product, as well as the principal organizing force, behind the classic form of human culture, the oral tradition. We are dominated by our stories, whether of magic, witches, devils, demons, great heroes, or genesis. They are the imaginative fodder of self-identity, morality, class, and authority. Stories can become so influential and so deeply rooted in the daily operation of the culture that they assume a special cognitive status, that of myth. Myths are standard versions of very old and widely shared stories. A mythic narrative surrounds great cultural heroes, such as Ulysses, Moses, and Christ. And mythic collections such as the Veda and Homeric myths can influence every phase of life in their societies. People model their lives on mythic figures. Western religion, through its domination of the mythic resources of Europe, was able to influence every date on the calendar, every major architectural and engineering project, every library, and every philosophical and literary work. It even financed the rise of science, often against its own better judgment. Advances in astronomy were pursued aggressively in seventeenth-century Europe mainly because the papacy needed more precise ways of determining the exact date of Easter. Similarly, the Crusades were fought, and new continents conquered, largely because of an idea fixed in a narrative framework of what the world was like or should be like. Stories of imaginary places, like the kingdom of Prester John, or of the coming of the Apocalypse have inspired people to give up their roots to live in danger and often to die.

Myths have driven the rise of every empire, from Islam, India, and China to Venice and London. Narrative traditions ruled the minds not only of poets and artists but of emperors and foot soldiers. Local narratives ruled every tribe and village, through their elders, chiefs, shamans, and bards. They have been a governing force in human thought and often move people to undertake the most incredible projects. They not only can spell out the project of a life but can provide a sense of group identification that drives people to attempt almost anything.

Human conscious capacity has been enormously expanded by this collective leap into the narrative domain. The process of mapping the form of any symbol onto meaning gives us much more than clusters of associations, or fast-moving maps of forms with discrete meanings, or ways of parsing and segmenting events. Words and grammars are merely the entry-level skills without which narrative traditions could not exist, but once they are acquired, they are secondary to the stories themselves. The cognitive impact of language can be measured primarily through evaluating the cognitive value of narrative skill, rather than vocabulary size or grammatical elegance. Narratives, especially shared life narratives, are the basis of autobiographical memory itself. Stories and myths can completely reshape our semantic spaces, leading to a consensual definition of a shared virtual reality that is the core of oral culture.

Language is secondary to mimesis in its early acquisition and does not emerge as an autonomous level of representation until children, somewhere around age four, become competent at telling their own stories and start developing an active autobiographical account of their own lives. This ability allows them to time-track their lives and to place various events in an all-encompassing model of their own experiences. This ability has introduced a new layer of culture into the traditions we inherited and has given us a new means of invading one another's minds. But language did not devalue or displace mimesis. It serves a different function in society. The mimetic body language, prosody, and ritual of life continue, but under the control of a new mythic mantle. Thus the storied king, with his divine right and his power protected by an armamentarium of validating myths, nevertheless had to rely on coronations, elaborate weddings, customs, such as bowing and curtsying, and all the formalities of a courtly hierarchy to consolidate and implement his power in daily life. We have retained everything that works: a narrative tradition, mimesis, and the episodic cognitive foundations we inherited. Each serves its own function in society. We still have craft, custom, skill, and ritual, all mimetic in their roots. But myth has come to dominate ritual, and ultimately narrative has come to dominate and surround mimesis. There is no doubt where the power lies in traditional societies. It lies with language and the common cultural myths it generates.

The creative tension generated by oral-mythic culture is not static. This is still evident after forty-five thousand years, and possibly more than one hundred thousand years, of linguistic evolution, marked by a frantic pace of linguistic invention. Groups of migrating human beings, as they moved over the Earth, spun off dozens of independent cultural and linguistic offshoots, many of which became isolated, and founded hundreds of languages. Those languages are constantly changing their particular mappings of form onto meaning, and this has led to incredible variation. The best-documented example is the history of the Indo-European languages, including all those related to Sanskrit, Persian, Gaelic, Latin, German, Russian, and Greek. This huge group of languages, spoken by a billion people, can ultimately be traced back to a single common ancestral tongue, called Proto-Indo-European, which existed less than eight thousand years ago. That ancient tongue, carried by early agriculturalists, spread from western Asia to Europe, northern Asia, and the Indian subcontinent. The myths of Proto-Indo-European also spread and underwent modification during the same migrations, although of course, particularities were added by each subculture, over several millennia. The result was incredible diversity, generated in a period of time that we can now see as a mere twinkling of an eye. Each culture was a spin-off of a rapidly evolving, multilayered, and distributed cognitive system. This pattern of incessant innovation illustrates the creative drive of language. Nothing could better illustrate its impact on culture. Cultures that have language are constantly redefining their collective knowledge networks and changing their maps of form onto meaning.

Traditional human culture is a complex blend of orality and mimesis. The term "orality" usually refers exclusively to those aspects of culture that are mediated by spoken language, or the oral tradition, which include all the natural products of speech. The oral tradition usually dominates the mimetic dimensions of culture, the gestures, skills, forms of social address, and habits of public spectacle that help define social life and become dominant influences in forming young minds. Together oral and mimetic traditions define self, tribe, and caste; express how life is to be lived; and specify, usually implicitly or obliquely, what is to be valued. Often these oral and mimetic traditions are summed up in a religion, a set of formal beliefs and rituals to which a society subscribes.

In the strictest definition of the term, oral cultures have no writing and no permanent external symbols of any kind, including sculpture, paintings, and monuments. There are still a few such societies on Earth, although they are rapidly becoming a thing of the past. However, when they first appeared, they constituted a revolutionary force. The amazingly rapid course of the second transition transformed human culture in less than a hundred thousand years. We transcended our mimetic foundations with powerful oral traditions that generated thousands of fully developed mythic civilizations. The ancient world was dominated by such traditions, and most of the modern world is not much different. Our myths still define us.

COLLECTIVITY OF MIND

Collectivity has thus become the essence of human reality. Although we may have the feeling that we do our cognitive work in isolation, we do our most important intellectual work as connected members of cultural networks. This gives our minds a corporate dimension that has been largely ignored until recently. The word "corporate" usually refers to institutional entities, such as banks and governments. This is not so anomalous a label as it might seem because corporations are unified in a cognitive sense, just like the bodies of living organisms. But unlike organisms, which are locked in on themselves, corporations can distribute their intellectual work over many minds and employ various external symbolic devices, such as writing systems and computers, to facilitate this distribution process. They can develop corporate perceptions, ideas, agendas, and even personalities. Individual minds are thus integrated into a corporate cognitive process, in which single individuals rarely play an indispensable role.

This gives us a good reference point for how we think as cultural beings. Our cultures invade us and set our agendas. Once we have internalized the symbolic conventions of a culture, we can never again be truly alone in semantic space, even if we were to withdraw to a hermitage or spend the rest of our lives in solitary confinement. Big Brother culture owns us because it gets to us early. As a result, we internalize its norms and habits at a very basic level. We have no choice in this. Culture influences what moves us, what we look for, and how we think for as long as we live. We work out the vectors of our lives in a space that is defined culturally. In some cases, this process involves a hierarchy of influences that are normally invisible to us. It is fairly easy to visualize the distributed work of great numbers of laborers, such as those who built the pyramids, invaded Gaul, or built thousands of Model T Fords. But it is difficult to imagine how that same metaphor applies to mental work. Nevertheless, the invisible mental labors of generations of scholars and composers, stock exchanges, research institutes, software sweatshops, and bureaucracies are also distributed, just like those of an assembly line. Distributed cognitive systems employ thousands of human beings for various collective agendas. Workers in such systems are, in their collective and professional identities, nodes in a distributed network. They may be active, intelligent people in their own right, absolutely convinced of their individuality, but they are nodes when they play their professional or corporate roles. Control lies elsewhere. Any management flowchart will testify to this.

This is especially true in those domains that are close to the heart of culture, such as government, law, and science, where the individual, on a good day, might render some small service to a vast system that has assembled its knowledge networks over a period of several millennia and is developing a global reach. Artists and writers are especially driven by their cultural-historical context. They emulate and modify the work of their predecessors, just as jurists, government officials, inventors, and scientists do. It is a rare idea, thought, hypothesis, or archetype that has not already been conceived and modified a thousand times, somewhere in the distributed webs of the human cognitive universe. The best an individual can hope for is a small degree of uniqueness, perhaps by becoming the conduit of new collisions of ideas or conjoining vectors on thoughts that have never before been brought together. The presence of culture, defined in this manner, is everywhere. In the domain of cognition, we seem to have shaken off Darwinian shackles for cultural ones.

The creative spark of cognition still depends on the individual conscious mind, but even this statement has to be qualified because creativity cannot be exploited, or even defined, without a cultural context. Success in any given culture is contingent on the successful incorporation (an awkward, but apt, term) of a "talented" individual by a culture that defines not only what talent is but what value should be assigned to it. Minds are called talented and creative if they are important generative nodes in a cultural system. Our historical juxtaposition in culture can explain much about the strange phenomenon we call "genius." Culture confers great power on anyone who can play the system. Geniuses travel inside the same knowledge vortex as everyone else, and the current state of that vortex may fix the possibilities available to each generation, but only geniuses can realize those possibilities. Under the right circumstances, the cognitive resources of an entire culture can become concentrated inside a single mind, and this can bring about an awesome concatenation of forces, like those giant hurricanes observed over the Caribbean that focus huge energies on a tiny geographic point. The whole of Elizabethan England somehow flowed into the mind of Shakespeare, as Victorian England flowed into that of Dickens. Shakespeare and Dickens gathered within their inner spaces the entire spectacle of their times and gave it tangible form. A different dimension of cultural knowledge came together in the minds of Newton, Darwin, and Einstein.

Conscious minds have always lived at the cognitive cutting edge of culture. This is true of everyone, not only of geniuses. The larger cognitive-cultural system embeds our minds into a distributed network, and it is this collective framework that directs the conscious process in certain creative individuals. After many centuries, this process has generated the complex web of habits, customs, and beliefs that define human culture. These are now unconscious, of course; in cultures, as in individuals, automatization is the other side of advanced consciousness. And the conscious mind is thus part of a larger fabric, much of which exists outside its grasp. Nevertheless, the essence of our human mentality is to harness ourselves to the tremendous collective organizing energy of culture. This is our obsession and our inevitable fate. For many, in an ironic twist on the notion of self-identity, culture has become a form of self-justification. We have created a collective organism that appears ominous at times. Our interlinked nervous systems, newly powerful in their electronic extensions, are now challenging the supremacy of the natural world.

But this masks the larger role of humanity in evolution. Through us, nervous systems in general have broken free of their autochthonous solipsism. We are the means whereby the physical universe has finally reflected back on itself and reaped the effects of our painfully slow evolutionary gestation. But ironically, our elaborate cultural games are subjecting our brains to forces that are far beyond our control, and the future direction of our civilization is not clear. We tend to ignore what is going on inside us and pay attention mostly to our cultural universe, rather than the natural world, which we now experience primarily through one of the primary filters of our new religion, science. We are a culturally bound species and live in a symbiosis with our collective creation. We seek culture, as birds seek the air. In return, culture shapes our minds, as a sculptor shapes clay.

8

The Triumph of Consciousness

Weiner remarked casually that Feynman's notes represented "a record of day-to-day work," and Feynman reacted sharply. "I actually did the work on paper," he said. "Well," said Weiner, "the work was done in your head, but the record of it is still here." "No, it's not a RECORD, not really. It's WORKING. You have to work on paper, and this is the paper. Okay?"

—FROM JAMES GLEICK'S BIOGRAPHY OF RICHARD FEYNMAN

I feel that painting is able to contain whatever one thinks and all that is. Painting can contain the politician in a Daumier, the insurgent in a Goya, the suppliant in a Masaccio. It is not the spoken idea alone, nor a legend, nor a simple use of intention that forms what I have called the biography of a painting. It is rather the wholeness of thinking and feeling within an individual.

—BEN SHAHN

The evolution of languages and oral cultures left our species with a formidable strategy for representing reality and elevated consciousness to a new plane. But the emergence of the human mind did not slow down following the arrival of our species. Rather it accelerated. We have traveled a great additional distance. Something about our mentality changed in the past few millennia, something that made us able to construct such exotic things as symphonies, philosophies, oil refineries, nuclear weapons, and robots. Do such achievements have implications for theories of consciousness? Many would deny that they do. They would claim that the parameters of mind were surely fixed long ago, when we emerged as a species, and that culture can add nothing to an equation written deeply into the human genome.

But that common belief does not stand up to scrutiny. The human mind has been drastically changed by culture. In modern culture, enculturation has become an even more formative influence on mental development than it was in the past. This may be a direct reflection of brain plasticity, rather than genetic change, but that does not in any way diminish the importance of the change from a purely cognitive standpoint. The human mind is so plastic in the way it carries out its cognitive business, individually and in groups, that the core configuration of skills that defines a mind actually varies significantly as a function of different kinds of culture. This is especially true of the most conscious domains of mind, such as those involved in formal thinking and representation.

Let me be very clear about what I mean here. I am not speaking of trivial cultural changes, such as variations in custom or language usage. These are by far the most common and have no proven cognitive impact. But there are other innovations that have a significant impact. The most important of these is literacy. Literacy skills change the functional organization of the brain and deeply influence how individuals and communities of literate individuals perform their cognitive work. Mass literacy has triggered two kinds of major cognitive reorganizations, one in individuals and the other in groups.

To become fully literate, the individual must acquire a host of neural demons that are completely absent from anyone who lacks literacy training. This involves massive restructuring. There is no equivalent in a preliterate mind to the circuits that hold the complex neural components of a reading vocabulary or the elaborate procedural habits of formal thinking. These are unnatural. They have to be hammered in by decades of intensive schooling, which change the functional uses of certain brain circuits and rewire the functional architecture of thought. This process can be very extensive. For example, multilingual people might have hundreds of thousands of lexical items, linked to a variety of grammars. Consider the impact of twenty or more years of schooling on the brain of someone who has acquired full symbolic literacy in several different technical, mathematical, scientific, and musical fields. These skills encumber neural resources on a vast scale and change how the person's mind carries out its work.

In such individuals, the demons of literacy reside in large neural networks that, complex as they are, form a conceptual entity that might be called the literacy brain. The literacy brain is a cultural add-on to the normal preliterate state of the brain. It determines a great deal about how the operations of the conscious mind are carried out. It affects the relative growth and synaptic richness of certain regions and structures of the brain. Even though the neural circuits of literacy are far from fully understood, we know that they engage both hemispheres, albeit asymmetrically, with more circuitry on the left than on the right in most people. Sometimes these networks occupy at least as much space as the networks of spoken language. They also break down in a parallel manner to disordered language, or aphasia, but independently of it. Damage to the literacy brain causes dyslexias (reading problems) and dysgraphias (writing and spelling problems) independently of other cognitive loss. This shows that there are several new physical systems involved in setting up a literacy brain, depending on the number and complexity of the symbolic skills the brain has acquired. These systems qualify as functional components in exactly the same way as the subsystems of spoken language, and they are unquestionably a product of enculturation, rather than genetic predisposition.

Figure 8.1

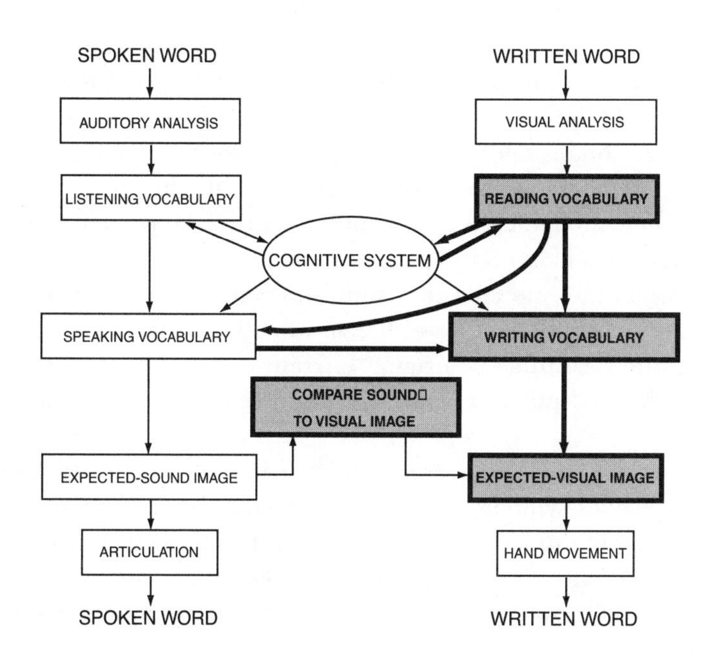

This diagram shows a bare outline of the literacy brain, which includes several immensely complex neuronal networks that must be wired together and interwoven with the rest of the cognitive system when a brain learns to read and write. Since literacy skills are usually piggybacked onto the speech system, the speech brain is also outlined in the figure, in a parallel column (left side) that traces the path of a spoken word as the brain processes it. The shaded boxes and thick black arrows highlight the special networks for written language. These must be added to the brain's basic network architecture when we learn to read or write. They include a reading vocabulary and a separate writing vocabulary, which does not always correspond to the reading one (this is usually the case in second languages). Reading and writing entail semi-independent networks, as we see in some special cases of dysgraphia, where patients lose the ability to write but can still read. The opposite syndrome is found in a rare class of acquired dyslexia where reading is completely lost while writing is intact. Additional networks are also needed in a fully-functional literate brain, including an important phonetic error-correction network that compares the sounds of words to their appropriate visual and tactile images. Taken together, these are the demons of literacy. Like the demons of speech, they make great initial demands on conscious processing, but eventually become automatic. Once they are fully automatic, they constitute a fast-track neural pathway by which visual symbols can access the highest representational networks of the mind.

The literacy brain did not evolve, in the Darwinian sense. Humans evolved as a species long before external symbols came into existence, and thus the brain could not have evolved specifically for their use. Moreover, literacy is neither natural nor universal. Humanity was illiterate until a few thousand years ago, and more than 90 percent of existing human languages have no indigenous form of literacy. Yet the children of all human cultures are able to acquire literacy if given a chance. This shows that the neural demons of literacy, in all their exquisite complexity, are entirely cultural in origin. Literate culture has capitalized on untapped cerebral potential and reprogrammed the human brain in its own image.

Literacy skills are a response to the invention of external symbols. As the technology of symbols evolved, literacy skills became more complex. But the effects of external symbols did not stop with the reorganization of the individual brain. They transformed the collective architecture of cognition and changed how the larger human community thinks and remembers. They also enabled many new forms of mental representation. There are now entire classes of cognitive work that cannot take place without external symbols. If we define symbolic technology very broadly, including everything from musical and mathematical notations to art, circuit diagrams, and maps, it is clear that most of our major cultural institutions and a high percentage of our cutting-edge work are completely dependent on symbolic technology.

Symbolic technology has changed the way we think, remember, and experience reality, individually and as a collectivity. In fact, it has triggered a third cognitive transition, equal in importance to the first two.

THE THIRD TRANSITION: THE INVENTION OF SYMBOLIC TECHNOLOGIES

Symbols can be internal or external to the brain. For example, the words of spoken language are internal symbols stored in the brains of speakers and listeners. The same words typed on a sheet of paper become external symbols, stored on the printed page, rather than in the brain. Symbolic technology is the enterprise of manufacturing and crafting external symbolic artifacts and devices. These have enabled us to build a vast cultural storehouse and an external symbolic storage system, which serves as a permanent group memory and includes such things as books, museums, measuring instruments, calendars, and computers. These are extensions of what archaeologists call material culture. But unlike most aspects of material culture, they are designed specifically to help us think, remember, and represent reality. They have gone through thousands of years of development, and the end result, in our time, is a plethora of very sophisticated technological devices that are interwoven with our minds. This book is an example of such a device. Quite independently of whether it says anything of significance, it is a tremendously advanced piece of technology that, under the right circumstances, might lead readers to think thoughts that were previously impossible for them to conceive. The same is true of paintings, maps, musical notations, and a huge number of cognitive machines, such as clocks and observatories. They revolutionize what we can do with our minds.

Symbolic technologies serve many different purposes, but they all share a common underlying principle. They liberate consciousness from the limitations of the brain's biological memory systems. The new physical media of symbolic technology have enormous advantages over brain-based memory media. One of their chief advantages, as we shall see, is that they are fully accessible by awareness. From the standpoint of evolution, this was a revolutionary development. It carried the great hominid escape from the nervous system to its logical conclusion. Because of the limitations of biological memory, conscious thought was enormously difficult when contained entirely inside the brain box. External storage changed this and gave thinkers new strategic options.

Preliterate cultures have functioned perfectly well for many millennia. But there was, and is, a price to pay. Such cultures cannot get around the limitations of the brain's working memory system, which is the final common path of all conscious mental operations. This places enormous constraints on thought, to say the least. Unaided by external symbols, preliterate cultures have only two means of constructing a cultural memory: narrative and mimesis. For example, to retain a strong cultural memory of a significant historical event, preliterate minds have to rehearse an oral narrative about the event, which could be learned by heart by virtually everyone, and to find a way to reenact the event regularly in public ritual. These traditions can be maintained across generations only by rigorous social enforcement. In such cultures, dominant ideas must therefore be preserved in the same manner, in metaphor and allegory, which tap the so-called narrative mode of thinking, a straightforward extension of storytelling. A moral law might be expressed as an allegory, or an ideal of manhood in a heroic myth. This preserves traditional ways of thought very effectively but limits the forms that representation can take.

The invention of permanent external symbols should have changed all that very quickly, in theory. But in fact, symbolic technology did not seem to have much impact when it was first invented. The mere existence of symbols and a literate class almost never results in rapid cognitive change, even today. The discovery of the cognitive potential latent in external symbols was very slow and could not be realized without massive social change. The first major systems of writing and counting, which appeared in Egypt and Sumer, consolidated some very old customs and ideas, rather than generate radically new ones. It took a great deal of time to realize the revolutionary possibilities inherent in those technologies, and very few societies were capable of the social changes needed to exploit them fully.

The earliest evidence of symbolic technology dates back at least forty thousand years. Microscopic analysis of very small carved objects from that period shows that they were inscribed deliberately, with the repeated use of special tools. Archaeologist Francisco D'Errico has written that the extreme care and craft with which such things were made support the idea that they were meaningful to their makers. One of the pioneers in this field, Alexander Marshack, suggested that some of these artifacts might have been simple calendars, used for tracking the phases of the moon during seasonal hunting expeditions. Their exact meanings have been lost to us; however, there seems little doubt that they were symbolic. Cave art, common all over the inhabited world thirty thousand years ago, also seems to have been symbolic. The magnificent images of ancient cave art are quite similar to those painted by some surviving Stone Age peoples, who invariably use them in relation to ritual, trance, and religion.

Writing was invented much later, long after the appearance of cave art.

It first emerged about ten thousand years ago, when inscribed tokens were developed to keep track of trade goods. Economic necessity led to the improvement of this technology and eventually to the first writing systems, which appeared in all centralized agricultural societies, such as those of Asia, northern Africa, and Central and South America. Other symbolic technologies were also developed for such practical purposes as navigation, construction, and measurement. Several of these came together in the great classical civilizations, which also produced the first evidence of literary work. Urban society could not have progressed without the massive use of external symbolic storage, which was necessary to centralize cognitive control in urban agglomerations. The first fully phonetic alphabet and improved notational systems for geometry and mathematics were developed in the city-states of ancient Greece. More important, Greek society also changed the way written texts were used. They became reflective instruments, in which thought itself could be exposed to systematic analysis. As many have suggested, written symbols allow us to decontextualize ideas and abstract them from the concrete situations from which they sprang. By achieving this, the thinker can extract general principles that might otherwise remain obscure.

From a contemporary perspective, symbolic technology seems to have taken an inordinately long time to develop and even longer to spread. This is understandable, because symbolic invention is never easy. How many people have actually invented a truly novel symbolic device or expression? Very few indeed. How many societies have actually exploited such symbols to the full? Almost none. Symbolic invention is usually the work of single minds, but its full exploitation is a collective enterprise. To exploit symbolic technology, a society needs both the tools and the procedural habits to use them effectively. Symbolic tools in themselves are not enough to engineer a cognitive revolution, and under some circumstances, they can even prevent one from happening. For instance, Chinese writing was so difficult to master that it actually hindered the spread of mass literacy, and Roman numerals made certain kinds of mathematics impossible to conceive. Until the right symbolic technology came along, certain kinds of thoughts simply could not be thought. But symbolic tools alone are not enough. Social change is also necessary before they can be exploited to the full, especially in their corporate applications. We need to inculcate countless invisible habits of mind, both individually and collectively, and weave them into an institutional fabric before we can discover the best uses of symbolic technology. This often requires massive social dislocation, and not every society can tolerate, let alone survive, such trauma.

Symbolic technology always originates in a very small creative nexus in the cultural-cognitive system, the interface between the conscious mind and the symbolic environment. This is where artists struggle with their easels, engineers with their drafting boards, and composers with their notations. Minds that use such devices are changed by that intimate process of interaction. Influence flows both ways, from thinker to symbol and from symbol to thinker. The symbol grows on the page or easel and then invades the mind and takes it over. This reciprocal flow can generate amazing changes in direction, and it is an infinitely open-ended process because it can carry on over many generations. It enables human culture to conquer time and space, its intellectual adventures having been permanently preserved for anyone with the codes to decipher them.

When combined, the conscious mind and its symbolic technologies generate a powerful chemistry. The brain-symbol interface is the birthplace of art, science, mathematics, and most of the great institutional structures that humans have built. Every advance in our intellectual enterprise can be traced to a symbolic innovation of some sort. Such innovations may be large or small, and most are probably trivial, but the process itself is revolutionary. Differential calculus, Boolean algebra, Shakespearean humanism, Impressionist light, and American political philosophy all are inventions of symbolically literate minds, achieved after many refinements and recorded in detail on artificial memory media. This applies to a wide range of symbolic innovations, including megalithic observatories, sextants, and the huge armamentarium of modern laboratory science. The symbolic efforts of humankind have been truly protean, and it is important not to narrow our definition of that technology down to mere writing. The great imaginative leaps of early geographers yielded better maps; that is, better external representations of space. The mathematical innovations of the Babylonians and Arabs were manifest in various new symbolic media. So were the calendars of the Aztecs, the clocks of medieval Europe, the navigational techniques of the great explorers, and the accounting innovations that made international banking possible. Such technologies have allowed us more than once to break the mold and think what was previously unthinkable.

A MIRROR OF CONSCIOUSNESS: THE EXTERNAL MEMORY FIELD

The power of symbolic technology can be attributed largely to its impact on the conscious process. External symbols are revolutionary because they transform the architecture of conscious mental activity. This can be seen most clearly in the way they are displayed to awareness. The region of symbolic display may be called the external memory field. This is where the brain-symbol interface begins and ends.

The external memory field includes everything in use at the moment and possibly some items not in use that might later become relevant. At present my external memory field includes a laptop computer, some printed manuscripts, a variety of books set out on two tables, and a bulletin board. These items are all easily available to my awareness. Any of them might be brought into the focus of attention and temporarily occupy the center of my consciousness (these two are not synonymous). For instance, I might now choose to narrow my external memory field to include only this line of print, viewed in the context of the previous paragraph, which I have just completed scrutinizing intensely. I might also choose to change the contents of this focus by zooming in or out and adding or subtracting items. Note that the items on display in my field are largely ideational in nature and have an enormous effect on the accuracy of what I am able to maintain in my working memory. The items of major interest are rarely mere lines, letters, words, or sentences. They are usually much larger ideational entities, such as hypotheses, graphs, arguments, conclusions, and themes.

Although this arrangement constitutes a very ordinary work environment in our highly literate society, it is an extraordinary historical development because *it changes the long-standing relationship of consciousness to its representations.* We can arrange ideas in the external memory field, where they can be examined and subjected to classification, comparison, and experimentation, just as physical objects can in a laboratory. In this way, externally displayed thoughts can be assembled into complex arguments much more easily than they can in biological memory. Images displayed in this field are vivid and enduring, unlike the fleeting ghosts of imagination. This enables us to see them clearly, play with them, and craft them into finished products, to a level of refinement that is impossible for an unaided brain. Thus the display characteristics of the external memory field expand the range of mental operations available to a conscious mind.

External symbols introduce more powerful memory media into human cognition. The biological memory records of the brain, known as engrams, differ from external symbols, or exograms, in most of their computational properties. Engrams are impermanent, small, hard to refine, impossible to display in awareness for any length of time, and difficult to locate and recall. In fact, to find a natural memory, we tend to rely on ancient associative principles, such as similarity and contiguity. In contrast, external symbols give us stable, permanent, virtually unlimited memory records that are infinitely reformattable and more easily displayed to awareness. Moreover, exograms are much easier to search, and we can recall them with a variety of retrieval methods. The availability of powerful external media increases the number of ways we can represent reality. Just as better memory media can transform what a computer can achieve, so they can revolutionize what a conscious mind can achieve. A cognitive change this revolutionary could never have occurred simply by evolving a larger brain. But it has been achieved, relatively quickly, with improved symbolic technology.

The external memory field creates a mirror world for consciousness. It reflects the architecture of biological memory back into the symbolic environment, and this mirror image is then reflected back into the brain. The conscious mind is thus sandwiched between two systems of representation, one stored inside the head and the other outside. An image might help clarify what this means and how the external memory field fundamentally changes the architecture of consciousness. Imagine the preliterate state of the conscious mind as a space (not literally, but in a metaphorical sense) with a simple structure, as shown below.

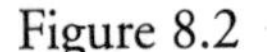

Figure 8.2

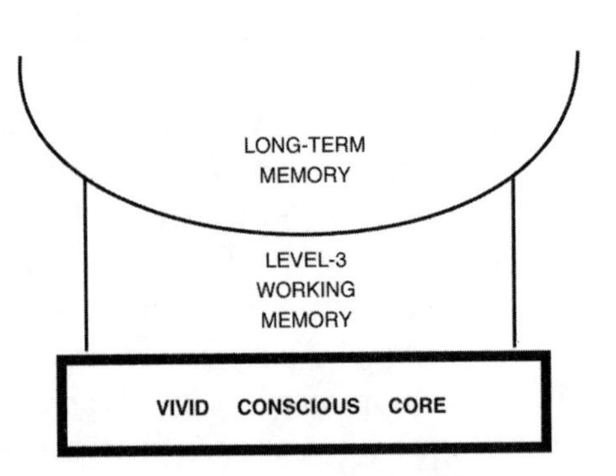

In this arrangement, sensory inputs fall into the vivid core of awareness, the focus of level-1 and level-2 processing. This core is surrounded by a less vivid conscious space, which is the broader realm of level-3 processing. This processing can endure over a much longer time span and incorporate many layers of ongoing activity. It also draws from the brain's long-term memory systems and recalls ideas or images into awareness as needed.

This is the canonical structure of the conscious mind, our "normal" state of mind. The high-turnover vivid core is supplemented by the many low-turnover, slower-moving networks running in the intermediate-term background of awareness. The "thinker" resides in the latter, although the high-turnover focus might dominate awareness in the short run. For example, while walking through the forest with my son, the vivid core of my awareness would probably be filled with sights, smells, and sounds. His voice and mine would trigger a second level of awareness, a barrage of ideas and images related to the conversation, which would inhabit my level-3 awareness and play out on more than one level. Consciousness would thus be influenced by both the vivid core and the background conversation.

This arrangement is drastically altered when I plug myself into a powerful symbolic device, such as a book or computer. When I confront such a device, my operational mental architecture is temporarily altered. Awareness now finds itself juxtaposed between two simultaneously present storage systems, one internal and biological, the other external and technological, each with long-term and short-term aspects. Two memory systems are thus available, whereas before there was only one. A mirror image of natural working memory is now available to awareness in the external memory field, as shown below.

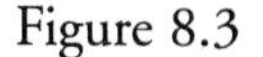
Figure 8.3

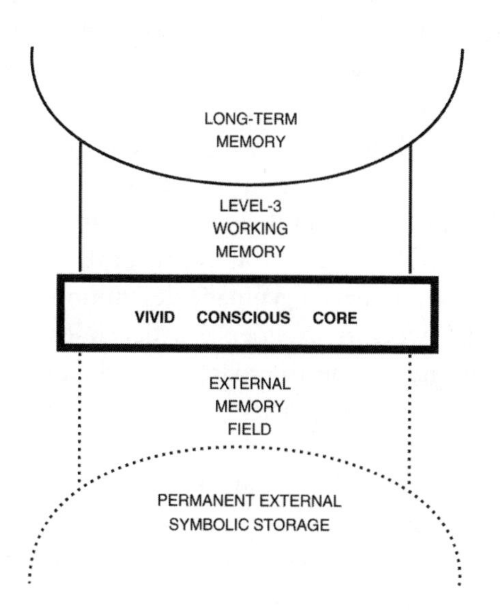

In this case, the conscious mind receives simultaneous displays from both working memory and the external memory field. Both displays remain distinct in the nervous system. The external memory field is not just another sector of working memory. It taps directly into the neural networks of literacy, located in brain regions that are distinct from those of working memory. Working memory and the external memory field thus complement each other, and this allows the brain to exploit their distinct storage and retrieval properties. This gives awareness a much richer structure. These two displays, the one externally and the other internally generated, are processed by the conscious mind at the same time and serve different functions in the same cognitive sequence, as when a literate thinker reads about something and at the same time reflects on what he is reading.

The vivid core of awareness, where it all started, still sits at the geographical center of the phenomenal universe, but not in a position of control. It is now between two much more powerful cognitive fields, both capable of condensing and focusing a host of culturally generated algorithms. Each has its own role to play in energizing and directing awareness. To maintain control, a fully literate consciousness must navigate a way through this complex of ideas and images, optimizing its use of resources.

This complex structure can work wonders. Recall the quote from physicist Richard Feynman on the first page of this chapter. He experienced his pen and his all-important sheet of paper as an integral part of the WORK of his mind, part of a tight loop from brain to hand to eye, and back. As a highly creative thinker he experienced his own thought process as a distributed arrangement that bridged two distinct memory fields, one outside, the other inside. The act of thinking actively links these two simultaneous memory locations, one biological, the other external, into a seamless algorithm. The conscious mind controls this process by "working" its own external memory field, as Feynman noticed. Ideas pour out directly onto a sheet of paper and reflect directly back into the brain, where they can be examined and modified, and so on. Thus the sheet becomes part of the process, intimately nestled in the thinker's awareness. Feynman's description is far from unique. All modern knowledge workers experience a dependency on paper or some other medium, as they refine the material in their own external memory fields and use its display properties to nudge, frame, and fine-tune their awareness.

If this sounds familiar, it should. It is surely obvious that the creative bob-and-weave boxing match between consciousness and spoken language is also played out with external symbols. It is the same conscious mind, after all, simply transposed to a new medium, that generates the domain of external symbolic storage. The mechanism the human mind uses to generate languages also equips it to generate external symbols. It is not the underlying creative mind that has changed. We are no more creative, in this sense, than our Stone Age forebears. It is the evolving change in the memory media themselves that have made such a huge difference in what the mind can finally achieve.

These two classes of memory media, biological and external, are also parallel in their long-term storage functions. The external memory field can draw ideas or images from an infinitude of external symbolic representations, in the same way that consciousness can draw ideas or images from long-term memory. But the external storage system far exceeds the capacity of biological memory, by many orders of magnitude. Thus the emergence of external symbols radically changed the total memory storage capacity of humans, as individuals and as a species.

This mirror arrangement also changes the reflective power of the conscious mind, because the external memory field gives working memory a much more solid display system for representations. Ordinarily, neither spoken words nor images can be displayed long enough in normal working memory, even in the intermediate term, to allow any prolonged or detailed reflection on them. Natural memory display is poor, lacks definition and detail, and is notoriously unreliable. The external memory field gives us sharper and more durable mental representations. This allows the conscious mind to reflect on thought itself and to evolve longer, more abstract, procedures that serve to verify and control the quality of its own actions. These procedures give us the basis for turning thinking into a more formal activity, subject to much stricter standards of revision. This results in a scaffolded cultural process that can accumulate and improve over time. It allows cultures to develop thought habits that can be communicated back to the individual and have a profound impact on how cognitive business is conducted. Given this, societies can absorb the efforts of previous generations at a faster pace, and the shared cognitive processes that result become easier to refine and reproduce. This leads to qualitative changes in what we can achieve with our limited mental resources. Endless repeated reflections on the same core material over successive generations can give a society a more powerful set of conscious procedures to follow; that is, much better ways of thinking.

But this has complicated the predicament of the conscious mind. Internal and external memory can now be played off against each other. Indeed, this has become one of the main activities of modern thinkers, who have to switch between internal and external representational systems and resolve tensions between them. This is the conscious mind in its new role as arbitrator. Multifocal awareness, forged in the furnace of oral-mythic culture, is stretched to the limit by the infinitely changeable, infinitely dense vagaries of symbolically literate culture. Its dance with symbols keeps pushing back the frontiers of thought, and the symbol-work itself, along with the cultural products that come from it, has assumed a tremendous presence in our lives.

The greatest of our symbolic representations can provide a life's work for generations of scholars. The effective and active use of the external memory field is a very difficult art in itself. Intelligent reading is a highly aggressive, active business. So is viewing art. Michelangelo himself could never have imagined the finished panorama of the Sistine Chapel that he himself painted. He created it in the intimacy of his own struggle with his external memory field, but the totality of the final result exceeds the reach of any imagination because it simply goes beyond the limits of basic capacity. It cannot be held in mind all at once. Our most challenging symbolic representations are like this. They deliberately exceed capacity. They demand very hard mental work to understand them. Such creations are seldom user-friendly. They consciously choose not to trivialize reality. Instead they challenge and stretch the mind of the conscious observer. Rather than serve up pap, they add power. The ideas, myths, styles, icons, and archetypes that Michelangelo summed up in his masterpiece do exactly this to the knowledgeable viewer. They challenge and provoke. They allow juxtaposition of all the central themes of Renaissance civilization at once, in a single field, served up to awareness for synthesis, ironic contrast, or whatever. The cleverest of viewers might gain insights that even the great artist himself would have missed. He would approve, heartily.

Multiple images such as those of the Sistine Chapel simply won't do their work on one's mind if it is passive, no matter how knowledgeable. They demand active thought and perception, and the outcome of such effort is always unpredictable. Viewing a complex painting requires rapid-fire attentional switching between the component elements of the picture, and the interpretation of the sum may not be predictable from the meanings of the component parts. Done well, such images release energy in specific places in the cognitive system and bring about unexpected fusions and fissions of meaning. Those reactions may be shared and recorded. They may also influence those of others, endlessly fine-tuning the shared semantic fields of culture. The skills involved in dealing with the most complex external representations of culture are like this, more like prizefighting than watching television, more active than passive.

Externally imposed conscious states have another interesting property that makes them stand out. They can be set up or abandoned very quickly. This allows a viewer to switch from one disposition to another with alarming speed. Biological memory cannot do this, at least not voluntarily. Hallucinogenic drugs can produce rapid flipping and switching, but we lose control. Crafted words and images allow us to achieve this same effect, under programmed control. This can do dramatic things to a conscious mind. It can speed it up and actually jerk it back and forth from one symbolically induced state to another. Arranged in the right order, a variety of symbolic technologies can lead a viewer to specified states of mind. This serial feature, the scripted aspect of external symbols, gives rise to novel cultural possibilities.

It makes it possible, for example, to create extremely powerful programmed experiences. Culture is now in the business of manufacturing virtual realities. The kinds of realities imposed artificially by culture can, in theory, outstrip natural experience in both intensity and variety. In modern society, the sequencing of experience by means of symbolic technology has developed into an exact science. It has gone far beyond the static image into the scripting of many images in series. This process is the basis of complex media such as film and the novel. Scriptwriters of various stripes have become the chief cognitive engineers of mass culture. They design many of our experiential trips into virtual reality.

The availability of external symbols thus makes culture infinitely more self-reflective and feeds back onto the texture of individual consciousness itself, making experience the object of its own deliberate capacities to plan. This might point to the ultimate triumph of consciousness, the control of consciousness itself, the scripting of individual awareness by culture to enhance our understanding of the universe. Unfortunately, the same shared technology can also be used to deceive, confuse, regiment, and even to prevent thought, as it was during the Second World War and behind the Iron Curtain. However it is used, it confers a terrifying power on the conscious process.

A CEREBRAL TROJAN HORSE

External symbols can always be translated into a series of internal physiological events. The external memory field gains its foothold on the mind through its ability to program brain activity. To enter the cognitive system, an external memory display must project into the brain of one or more conscious recipients. Otherwise, it is a mere lifeless object. It acquires psychological reality only by gaining control over a set of automatic demons in someone's brain. This engagement seems to occur initially in the parietal and temporal lobes of the cortex, where high-level perceptual analysis is carried out. If the literate viewer has the appropriate demons, there will also be a higher-level cognitive response distributed throughout many areas of the brain, which opens the doors of semantic memory, calls ideas and scenarios into awareness, and juxtaposes experience. We are not yet sure how anything this complex translates into neural activity on a microscopic level, but it does. It harnesses the basic executive brain systems. In addition, it engages huge amounts of memory. Thus the external memory field provides a fast channel from the cultural universe to the control centers of the brain. It is a shortcut from the engineered symbolic universe into the deepest, most complex networks of the mind. It externalizes control and allows the outside world to pull our mental strings directly. There are gains in this, of course. But there are also losses.

The external memory field is really a sort of cultural Trojan Horse into the brain, a device that invades the innermost personal spaces of the mind. It can play our cognitive instrument, directing our minds to predetermined end states along a set course. This has roots in the pedagogy of oral culture, where words and gestures serve a similar purpose, deliberately leading audiences to achieve certain end states. At first glance there might appear to be nothing new here. But this is not so. Although the external memory field may not feel any different from spoken language, and although it is no more directive than, say, a mother teaching a child, its cognitive properties are very different. Temporarily it translates all the advantages of external storage media—permanence, accessibility, refinement—directly to the brain. These advantages are conveyed to the brain's internal networks and can be maintained as long as the algorithms involved unfold in the right order. This magnifies the mind's cognitive power and amplifies the impact of representational objects.

To an extraordinary degree, this makes the human mind externally programmable. In symbolic technology, consciousness has gained a means to reflect its own activities back on itself and program its own operations. It can redirect the narratives and images of culture back into the flow of individual awareness, where it can take advantage of the tremendous resolving power of our perceptual apparatus. Through the external memory field, external symbolic devices create stable, vivid displays of many things that were previously undisplayable. Ideas, theories, strategic plans, even imaginary events are left suspended in the mind, to be processed as we please. Or to be processed according to the dictates of platoons of cognitive engineers. The essence of the modern condition is the proliferation of deliberately engineered sources of experience. Modern conscious experience has drifted very far from the natural streams of events that we imagine to be the normal source of experience. We now have entire industries that deliberately design and manufacture experience.

One might argue that it has always been thus. Perhaps. Certainly, tribal society is permeated by religion, which provides a formidable means of controlling how life is experienced and what kinds of thoughts and images are acceptable. In traditional urban societies, religion sums up the whole hierarchy of expression and ideas offered by that society and provides a framework within which life is experienced. However, it is doubtful whether traditional religion as a whole could be characterized as a deliberately designed means of controlling experience. It is more often a kind of cultural attic, containing the major ideas and images that influence a people's history. Everything, from the architecture of such societies to their costumes, music, habits of prayer, and formal rituals, has an impact on experience, as it sets the main emotional markers in the cultural landscape. But these are usually not so much designed as they are spontaneous outgrowths of tradition and the popular will.

Modern society is different. It has hundreds of very powerful external symbolic technologies that provide a medium for collaborative cognitive work, and these are used routinely to design the cultural environment in a more deliberate manner. Entire industries do this for individual consumers, following the directives of advertisers and programmers. They offer more choice than ever before, and many more alternatives than those offered by traditional society, but the entire menu is highly contrived. Symbolic devices are absolutely essential to conceive, set up, and execute such schemes. A proscenium stage, or a large-screen monitor, or a television network can coordinate experience on a scale that was previously impossible to contemplate. These technologies are designed to induce a common end result, a conscious state of great subtlety, with textures and complexities of raw feeling that are almost impossible to describe.

One of the most powerful examples of such technology is grand opera. Until the last century this was probably the ultimate example of a programmed experience tied to various symbolic technologies arranged in a distributed cognitive system. It may go against the grain to subject this stupendous art form to a dry cognitive reduction, but it is illuminating to do so. Any great opera house, say La Scala in Milan or the Met in New York, is really a carefully designed machine for orchestrating experience. This is its only purpose, and to this end the design of an opera house incorporates every level of human cultural and cognitive evolution. Its symbolism begins with the location of the house itself, its scale and architecture, and its significant interior spaces, which communicate a particular ideal, usually rich in semiotic devices. The symbolism of this environment taps our mimetic vein; it extends down to the arrangement of the seats and boxes and the dress and manners of the huge collectivity in attendance, from audience to conductor. Everything, from the placing of the musicians to the statuary on the walls, has been scripted to the last detail toward a single-minded cognitive purpose.

On any given evening this gigantic machine will be operated by groups of minds whose sole objective is to design the audience's conscious experience. They will have previously hammered out their ideas for months or even years, using symbols at every stage of the process, often to create larger megasymbols, as when a small musical theme emerges, grows, and eventually invades the entire score or when a simple idea of place and time expands to determine the entire style of a production. An actual performance exposes one of the most distinctive possibilities of having an external memory field, its capacity for symbolically mediated coordination on a grand scale. Hundreds of musicians are assembled, all with sheet music in the center of their own personal external memory fields (the audience can only hope they have the right demons to read them). They cannot hear one another very well and therefore cannot judge the overall sound of the orchestra or follow the singers very well. The conductor is there to solve that problem and guide the orchestra toward an ideal interpretation. To that end he coordinates the whole ensemble, using a very peculiar set of mimetic signals: grimaces, postures, facial tics, stares, hops, and baton gestures, which differ vastly between conductors. The singers and dancers are rehearsed separately in advance, then brought together with the orchestra, and the whole production is then combined with lighting, set design, costumes, and so on.

There is a huge distributed network operating behind the scenes in the opera house, coordinated with every symbolic trick in the human repertoire: mimesis, spoken language, and a plethora of external symbols. External symbols have crept into the creative process at every step. No performance this complex could exist without them. Musical and dance notations, sketches for set and costume design, even the current vernacular of the theater, and the audience's known taste in this regard all can influence the final result. Other notations come into play, for choreography, stage management, and even the financial, managerial, and marketing aspects of the production. The audience has been extensively groomed for the experience by years of enculturation with similar symbols to scaffold them up to a certain level of sophistication.

When the actual performance finally begins, the whole point, weird as this would undoubtedly sound to someone unfamiliar with this art form, is to generate an extraordinary state of mind, of a certain accepted kind. That is all. That is how it will be judged. There are rigorous standards by which this completely artificial and otherwise impossible experience will be judged. No matter what the desired result, whether it is tragic exhaustion or heroic exaltation, it must temporarily herd the experience of thousands of people in the desired direction, using this utterly fantastic collective apparatus. Building, conductor, designer, musician, audience: All are wired together in a distributed cognitive system, embedded in a largely invisible network of symbolic technology, most of which is not evident in the actual performance. The objective of this exceedingly irrational, but prototypically human, process is the final catharsis or illumination. The consciousness of the audience will have been teased into a frenzy by clever cognitive engineers using every trick in the human cognitive book.

If we take the long view of Darwin, this is all very peculiar. What an extraordinary aberration we are! Try to picture a gigantic assembly of modified apes, locked into an experience machine of their own design, endlessly debating the virtues and vices of how to use it and inventing whole industries for that sole purpose! After two million years of evolution, the existence of elaborate experience industries testifies to our infinite strangeness. They are part of a closed-circuit cultural system that magnifies imagination and inadvertently emphasizes our loneliness, in a universe of empty celestial mechanics. They also celebrate our potential as a cognizing species. Perhaps good art is the ultimate triumph of consciousness.

Opera may have been the product of another century, but film is the ultimate contemporary art. With all the advantages of recent technology at its disposal, it is more scrupulously planned and scripted than any live opera performance could ever be. Film is so abstract in its working machinery that its actual operational structure is much more difficult to analyze, or reduce to a cognitive diagram. Anything goes. Scenes can be filmed out of order or thrown out, as needed. Performances can be repeated an infinite number of times. Simulation can replace actual live acting. Film can press every button and exploit every nook and cranny of the human psyche for its objectives. There is no adequate code to bring together all the strings of a production such as the remarkable film *Shallow Grave*, which plays the minds of its audience in an apparently unlimited number of ways. It bypasses most of our traditional categories for classifying symbolic technology and taps the whole human hierarchy of representation, from our ancient episodic origins to our traditional mimetic and narrative selves to various high-tech–driven combinations of recent symbolic technology. The twenty-first century will probably develop further in this direction until it can produce distributed interactive technologies with even more powerful manufactured experiences. It is impossible to predict where this will go, or to imagine the kinds of experiences that will soon become possible, with still-unimagined media. But the direction of the trend is unlikely to change.

Film and opera are presented under special conditions, often in a special place, and experienced as an interruption of the normal stream of consciousness. This may have given the reader the impression that symbolic technology has control of our conscious agenda only under special conditions. This is not true. In an urban environment, virtually every thought and action is framed by external symbolic devices. This includes some that are extremely down-to-earth, such as advertising circulars, traffic signs, and calendars, and others that are abstract and distant, such as government tax policies and oil-pricing legislation. The latter are held in permanent codices stored far away, but they are nevertheless very real in their day-to-day impact.

In submitting my every action to this degree of external symbolic control, I am typical of my generation and profession. Most professionals and managers are servants of the global external symbolic storage system. They serve it in much the same way that assembly line workers serve large industrial corporations. Many other occupations also make extensive use of the external symbolic environment, and entire careers are pursued within the parameters set by the system, dedicated to adding, refining, or organizing it or (in the case of historical scholarship) reinterpreting and rediscovering some of its secrets. Lives are lived out in the service of the system, territories are fiercely contested, and success is measured, in large part, by how far an individual manages to penetrate its mysteries. We are truly children of the third transition.

MULTILAYERED CULTURES, MULTILAYERED DOMAINS OF AWARENESS

The fate of human consciousness is thus tied to all the stages of human cultural evolution. Each major transition has increased the scope of basic awareness by looping it through a new kind of public process. We now have three distinct layers of cultural representations, each of which has a magnifier effect on the previous layers. This has increased the number and variety of conscious modes. However, the core brain structures involved are exactly the same. On the basis of what we know of our brain's evolution, we must conclude that the raw feel of being human is probably not qualitatively different from the raw feel of being any kind of primate. The same fundamental brain functions negotiate all primate experience: binding, short-term memory, and intermediate-term governance.

But if the raw feel is the same, consciousness has gained far more control over mind and action in humans than in any other species. It has also enriched the content of individual awareness, through its immersion in communities of mind. The differences between us and other primates can be attributed largely to deep enculturation and the resultant reprogramming of conscious experience. Our unique relationship with culture gives us a hybrid awareness and several channels for transmitting cultural influence. We are married to culture and fated to play out its algorithms in our conscious acts. The course of our evolution has been marked by an almost incredible series of intellectual revolutions, whereby conscious capacity has been modified to cope with the overwhelming presence of culture. The collective matrices of culture have exploded, impressing their memory records into our rapidly evolving brains. As a species we have exploited the conscious resources of the individual mind to build a new collective process. This has resulted in a multilayered cultural framework for modern consciousness, which might be conceptualized as a set of concentric rings, as shown below.

Figure 8.4

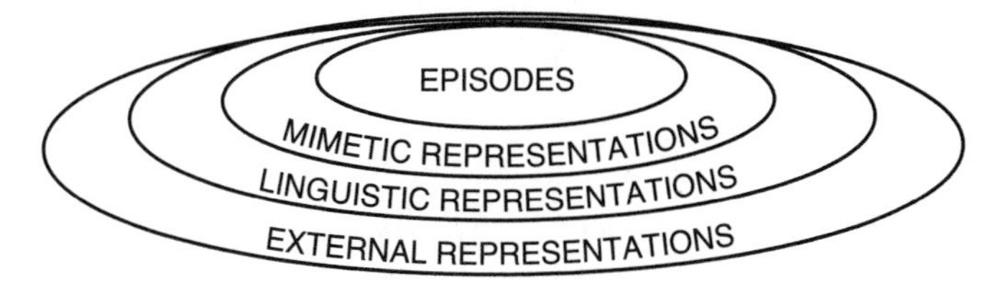

The center ring is the episodic core we inherited as members of the primate family, the bedrock experience base of a precultural mind. At this level consciousness consists of a stream of discrete events, served up as the natural units of experience. Surrounding that central ring are three culturally defined levels of consciousness, each connected to a corresponding cultural matrix, or pattern of convention. The mimetic matrix is the theatrical domain of human life, the actor's realm, with its attitudes, gestures, postures, stances, and unspoken nuances. This matrix defines both actor and audience and sets out the basic rules for all communication and expression. The linguistic matrix is a highly precise and efficient system of knowledge representation, encoding countless stories, myths, and traditions. The external matrix is an even more precise collective system, which provides more powerful media for formulating and displaying knowledge as well as for operating institutions, governments, and formal symbolic systems of thought.

All three of these domains are irrevocably tied to changes in the nature of awareness, especially self-awareness. Mimetic skill is all about self-representation and self-definition, but mimetic style has meaning only in relation to the actions of others. Language is a powerful means of constructing autobiographical memories, but our sense of self takes on meaning only within a shared oral tradition. External symbols define us as individuals in a thousand ways, but they too acquire meaning only in the context of a collective sense of social structure.

The nature and range of human conscious experience are no longer a biological given. Rather they depend on a somewhat unpredictable chemistry of brain and culture, whereby the processes of mind can be endlessly rewritten and rearranged by cultural forces. A mind isolated from culture is deprived of the outer layers of our representational map, despite having the brain mechanisms required to support them. Culture must provide the necessary demons before consciousness can exploit them. We can achieve full awareness only if culture is able to write its key algorithms onto our minds. Evolution has nudged us in this direction, interlocking brain and culture, toward the absorption of individuals by communities of mind. The human conscious process is a specialized adaptation for navigating the turbulent waters of culture as well as the primary channel through which cultural influence can be transmitted back to us.

THE ESSENTIAL UNITY OF THE CONSCIOUS HIERARCHY

Consciousness has served as the engine of our evolutionary adventure, and still does. Each stage in human evolution was constructed by expanding the same set of foundation skills and reprogramming the mind as we evolved more and more elaborate cultural networks. This enriched the human intellectual universe without sacrificing the basic integrity of cognition and awareness.

Human conscious capacity has thus been transformed in its function into an integrative device. It is the final common path of the labyrinthine cognitive-cultural system we evolved. Ultimately the entire distributed network relies on a set of simple operations, performed one at a time, in a specified order. Out of simplicity, complexity. The capacity of the executive brain system thus sets a limit on the kinds of operations that a mind can assimilate from culture and on the complexity that a cultural system can tolerate. Human conscious capacity must be regarded as a distributed resource that limits the rate at which culture can accumulate knowledge and determines what kinds of representational systems a culture can successfully construct and maintain.

The unity of this sprawling cognitive-cultural system, which has been a central consideration in constructing this theory, can be seen in the flow of control through the three evolved layers of human mind and culture. The core episodic layer in Figure 8.4 is inherently other-directed. It observes the world from an unselfconscious perspective. In contrast, the mimetic layer is acutely self-conscious; its own actions are subjected to relentless comparison and review. The mimetic imagination is a dramatic actor. It gives culture shape, by using essentially the same set of conscious operations as an episodic mind, in a wider theater of action.

The narrative layer adds greatly to this cognitive-cultural system. It gives ideas a certain autonomy from personal experience and creates the possibility of abstract beliefs and public discourse. But narrative constructions can govern the larger cognitive-cultural system only if they are successful in dominating the mimetic layer of the system. Narrative imagination wins control by altering the mimetic templates of culture and bringing them in line with myth. The chain of command is clear: Words and stories will dominate only if they successfully influence role playing, skill, and style. Myth gains power only when it directly controls the mimetic imagination. Mythmaking has become an important industry in our cultural hierarchy, and it usually emerges victorious. But its victories never destroy the unity of conscious experience because as we have seen, language floats on the surface of a very deep cognitive ocean. It has no autonomy. It cannot shatter the basic unity of experience and imagination because like episodic and mimetic cognition, it is rooted there.

External symbolic storage significantly alters the powers of our larger cognitive-cultural system by creating conditions within which formal theoretic systems can emerge. But symbolic technologies do not shatter the basic unity of conscious experience either. Symbols and theoretic systems are subject to the same centrifugal forces as every other medium of symbolic thought, within the larger system. Theoretic representations can win governance over culture only if they successfully operate the levers of the hierarchy and assume direct control over narrative, mimetic, and episodic imagination. Theoretic skills are important for survival, but in isolation they never endure. They will govern human culture only if they can dominate the entire cognitive-cultural hierarchy. The governing mode must attain the peak of the collective system, the capstone of the cognitive pyramid. There is no way to short-circuit this elaborate social-cognitive mechanism, and no reason to do so. It works wonderfully well.

The unity of this system is closely related to the question of teleology or purpose. Purpose implies conscious deliberation. There is no such thing as unconscious purpose or unconscious will. There may be unconscious motives and drives, but purpose, by definition, implies conscious design. However, its proven reach is very limited. Human beings are capable of micromanaging very small events, such as wars, fish stocks, and small economies. If we were able to control the Earth's climate, we might make a credible claim that purpose was now a larger player in the universe, but this would still amount to only a small victory in the context of a universe of chance.

Although purpose is anchored in consciousness, it can never be truly attributed to a single conscious mind, in isolation. It is the conscious mind *in culture* that constitutes the source of teleology in the affairs of the human world. There might be a race of beings somewhere out there that is capable of conceiving purpose in a truly solipsistic manner. But human purpose always has a cultural dimension and is inherently distributed in its origins. Ours is a collective teleology, and its creative engine is a conscious mind that has assimilated the algorithms of a culture and is thus a vehicle through which the collective cognitive-cultural hierarchy can act.

CODA

The human brain is the only brain in the biosphere whose potential cannot be realized on its own. It needs to become part of a network before its design features can be expressed. Since we are living beings, the networks we create are complex, fuzzy, and multilayered, rather than lean and mean, or driven solely by the needs of symbolic communication. This makes our networks radically different from those that have been invented for nonliving entities, such as computers. The cognitive infrastructure of human culture includes many things that we do not normally call symbolic, such as patterns of public action, the built environment, and conventional expressions of emotion. These things serve a cognitive purpose, because they convey a great deal about intention, bonding, affiliation, attachment, and hierarchy. They provide structure. By eliminating the need for the explicit encoding of many important parameters of collective cognition, and by embedding layer upon layer of tacit or implicit knowledge in a cultural network, they make it feasible to evolve in a distributed fashion the kinds of highly abstract symbolic systems that we now take for granted. The result is that we are plugged-in, as no other species before us. We depend heavily on culture for our development as conscious beings. And by exploiting this connection to the full, we have outdistanced our mammalian ancestors. We have evolved a far more autonomous awareness, capable of incredible flights of imagination and invention. Consciousness has conquered the mechanical world of the automaton. It has triumphed.

But there has been a cost. Without culture, our world-models, those highly personal and idiosyncratic visions of current reality that define all conscious experience, will inevitably shrivel. If we line up the key features of the many different kinds of minds that coexist with us on Earth and rank the breadth and complexity of their world-models, we see how deeply we depend on our cultural hookup. Table 8.1 lists the size and complexity of the world-models of various species, and indicates approximately how long these models can endure in the mind, independently of external stimulation. This

TABLE 8.1

World-models that typify various levels of conscious capacity

Level	Predominant type of world-model
3rd-order hybrid	Shared theoretic world-models, and external symbolic networks
2nd-order hybrid	Shared narrative world-models, and autobiographical self-awareness
1st-order hybrid	Shared mimetic world-models that incorporate the physical self
Level-3 basic	Complex social world-models held in extended working memory
Level-2 basic	Complex events that can be held in short-term memory briefly
Level-1 basic	Simple perceptual events, integrated across time but ephemeral
Pre-conscious	Very simple perceptual objects, automatically bound, transient

gives us a rough scale, which is less a metric than a hierarchical chronology. It shows the approximate order in which more abstract, and better differentiated, world-models appeared in the biosphere.

This is a graded scale in which the steps do not represent a true qualitative change, or a progression of any kind, because there are so many species that are idiosyncratic in one way or another. One might reasonably ask, If consciousness has so many fine shadings, why deny it to the organisms at the bottom of the hierarchy, the so-called preconscious organisms that are ruled by innate, automatic responses? Might they not have some kind of elemental awareness, passively-induced modes of experience rather like Condillac's Statue on day 1? I have no compelling answer to this, except to fall back on Chapter 4, where we saw that there is an accepted convention that consciousness must have some autonomy from the physical world. Maybe that is an arbitrary stance, and consciousness should be attributed even to the most rudimentary nervous system. I cannot deny that possibility. But such an admission would not change much.

There is a more serious problem with this scale, however. It occurs at the other extreme of it, where we see the three levels of brain-culture interactions that are special to humanity. This ordering of the levels of consciousness seems to imply that an awareness capable of holding theoretic models is somehow the highest form of consciousness. I cannot emphasize too strongly that I intend no such meaning. This scale says nothing about such value judgments. While it is true that theoretic models appear late in the chronology of consciousness, I am not interested in attributing value, intellectual or otherwise, to that fact. Theoretic minds certainly have greater powers of abstraction than minds that are locked into a purely mythic stance—hard-won powers, bought at a great historical cost. But there are ways in which mythic knowledge representations reign supreme over theoretic ones, just as there are realms where a mimetic stance does. Myth and mimesis rule politics and art, as well as the entire domain of interpersonal relationships. No mere theory can improve on good art, effective body language, or strong tribal identification (even the tribal identifications of logicians or theoreticians). Nevertheless, a powerful theory can take the conscious mind on a voyage to Mount Olympus as no other kind of representation can. For that brief moment when we grasp an elegant theory fully, in all its implications, we are granted a glimpse of what it might have been like to be a god.

As our latest speculations about awareness pour onto the pages of more and more books and scientific papers, they will change some of our old ideas about awareness. But the ideas were not really our own in the first place; that is, they do not belong to any of us in a personal way. They are products of a collective process that defines the peaks and valleys of a virtual landscape of meaning. Ideas are under constant revision by a collective process that often masquerades as a highly personal quest. If anything in this book can be construed as truly "my own" or acknowledged in some way as "my" achievement, it is only in the sense that I might have successfully become a temporary vortex within the culture, a point of convergence whereby certain forces have become concentrated in my consciousness temporarily, before returning, transformed, to the collective matrices whence they came.

Which is not to say that I would have it any other way. Fate has given us this hybrid nature, by which we are joined to communities of our own invention. We remain distinct from them but are never fully autonomous. The human brain is a poor thing on its own, an inarticulate, undifferentiated, metaphorizing beast like any other. But joined to a community of its fellows, it has this remarkable capacity to create a community of mind, acquire symbolizing powers, and vastly expand the range of its own awareness, in proportion to the depth of its enculturation. If this appears strange to us, this is surely only a reflection of our conventional notions of, among other things, strangeness. We have lived comfortably with the myth of the isolated mind throughout most of our history. We like to think of ourselves as self-complete little monads dwelling inside our sealed biological containers, peering out at the world from the safe haven of consciousness.

But we are edging closer to the truth. We are collective creatures, even to the texture of our awareness. When we face that fact, the irrational beast that drives our cognitive enterprise will finally be able to see itself as it really is, as part of a collective process that has freed us from our solipsistic prison, if not from our materiality. The triumph of consciousness will be complete when it can finally reflect on the collective process itself and see only itself, in the mirror of its own reflection.

Notes

PROLOGUE

Page

xi Borges, 1999.

xiv Definition of culture. This is a complex topic, as shown in Kuper, 1994, 1999. Our concern here is not to decide how to define culture for once and for all, but rather to focus on its cognitive function, and its possible evolutionary role, as a distributed system of "knowledge" for the human species. The cultural system stores much more than "information" as it is usually defined. It stores both tacit and explicit knowledge. Even such things as art, ritual, and the built environment can store and transmit important cultural knowledge. Plotkin's theory of multilevel evolution also defines culture as a storehouse of replicative information for the human species (Plotkin, 1993, 1998).

1. CONSCIOUSNESS IN EVOLUTION

Page

1 Spencer, 1855. 1970 edition.

2 The irrelevance of consciousness. See Pinker, 1994, 1997.

4 Memes. See Dawkins, 1990; and Dennett, 1995, pp. 342–70.

4 Brave New Mind. Peter Dodwell has put a very different spin on this expression, in a recent book of the same title (Dodwell, 2000).

5 Agnosia. See Sacks, 1970; also Milner and Goodale, 1992; and Weiskrantz, 1997.

6 Unconscious perception. See Marcel, 1988; and Mack and Rock, 1998.

7 Language acquisition. See Jusczyk, 1997.

8 The demons of culture. This is another complex and controversial topic. Tooby and Cosmides (1992) have constructed the best known defense of what they call the evolutionary "modules" that greatly constrain human cultural patterns. Their viewpoint is obviously very different from the theory expressed here, since it downplays the role of consciousness. However, the two theories are not completely incompatible. The area of disagreement is largely confined to higher cognition and language, and seems to reflect a difference in how a neurocognitive "module" should be defined.

9 Physical origins of consciousness. See Penrose, 1990, 1995; and Crick, 1994.

2. THE PARADOX OF CONSCIOUSNESS

Page

13 Marcus Aurelius, 170.

16 Minimalist. To be clear, the use of the term has nothing to do with the "minimalist" program in linguistics.

21 Sensory deprivation. See Heron, 1967.

22 Spatial limits to consciousness. See James, 1890; and Miller, 1956.

26 The unconscious context of knowledge. See Baars, 1988, 1997.

28 Consciousness as an epiphenomenon. See Churchland, 1986; and Rey, 1997.

29 The Cartesian Theater. See Dennett, 1991.

29 The Society of Mind. See Minsky, 1986.

30 Classic AI theory. See Simon, 1969.

30 Darwin Machines. See Calvin, 1987.

31 Selfhood. See Rorty, 1989.

31 Humphrey, 1983, p. 97; p. 98.

31 Raw feeling. See Kirk, 1994.

32 Emergent complexity. See Sinnott, 1961; and Mayr, 1988.

35 Primacy of perception. See Berkeley, 1710.

35 Self-consciousness as a cultural invention. See Jaynes, 1976.

35 The autonomy of linguistic awareness. See Rorty, 1989, p. 21.

35 Experience as essentially linguistic. See Gadamer, 1976; and Norris, 1993, p. 289.

36 Memes and representational awareness. See Dennett, 1991.

36 Linguistic modularity. See Fodor, 1983; and Chomsky, 1984, 1995.

39 Dennett's dangerous idea. See Dennett, 1991, 1995.

39 The most thorough point-by-point response to Dennett's many examples is Owen Flanagan's book *Consciousness Reconsidered* (1992).

40 The proprium. See Allport, 1961, 1966.

41 McGurk effect. See McGurk and MacDonald, 1976.

41 Color Phi. See Kolers and Von Grünau, 1976.

42 Cyclopean perception. See Julesz, 1995.

42 Bidwell's phenomenon. See Bidwell, 1899.

43 Delayed awareness of sensory input. See Libet, 1985a, 1985b, 1993.

3. THE GOVERNOR OF MENTAL LIFE

Page

46 James, 1890.

47 The Cognitive Revolution. See Gardner, 1985.

50 Width and depth of working memory. Alan Baddeley's concept of working memory (Baddeley, 1986) includes a central executive component that might presumably account for larger time spans than fifteen or twenty seconds, but he is rather vague on this point.

52 Limitations of standard human paradigms in psychology. See Neisser, 1982.
52 Long-term working memory. See Ericsson and Kintsch, 1995.
58 Unconscious learning and perception. See Mack and Rock, 1998; and Marcel, 1988.
58 Unconscious learning of grammar has been claimed by Reber (1996), who holds that many kinds of learning are tacit, or implicit, and therefore unconscious. However, in the broad definition of consciousness used here, many of his cases might be used as excellent examples of conscious learning. Tacit knowledge is not necessarily acquired unconsciously. By the definition used here, when learning relies on working memory, selective attention, and metacognitive supervision, it must be labeled as conscious, or controlled, learning. This rule applies to any cognitive process that requires the allocation of conscious capacity, whether or not the subject explicitly realizes or recalls what was learned.
61 Carroll, 1865.
61 Carlyle, 1857,
61 Mind reading. Baron-Cohen (1999) has suggested that our capacity for mindreading might be an evolved module of the human brain, possibly based on a special capacity for sharing attention. However, enculturated apes can also share attention (Savage-Rumbaugh et al., 1998), and even wild apes are able to practice deception. This militates against the idea that human mindreading evolved as a uniquely human module. However, the amodal nature of this skill in humans, and its reliance on the executive brain system, suggests that what really happened was that humans amplified primate mindreading skills by connecting them to a powerful amodal executive brain system with an expanded range of influence.
62 The disciplined use of phenomenology. See also Varela and Shear, 1999.
67 Physical self-familiarity. See Varela et al., 1995; Lakoff and Johnson, 1999; and Damasio, 1994, 1999.
68 Anosognosia. See Damasio, 1994, p. 237; p. 236.
69 Paroxysmal aphasia. See Lecours and Joanette, 1980; also Donald 1991.
70 Metacognitive self-supervision. This idea of consciousness is discussed in detail in Koriat, 2000.
71 See Luria, 1987.
72 Ibid., p. 47.
73 Ibid., pp. 62–64.
74 Ibid., p. 69; pp. 80–81.
75 Ibid., p. 158.
76 Zasetsky's mirror twins. See Luria, 1966.
76 Frontal lobe injury. See Stuss and Benson, 1986.
79 James, 1886, p.164.
83 Stendhal, 1830, pp. 314–15.
85 Balzac's project. See Robb, 1994.
86 Humphrey's stand on the social function of consciousness. See Humphrey, 1983.
87 Helprin, 1991, p. 2.
88 Joyce, 1937.

4. THE CONSCIOUSNESS CLUB

Page

92 Griffin, 1992; and Hamilton, 1859.

93 Mysterianism. See McGinn, 1991.

93 The Hard Problem. See Chalmers, 1996.

95 Exaptation. See Gould and Vrba, 1981.

95 Evolutionary determinism. See Huxley, 1953.

97 Total human biomass. See Wilson, 1978.

98 The life essence. See Bergson, 1959.

101 Cortical columns. See Szentágothai, 1969; Szentágothai et al., 1977; and Mountcastle, 1998.

102 Cortical architecture. See Nauta, 1993; and Nauta and Feirtag, 1986.

106 Adequacy of the human brain. Owen Flanagan (1992) has made this same point in convincing fashion, as have many others, including Churchland (1986).

107 Propriospinal and corticospinal tracts. See Lemon et al., 1999.

108 Song nuclei in bird brains. See Nottebohm, 1984.

109 Legend for abbreviations in Figure 4.1. Shaded ovals: OPT. T = optic tectum; GL. P = globus pallidus; HAB = habenula; HTh = hypothalmus; IPN = interpeduncular nucleus; RF = reticular formation; CER = archicerebellum; VEST = vestibular nuclei. Unshaded ovals: S. NIGRA = substantia nigra; SUBth = subthalmus; NEO-CER = neocerebellum; RED N. = red nuclei; I. OLIVE = inferior olive.

111 The Great Module Hunt. See Broca, 1861; Brain, 1973; and Wernicke, 1874; also Geschwind, 1965, 1970.

112 Primate brain size. See Passingham, 1982.

112 Expansion of the frontal cortex. See Deacon, 1997.

114 Jumping spiders. See Foelix, 1996.

120 Bonobo-human cultures. See Savage-Rumbaugh, 1986; Savage-Rumbaugh and Lewin, 1994; and Rumbaugh, 1974.

121 Language mediation. See Nelson, 1996.

125 Mental autonomy. See Hebb, 1961, 1963.

126 Consciousness in animals. See Griffin, 1992.

128 Brainy monkeys. See Milton, 1980.

130 Deception in primates. There seems to be no doubt that it is deliberate, in some species. Based on the database collected by Byrne and Whiten (1988) and other evidence, Hauser (1996) and others have concluded that primates who live in large social groups (such as chimpanzees and macaques) are more skilled at this than monogamous species, such as gibbons.

132 Resourcefulness of mammals. Mark Hauser gives some excellent examples of this kind of cleverness in various species in his recent book, *Wild Minds* (2000). However, there are limits on animal cleverness, as shown in their tendency to use tools in a very narrow manner, usually applying them to only one task. In contrast, when human beings invent a tool, they tend to find many uses for it.

Mithen (1996) has suggested that it is this flexibility rather than the mere fact of tool use that marks humans' practical intelligence. Again, this illustrates the amodal, domain-general mind-set of humans.

134 The homunculus. See James, 1890; and Allport, 1961, 1966.

135 The physical self. See Lakoff and Johnson, 1980, 1999.

138 The zone of proximal development. See Vygotsky, 1978.

139 Table 4.1 is discussed in detail in Donald, 1998a and 1998b.

139 The bailout option. See Smith, 1992.

140 Multitasking. See Savage-Rumbaugh, 1991.

140 Self-reminding in apes. See de Waal, 1989; and Savage-Rumbaugh et al., 1998.

141 Autocuing in apes. See Fouts and Mills, 1997.

141 Autocuing in humans, also known as explicit recall. This ability is heavily dependent on language. It comes from what Karmiloff-Smith (1992) calls the redescription or "explicitation" of reality, an idea closely related to Gerald Edelman's (1989) earlier notion of "recategorical" memory. However, this high-level form of retrieval must ultimately depend upon a more fundamental capacity based on imagery.

141 Self-recognition in apes. See Gallup, 1970, 1982.

142 Imitation in apes. See Byrne, 1995; and Byrne and Russon, 1998.

143 Mindreading. See Baron-Cohen, 1995; Byrne and Whiten, 1988, 1997; Tomasello, 1999.

144 Pedagogy. See Premack, 1976, 1987, 1993.

144 Bonobo pedagogy. See Savage-Rumbaugh et al., 1998.

145 Critical review of ape language. See Terrace and Bever, 1980; Lieberman, 1991; Fouts and Mills, 1997; and Hauser, 2000.

145 Protogestures. See Wrangham, 1994.

5. THREE LEVELS OF BASIC AWARENESS

Page

149 Wittgenstein, 1980; and Young, 1960.

150 Variations in ape culture. See Wrangham et al., 1994.

150 Saussure, 1983.

151 Biological superorganisms. See Wilson, 1971.

152 Ecological theories and distributed cognition. See Gibson, 1950, 1979.

153 The great computational divide is described in a similar way by O'Brien and Opie (1999), but they still try to maintain a solipsistic approach to human higher cognition.

159 The octopus brain. See Young, 1960; and Hanlon, 1996.

163 Explicit memory. Of course, images can also be recalled, but brain-imaging studies suggest that this depends on much more than the activation of the primary sensory cortex. Imagery seems to engage the primary regions centrifugal-

ly, via the secondary and tertiary parietal and frontal regions. See Kosslyn, 1994; Posner and Raichle, 1994.

163 Memory and the cortex. See Damasio, 1994, 1999.

164 Language in deaf children. Goldwin-Meadow and Mylander (1998) have shown that without adult help, deaf children can create a communication system that has some of the properties of language (it appears to be a largely mimetic system). However, their subjects were neither sensory-deprived nor isolated from culture.

165 Tertiary cortex. Luria (1966) based his neuro-cognitive model of the brain on this classification of the cortex. There is an excellent presentation of primary, secondary, and tertiary cortex in Kolb and Whishaw, 1990.

167 Abstract attitude. See Goldstein, 1948.

168 Prefrontal cortex in human and monkey working memory systems. This is a very active area of research. For an introduction, see Goldman-Rakic, 1992.

169 Consciousness and brain waves. See Magoun, 1965; and Moruzzi and Magoun, 1949.

169 Consciousness and brain activation. See Jasper, 1958, and Lindsley, 1961.

169 The arousal switch. See Lindsley et al., 1950.

171 Cell assemblies. See Hebb, 1949; and Milner, 1957.

171 Retinal rivalry. Edelman and Tononi (2000) have shown that he neuromagnetic activity of cortical networks can be amplified 50 to 85 percent by conscious attention in a rivalry situation.

172 The junctional microstructure. See Pribram, 1971.

173 The clock not ticking effect. See Picton and Hillyard, 1974.

174 Electrical signs of selective attention. See Hillyard et al., 1973; Donald, 1982, 1983; Näätänen, 1992; Parasuraman, 1998.

174 Electrical signs of stimulus evaluation. See Kutas et al., 1977.

175 The executive system. See Posner et al., 1997.

176 Brain imaging and conscious effort. See Posner and Raichle, 1994.

181 Stages of vision. See Marr, 1982.

182 Gamma waves. German neurobiologist Wolf Singer (1994) was one of the first to observe these 40-Hz or gamma oscillations. He suspected that they might be a binding mechanism because common rhythmic activity could conceivably link together several disparate neural circuits that were active at the same time. These rhythms have been demonstrated most clearly in the visual system, where the color, shape, and pattern of movement of a stimulus will typically activate three quite disparate regions of the visual brain, which will share exactly the same temporal oscillation pattern if they are bound together. See also Crick, 1994; Crick and Koch, 1995; Engel et al., 1999; Gold, 1999; and Sauvé, 1999.

182 The notion that attention is necessary for neural binding converges nicely with an idea proposed in a recent monograph by Arien Mack and the late Irvin Rock (1998), a phenomenon they have called inattentional blindness. Their message is simply that nothing is perceived consciously without attention. They chal-

lenge the traditional Gestalt notion that basic phenomena are perceived automatically, claiming that even the division of the visual array into figure and ground, or the perception of large groupings as unified fields, require attention. This suggests that the link between binding and attention may be even stronger than Crick and Koch (1995) proposed.

182 Stages at which consciousness intervenes in perception. The classic theoretical proposal is by Treisman and Gelade (1980). Jackendoff (1985) has made an important effort to integrate this viewpoint with linguistic research.

183 Binding theory. Nancy Woolf (1999), a neurochemist, has made the interesting suggestion that an anatomical complex called the cholinergic forebrain system, which radiates out into the whole cerebral cortex, might be in control of the binding process. It feeds axons into every region of the cerebral cortex, so that, in theory, it can reach every cortical column and therefore has the wiring needed to weld local circuits together, even within single cortical columns. This system is already known to be associated with mental imagery. Stimulating it with drugs triggers dreaming in a sleeping person and vivid imagery in one who is awake. According to Woolf, the cholinergic forebrain system triggers imagery by adjusting microscopic structures, called microtubules, that make up the "skeletons" of nerve cells. It does this through its effect on a specific protein that alters the internal geometry of neurons that are simultaneously active in a given cell assembly. This altered geometry is itself a binding mechanism not unlike the one proposed by physicist Roger Penrose (1990, 1995). It aligns or attunes the cells in the assembly, creating microtubular coherence across the circuit, an effect that is not unlike rhythmic binding, in principle. One limitation of this particular mechanism is that it appears to be too local to account for some of the best documented rhythmic binding phenomena, in which bound circuits extend over considerable distances in the brain.

183 The Chinese Room. See Searle, 1997; and Block, 1978, 1997.

184 Binding of local sensory circuits. There is evidence that gamma-range neural coherence may not be as local as formerly thought, and may integrate very widely distributed neural circuits. See Rodriguez et al., 1999; and Miltner et al., 1999.

187 Spatial attention network. See Posner et al., 1997. This group places the main control center for attention in the anterior cingulate gyrus (a controversial position that probably exaggerates the role of that structure). Together with the prefrontal cortex, the midbrain, and the thalamus, the anterior cingulate gyrus forms the core an "executive attention network" that controls the working memory regions of the prefrontal cortex and enhances feature recognition in the secondary sensory cortex.

188 Figure 5.4. Modified from McIntosh et al., 1994. This figure shows only one of the many possible ways in which networks could be rearranged. In this case, roughly the same brain regions are involved in performing judgments of two visual tasks: object recognition and spatial location. Thus when the subject switches from one task to the other, network flow is redirected within the same overall neural architecture. But networks can be rearranged in more drastic

ways, switching over to a completely different architecture; for instance, when switching from visual to auditory search, or from reading to talking. Network rearrangement also occurs on a more microscopic level, but we cannot monitor such small changes on-line with present technology.

191 The thalamus as an amplifier. See LaBerge, 1995, 1997.

192 ERTAS. See Baars, 1986, 1988, 1997; Newman, 1995; Newman and Baars, 1993; and Newman and Grace, 1999. According to Baars, the functions of the brain fall into three general classes. The first, and most prevalent, is modular. It includes automatic systems, such as vision, hearing, emotions, and basic action patterns, which are built in and automatic. These are localized in self-contained, encapsulated modules that normally operate outside consciousness. The second class is long-term memory, which is concerned with providing background, or context, for all perceptions. Our permanent memory networks are usually dormant, but they are called up to consciousness when relevant to an ongoing pattern of mental activity. They provide the meaningful background knowledge for all action and thought, including the largest contextual structure of all, the self. The self-system plays an important role in human executive function but it is not part of the executive system per se. The third class includes all executive brain functions, including the ERTAS network.

194 The neocerebellum and learning. See Thatch, 1998; Kawato, 1999; and Imamizu et al., 2000.

196 Hominid expansion of prefrontal cortex. Boston neuroscientist Terry Deacon (1997) has mapped out the evolutionary evidence for these prefrontal changes in detail, arguing that during brain development, they expanded their anatomical range by generating more source cells in the maturing brain and establishing more functional synapses than their competitors.

197 The cerebellum as an optimizer. See Leiner et al., 1993. William Calvin (1983, 1996) has proposed that humans needed greater brain capacity to carry out the complex computations necessary to execute high-speed ballistic skills, such as throwing. Ballistic movements, like bullets, cannot be controlled or corrected once they have been initiated. They occur too fast to correct, as when we start to swing a golf club or a baseball bat and reach a point of no return. The elementary units of speech, phonemes, are ballistic movements. Each phoneme trips off the tongue at an incredible speed. This requires the instantaneous calculation of very precise movement vectors, and the human capacity for learned ballistic skill hierarchies might be related to the expansion of the hominid neocerebellum.

197 The supervisory system and the prefrontal cortex. See Norman and Shallice, 1986; and Shallice, 1988.

198 Long-term working memory and the frontal lobes. See Stuss and Benson, 1986; Ericksson and Kintsch, 1995; and Engle et al., 1999.

199 Slower processes and intermediate-term working memory. Tulving (1983) proposed the concept of "active semantic memory," a primed region of long-term memory. However, its on-line updating function rules against labeling it as "long-

term" memory. Complex processing that is on-line and current is, by my definition, conscious.

201 Episodic cognition. As it is used here, the term refers primarily to event-perception, rather than to episodic recall as classically defined by Tulving (1983), which usually involves explicit retrieval. However, my use of the term is not incompatible with the latter definition. In a multiple review of his book on episodic memory, Tulving (1984) conceded that some animals probably have episodic memories, but cannot access them explicitly, as humans can. In other words, episodic recognition memory must have evolved in animals long before humans evolved the neural machinery of explicit episodic recall, as I proposed in my first book (Donald, 1991).

203 Sherman and Austin. See Rumbaugh, 1974.

204 The evolution of phonology. See Levelt, 1989; Lieberman, 1984, 1991; and Donald, 1991. Also see Armstrong et al., 1995; and Studdert-Kennedy, 1998.

6. CONDILLAC'S STATUE

Page

205 Condillac, 1754.

210 Plasticity of the sensory brain. See Merzenich et al., 1987; also Neville and Bavelier, 1999.

210 The Baldwin effect. See Baldwin, 1896; and Dennett, 1995, pp. 77–80.

210 The runaway brain. See Wills, 1993.

211 Evolving plasticity. Mayr (1974, 1982, 1988, 1991) has written extensively on what he calls "open" and "closed" programs in evolution. Open programs allow for modifications during the lifespan of the organism, while closed ones do not, and can only be modified over many generations. Most of the cognition-related bioprograms of humans tend to be open, according to Mayr's definition. This is more efficiently explained in terms of a single domain-general adaptation that has a very wide cognitive influence, than in terms of many single, specialized adaptations to isolated cognitive modules.

214 Condillac as the French interpreter of Locke. Of course, this just reflects our customary Anglo-Saxon narcissism. To the French, he is one of the immortals, rightly regarded as a force in his own right.

214 The authorship of Condillac's *Treatise*. Condillac did not devise the Statue completely on his own. He acknowledged that the idea had been suggested by a student and associate, Mlle. Ferrand, who died before the book was published. She played such a central role in its writing that Condillac wrote an extensive tribute to her in his Preface.

214 Leibniz's Nativist views. See Leibniz, 1714.

216 Constructivism. See Bruner, 1990; Nelson, 1996; Bates et al., 1991; Karmiloff-Smith, 1992; Elman, 1996; Quartz and Sejnowski, 1994; and Tomasello, 1999.

218 Self-reinforcement as a basic property of neural networks. See Becker and Hinton, 1992.

219 Condillac, 1754, p. xxxvii.

219 Ibid., pp. 3–4.

220 Ibid., p. 6.

220 Ibid., pp. 6–7.

221 Ibid., pp. 9–10.

221 Ibid., p. 10.

221 Ibid., p. 11.

223 Ibid., p. 19.

224 Ibid., p. 20.

225 Ibid., p. 47.

227 Mandler's dictum. See Mandler, 1959, 1985, 1997.

229 Attention and language learning. See Jusczyk, 1997.

230 Scaffolded hierarchy of conscious control. See Zelazo (2000).

233 Helen Keller. The most comprehensive biography is that by Lash, 1980.

235 Helen's credibility. Most medical case histories are accepted at face value, but Helen faced challenges from many quarters, perhaps because of the harsh nature of the society in which she grew up. When she first became a public sensation, she was accused of merely borrowing stock phrases that she did not understand. In a sense, she was accused of being a sort of overblown parrot. The ten-year-old Helen was subjected to the notorious "Frost King" episode, named after a story that she had supposedly plagiarized. She was later vindicated, and the charges appear merely ridiculous today. Such criticism would have crushed most people. They seemed instead to drive her to higher levels of accomplishment.

235 Admission to Radcliffe. This whole adventure has an absurd, Monty Pythonesque quality to it. But there can be no doubt of the authenticity of Helen's entrance examination to Radcliffe, and this publicly watched event took place fairly early in her life. If she was capable of this level of achievement then, there is no reason to question her later victories. Her resourcefulness in meeting academic challenges was legendary. A good example was her approach to learning geometry. Her sense of three-dimensional space was not good, and she had difficulty walking and moving about. Yet she studied many subjects, including geometry, that require an abstract conception of space. Normally this depends heavily on vision for its development, but Helen learned about geometric shapes and solutions by using a pegboard with strings, tracing out patterns with her fingers, and using bendable pieces of wire to fashion new ones. She had some trouble mastering geometric notation but eventually managed to understand it and excelled at her exams. She also learned several foreign languages, a skill normally reliant on the sense of hearing. Indeed, without hearing or any sound reference for the printed symbols, what precisely is a foreign language? While learning Latin prosody, she developed her own system of signs for its various metric schemes. Moreover, as if to drive home the point that she could think in words as well as anyone, she was first in her Radcliffe English class. The magni-

tude of that achievement is truly astounding when we consider her cumbersome reading routine.

236 Neurological comparison of Helen Keller and Laura Bridgman. See Tilney, 1929.

237 Active touch. See Keller, 1908.

238 Sullivan, 1891, p. 3.

242 Keller, 1902, p. 98.

245 Keller, 1902, p. 316.

246 The experience of an epiphany that initiates and accelerates language acquisition is not unique to Helen Keller. This experience is not typically remembered by neurologically normal children because their language capacity matures before the development of autobiographical memory. But there are similar reports from late language learners who are deaf. Susan Schaller (1991) described the case of Ildefonso, a congenitally deaf man who did not acquire language until the age of twenty-seven. He had been raised in rural Mexico by hearing parents and had no sign language as a child. Initially he did not understand what Schaller was trying to achieve when she started teaching him to sign. They went through an arduous period for several months, during which Ildefonso did exactly what Helen had done, mimicking the hand signs of his instructor without realizing that they represented anything. He simply did not seem to grasp the referential principle of symbols. One day he had a flash of insight, followed by excitement and a rush of new signs. He learned dozens of signs within a day. Then he slowed down, and it took several more years before he acquired grammar and was able to formulate entire sentences. It took him even longer to become fairly fluent. For both Helen and Ildefonso, full-language acquisition was not as simple as learning a few signs for objects and names for actions. Acquiring the ability to use language in more complex ways, to convey ideas in sentences, took much longer: more than a year in Helen's case and several in Ildefonso's.

250 The Outside-Inside principle. See Vygotsky, 1978, pp. 52–57.

251 The origins of symbolizing algorithms. Gabora (1998) has made some useful suggestions about the leap from a solipsistic to a distributed system, based on a computational model of cultural evolution.

7. THE FIRST HYBRID MINDS ON EARTH

Page

252 Flaubert, 1857; and Lakoff and Johnson, 1980.

253 Social theories of language origins. There are too many to cite here in detail. A representative sample can be found in recent books edited by Hurford, Studdert-Kennedy, and Knight (1998); Jablonski and Aiello (1998); and Corballis and Lea (1999). I also strongly recommend recent works by Dunbar (1996, 1998); and Tomasello (1999), which make a strong case, from two very different research directions, for the cultural origins of human higher cognition.

253 The evolution of speech in relation to language. This is beyond the scope of the present discussion. In my 1991 book I proposed that the vocal capacities underlying human speech evolved in two stages. The first stage came long ago, more than two million years before the present. It involved the evolution of deliberate voice modulation, as part of a broader capacity for body expression called vocomimesis. This later became prosody, the emotional envelope surrounding speech sounds. The second stage came only with our species and involved the evolution of phonology, the complex special capacity that makes up syllables and words. The latter could be called a modular capacity, but only as a subset of a larger cognitive adaptation that was nonmodular, and culturally driven. A variant on this two-stage theory has been proposed by Deacon (1997).

254 Communities of mind. Does this idea imply group selection, as defined by Sober and Wilson (1998)? This is not a simple question to answer, especially by someone whose specialized training is not in this area. Mindsharing communities may have evolved along classic selfish-gene lines, but group selection would be a much more powerful explanatory framework for the rapid emergence of such communities, once archaic hominids had started to move toward a collective strategy. Understandably, Sober and Wilson have focused on the central problem of their field, human altruism; but the fact of human mindsharing is surely an equally salient reality in human life, and presents a somewhat similar dilemma to traditional selection theory.

255 Finding the codes of culture. Paul Bloom (2000) has made the telling point that children, and even infants, usually discover the meanings of words by watching others talk; that is, by tracking their intentions, rather than by listening to child-directed speech. This argues for the existence of very abstract social sensitivities, and an extraordinary capacity for remembering exactly where to direct attention, in infants less than two years old. See also Nelson and Shaw (2000) for a discussion of how socially shared symbolic systems emerge.

259 The stages of human cultural evolution. See Donald, 1991, 1993a, 1993b.

263 The conscious control of mimetic action. This is a domain-general capacity. It is not confined to any specific kind of action. Mimesis can take its inputs from various sources, and implement its output in any set of voluntary muscles. Humans have excellent supramodal capacities; thus I might read the letter *a* with my eyes, or my fingers, or, for that matter, my nose, and I can write the letter *a* with my finger, my wrist, my elbow, my lips, my head, or even my legs. This means that my abstract perceptual template for the latter *a* can be mapped onto a motor program so abstract that it can be executed in virtually any subset of my voluntary musculature.

263 The importance of pretend play in the early development of symbolic processes. See Harris, 1998. For more background, see Greenfield and Smith, 1976; Trevarthen, 1979, 1980; Bruner, 1983; and Suddendorf, 1999.

264 Imitation. See Tomasello, 1999, for its relationship to social cognition. See Meltzoff and Moore, 1997, for a theory of infant imitation that has key simi-

larities to the model of mimesis I have presented here. See also Donald, 1991, 1993, 1999.

264 Consciousness and imitation. Georgieff and Jeannerod (1998) have pointed out some of the perils of assuming that mimesis is a conscious process. Proprioceptive signals that are essential for accurate imitation tend to fade quickly, and are not available to consciousness for very long. Nevertheless, the larger schema of self-in-its-environment that underlies mimesis is available to consciousness in the intermediate-term. This model requires the subject to integrate first- and third-person information, as Meltzoff and Moore (1997) have shown.

264 Mimetic skill. My ideas about skill are somewhat similar to those of New Zealand psychologist Michael Corballis (1991). However, his notion of "generative praxis" is computational, in that it breaks action into irreducible components and recombines these components. Thus a toolmaking sequence would be reducible to a few elements that could be combined in various ways. In my view, mimesis does not work in this way; it is holistic. The inspiration for Corballis's concept, and for Harvard psychologist Stephen Kosslyn's (1988) closely related ideas about visual scene analysis, is the modern computational habit of breaking down every event into a set of explicit symbols. But the mammalian brain does not seem to do this on its own. It is impossible to reduce body language, or mime, to discrete symbolic elements.

265 Infant gestures. See Acredolo and Goodwyn, 1988; and Goldin-Meadow et al., 1996, 1998, 1999.

265 Evidence for mimetic culture in children. See Nelson, 1996.

266 Pretend play and mimesis. See Harris, 1998.

267 Toolmaking without language. Even the simplest stone toolmaking industries could not have come into existence without consensual patterns of group behavior. Current archaeological findings are pushing back many of the dates we previously assigned to hominid cognitive achievement. As a result, it is becoming more and more likely that the elaborate structure of archaic hominid group activity must have been constructed, and regulated, by some means other than language. Ancient mimetic scenarios are not difficult to imagine since we still see similar patterns, similarly maintained, in modern society. Imagine the following series of archaic collective tableaux, and how they might have been transmitted across generations: (1) discovering and communicating the location of a good source of stone; (2) transporting the material to a home base; (3) storing the material and protecting it; (4) teaching many individuals how to chip flakes away from the core and practicing the skill; (5) practicing the making of actual tools; (6) hunting, killing, and butchering game, using the tools; (7) driving off a food raid from a neighboring tribe, using tools; and so on. These are entirely mimetic scenarios, with mimetic custom providing a cognitive frame for the organized behavior of the group and mimetic rehearsal enabling the performances of individual actors in the drama.

268 Evolution of sign language. Armstrong et al. (1995) have proposed that gesture evolved first into sign, then into speech.

268 The particulate principle and its relationship to mimetic skill. See Studdert-Kennedy, 1998.

271 The leveraged takeover of the prefrontal cortex. Deacon (1997) proposed this as the evolutionary basis of language, but I view it as the basis of mimetic skill.

271 Mimesis and the frontal cortex. The most exciting recent discovery is probably that of "mirror neurons" in the premotor cortex (Rizzolatti et al., 1996, 1998). Mirror neurons get very excited when we perform a specific action, say, arm flexion. But they also get excited when we watch someone else perform the same action. They, or rather the networks of which they are part, can draw equivalences between our own actions and those of other people. They cannot achieve this by visual comparison since other people's actions look completely different from our own. They can do it only by mapping the other person's actions onto a body map, a kinematic model of the self, which must be mapped onto a higher-order map, an internal model of models that places one's own actions in the context of a larger world-model. This is direct evidence of a third-party model view of the self (see Georgieff and Jeannerod, 1998). Since it exists in primates, it could have served as the predecessor of the human mimetic controller, but not as the sole foundation of human mimesis.

272 Rhythmic dance and mimetic skill. Dance scenarios might better be called minidramas because they usually have a heavy emotional overlay. Emotional public actions are common in apes and serve to maintain both a dominance hierarchy and a system of alliances. Mimetic dance builds on this primate foundation. Many dance scenarios are elaborate variants of a basic primate expressive pattern, with the crucial additional property that mimetic skill allows us to refine and vary the form of such scenarios deliberately.

273 Self-representation and awareness. See especially Varela et al., 1995; Damasio, 1999; and Lakoff and Johnson, 1999.

275 Flaubert's complaints about his reading public. Attributed to him by Julian Barnes, 1985

279 Jackendoff's views on the relationship between language and other domains, such as visual space, music, and thought are briefly summarized in his book *Patterns in the Mind*, 1994.

280 Criticism of educational Constructivism. See Pinker, 1994.

280 Constructivist manifestos. See Elman et al., 1996; and Quartz and Sejnowski, 1994.

280 Cultural storage in evolution. See Plotkin, 1993, 1998.

281 Language evolution, creoles, and pidgins. See Bickerton, 1981, 1990, 1995.

282 Universal Grammar. Newmeyer (1998) suggests that it is better seen as a descriptive system, rather than a biological mechanism for language. Computer simulations of language evolution support the notion that grammar is the historial product of groups of brains striving for optimal clarity in communicating. See Batali, 1998; and Kirby, 1998.

282 Metaphoric nature of thought. See Jaynes, 1976.

283 Metaphor in language. See Lakoff and Johnson, 1980, 1999; also Langer, 1967.

283 Mimetic origins of linguistic universals. See Armstrong et al., 1995; Bates et al., 1991; Hurford et al., 1998; and Corballis and Lea, 1999.

284 Social metaorganisms. See Wilson, 1971.

284 Co-evolution of brain and language. Deacon (1997) proposed a theory of language evolution that tries to finesse the need for a mimetic preadaptation. He argues that hominids made the transition from primate "indexical" representation (as shown in operant conditioning) to fully symbolic representation (as shown in language) in one continuous evolutionary progression. This idea does not account for the metaphoric nature of language, the existence of mimetic skill, or the fact that language normally develops in a mimetic framework, and operates by metaphoric or mimetic principles. Moreover, it cannot scale the wall set up by the primate zone of proximal evolution.

291 Sentences as the primary units of language. See Croce, 1992.

291 Hunting for meaning. Psychologist Michael Tomasello (1999) has analyzed in detail the process that takes place as the typical three-year-old child starts to construct hypotheses about what other people intend to communicate when they speak. The child initially relies on gesture, pointing, and other mimetic indicators to know where to look for meaning. Otherwise, the meaning is lost. At age three, children are discovering the meanings of nine new words a day, which is a full-time job, even for Sherlock Holmes! The gestural cues given by adults, as agents of culture, must be precise and unambiguous, or the meanings of words will not be discovered by the child. At this critical early stage, the attention mechanism must learn where to attend, how to find cues, and what to remember. Once this is learned, the discovery of words becomes easier and faster.

291 Mimetic disambiguation. See comments by Studdert-Kennedy (1998) for the possible role vocomimesis might have played in the evolution of the lowest rung in his proposed "particulate" hierarchy of language, morphophonology. Peter MacNeilage (1998a, 1998b, 1999) has suggested that the unique human ability to produce speech sounds is the direct result of the expanded prefrontal cortex, which gave us better voluntary control over the vocal apparatus by bringing two hitherto independent action systems under unified control. The two distinct neural systems exist in primates, one for lip smacking and vocalizing (a voluntary pathway), the other for mastication and swallowing (an involuntary pathway). By bringing these two systems under the unified control of the prefrontal cortex and conscious regulation, humans evolved a coordinated capacity for producing high-speed streams of articulated sound.

292 The road to narrative from vocomimesis. A useful literary perspective on the blending of mimesis into linguistic narrative was developed by Mark Turner (1996).

293 The shift from implicit to explicit expression in development. Karmiloff-Smith (1992) calls this the "explicitation" of the child's representational systems.

296 Mythic motives for evolving theoretic culture. The exact date of Easter falls on the Sunday after the first full moon after the vernal equinox. This date occurs

at different times in different places, and is difficult to standardize. The Vatican funded a massive scientific effort to fix the exact time of Easter, to the fraction of a second, by converting a number of cathedrals into state-of-the-art solar observations. Many of these curious observatories, called "meridiana," were in the forefront of astronomy, and still exist. See Heilbron, 1999.

296 Myth and the cultural matrix. Jordan Peterson (1999) has written a fascinating contemporary review of this topic. He shows that myth and archetype mediate the transition from childhood to adulthood in human beings. In effect, the conserved wisdom of the group, preserved largely in myth and allegory, has always been a powerful collective cognitive resource, one that serves to maximize individual potential in an efficient manner.

297 Rapid spread of Indo-European language and culture. See Renfrew, 1987.

8. THE TRIUMPH OF CONSCIOUSNESS

Page

301 Gleick, 1992; and Shahn, 1957.

303 Figure 8.1. This is a global, generic diagram that summarizes how reading and writing are represented in the brain, as revealed in a variety of dissociable neuropsychological deficits. Any of the functional boxes or arrows (pathways connecting boxes) in the diagram can be damaged independently of the others. See reviews by Coltheart et al., 1980; Shallice, 1988; and McCarthy and Warrington, 1990.

303 The literacy brain. I have suggested (1991) that there are at least three dissociable high-level visual interpretative paths involved in the literacy brain. The most basic pathway is the "pictorial" route, which is used to interpret pictograms and visual metaphors. Even at this primarily visual level there are numerous interpretative (mostly metaphoric) conventions to master. The second is the "ideographic" path, which is sometimes called the direct visual-semantic path in studies of reading. It maps visual symbols directly onto ideas, as in the case of Chinese ideographic writing and Roman numerals. The third is the "phonetic" path, which might also be called the spelling path. This neural network maps written forms, such as alphabetic letters, onto the sounds of speech. These three learned paths, or demons, emerged in different historical periods and are functionally independent of one another in the nervous system. Each path supports a distinct visual input lexicon containing thousands of recognizable forms.

304 Cultural reprogramming and literacy. Edelman's principles of reentrance and competitive growth selection might help explain the radical effect of literacy training on the individual brain. Edelman (1987) originally conceived of neural reentrance in the context of a solipsistic brain model, but his principle should also apply to a wider distributed network. Following the logic of Chapter 6, the cultural network sets up and delivers key experiential algorithms that shape the pattern of re-entrant

activation during development. In this way, the cultural matrix ultimately sets up the neural architectures needed for literacy. The mediator of this interaction is conscious capacity, but the source of the programming is cultural.

306 Cognition and the prehistory of external symbols. See Marshack, 1972; D'Errico, 1995; Mithen, 1996; and a recent volume by Renfrew and Scarre, 1998.

307 The slow adoption of symbolic technology. In 1991 I summarized the historical trend in the teaching of metalinguistic thought skills in Western academies over the past two thousand years. They moved from an early emphasis on oral and narrative skills toward modern visuosymbolic and paradigmatic skills. Denny (1991) has suggested that the major thought pattern attributable to literacy is a property called decontextualization, and Olson (1991, 1998) has stated that writing allowed the "objectification" of language and consequently the development of formal thought skills. These proposals are compatible with my own suggestion that literacy allowed the thought process itself to be displayed in public and subjected to iterative refinement, like any other artifact.

307 The interface between the conscious mind and external symbols. One brain, interacting with a set of symbols, constitutes a distributed cognitive system in itself. Mathematical thought, for instance, involves a flow of information between internal and external representations. Zhang and Norman (1995) have analyzed various numeration systems in this context.

308 Symbolic technology and the birth of science. Our symbiosis with symbols supported the growth of a novel, semiautonomous realm of human culture, based largely on an institutionalized literate elite. The algorithms of scientific or "paradigmatic" thought (see Bruner, 1990) have been cultivated gradually over thousands of years of experience with symbolically driven cultures. Theoretic skills include a wide range of thought algorithms that are not innate and are inconceivable outside the context of a highly symbol-dependent society. I call this third stage theoretic culture. This level of culture is dominated by a relatively small elite with special literacy-dependent skills. Its principal instruments of control are codified laws, economic and bureaucratic management, and formal scientific and cultural institutions. These have emerged as the governing level of cognition in modern society.

309 More powerful memory media. The transformative effects of symbols can be demonstrated very simply, by alluding to the computer metaphor once again. There are many ways to generate a capacity change in a computational system. One might increase the speed of the processors, for example. Or increase the amount of raw memory available. Or increase the size of a central processing component, such as the accumulator (a central point in digital computers where all calculations must be done). Symbolic technologies have achieved all these things.

316 Externally programmable minds. Recent research on child development shows that the cognitive enculturation of modern children is highly complex. Their introduction to formal cognitive operations is through the educational system,

as they are led through a tangled web of representational modes and complex institutionalized algorithms, all of which are tied to symbolic literacy (see, for example, Nelson, 1996; and Karmiloff-Smith, 1992). In effect, children's minds are being fashioned into multilayered cognitive systems, and thus they carry within themselves, as individuals, the evolutionary heritage of the past few million years.

317 Collaborative cognitive work. Cognitive studies of the modern workplace (for instance, Hutchins, 1995; Kitajima and Polson, 1997; Suchman, 1987; and Olson and Olson, 1991) examine the way that electronically distributed knowledge representation and computer-coordinated planning and problem solving are affecting the relative roles of individual minds and external memory devices in this collective enterprise. There is no doubt that these roles are changing as the technology supporting distributed networks evolves and that the nature of cognitive work itself is being transformed in the process.

References

Acredolo, L. P., and Goodwyn, S. W. 1988. Symbolic gesturing in normal infants. *Child development.* *59*, 450–66.

Allport, G. W. 1961. *Pattern and growth in personality.* New York: Holt, Rinehart and Winston.

———. 1966. *Becoming: Basic considerations for a psychology of personality.* New Haven: Yale University Press.

Arbib, M. A., Érdi, P., and Szentágothai, J. 1998. *Neural organization: Structure, function, and dynamics.* Cambridge, Mass.: MIT Press.

Armstrong, D. F., Stokoe, W. C., and Wilcox, S. E. 1995. *Gesture and the nature of language.* New York: Cambridge University Press.

Ashworth, P. D. 1979. *Social interaction and consciousness.* New York: John Wiley & Sons.

Baars, B. J. 1986. *The cognitive revolution in psychology.* New York: Guilford Press.

———. 1988. *A cognitive theory of consciousness.* New York: Cambridge University Press.

———. 1997. *In the theater of consciousness: The workspace of the mind.* New York: Oxford University Press.

Baddeley, A. 1986. *Working memory.* Oxford: Clarendon Press.

Baldwin, J. M. 1896. A new factor in evolution. *American naturalist.* *30*, 441–51, 536–53.

Barnes, Julian. 1985. *Flaubert's parrot.* London: Pan Books.

Baron-Cohen, S. 1995. *Mindblindedness and the language of the eyes.* Cambridge, Mass.: MIT Press.

———. 1999. The evolution of a theory of mind. In M. C. Corballis and S. E. G. Lea (eds.), *The descent of mind: Psychological perspective on hominid evolution.* New York: Oxford University Press. 261–77.

Bartlett, F. C. 1932. *Remembering: A study in experimental and social psychology.* Cambridge, U.K.: University Press.

———. 1958. *Thinking: An experimental and social study.* London: Allen & Unwin.

Başar, E., Başar-Eroğlu, C., Karakaş, S., and Schürmann, M. 1999. Are cognitive processes manifested in event-related gamma, alpha, theta, and delta oscillations in the EEG? *Neuroscience letters.* *259*, 165–68.

Batali, J. 1998. Computational simulations of the emergence of grammar. In J. R. Hurford, M. Studdert-Kennedy, and C. Knight (eds.), *Approaches to the evolution of language: Social and cognitive bases.* New York: Cambridge University Press. 405–26.

Bates, E., Thal, D., and Marchman, V. 1991. Symbols and syntax: A Darwinian approach to language development. In N. A. Krasnegor, D. M. Rumbaugh, R. L. Schiefelbusch, and M. Studdert-Kennedy (eds.), *Biological and behavioral determinants of language development.* Hillsdale, N.J.: Lawrence Erlbaum Associates.

Becker, S., and Hinton, G. E. 1992. Self-organizing neural network that discovers surfaces in random-dot stereograms. *Nature. 355*, 161–63.

Bergson, H. 1959. *L'évolution créatrice.* Paris: Presses Universitaires de France.

Berkeley, G. 1710 (1952). A treatise concerning the principles of human knowledge. In T. Jessop (ed.), *Berkeley: Philosophical writings.* London: Nelson.

Bickerton, D. 1981. *Roots of language.* Ann Arbor, Mich.: Karoma Publishers.

———. 1990. *Language and species.* Chicago: University of Chicago Press.

———. 1995. *Language and human behavior.* Seattle: University of Washington Press.

Bidwell, S. 1899. *Curiosities of light and sight.* London: Swan.

Block, N. 1978. Troubles with functionalism. In W. Savage (ed.), *Perception and cognition: Issues in the foundations of psychology.* Minnesota Studies in the Philosophy of Science. *9*, 261–326.

———. 1995. On a confusion about a function of consciousness. *Behavioral and brain sciences. 18*, 227–47.

———. 1997. *The nature of consciousness: Philosophical debates.* Cambridge, Mass.: MIT Press.

Bloom, P. 2000. *How children understand the meanings of words.* Cambridge, Mass.: MIT Press.

Boden, M. 1977. *Artificial intelligence and natural man.* New York: Basic Books.

Borges, J. L. 1999. *Selected non-fictions.* E. Weinberger (ed.). Toronto: Viking.

Brain, W. R. 1969. *Diseases of the nervous system.* 7th ed. London: Oxford University Press.

———. 1973. *Clinical neurology.* 4th ed. London: Oxford University Press.

Broca, P.-P. 1861. Remarques sur le siège de la faculté de la parole articulée, suivies d'une observation d'aphémie (perte de parole). *Bulletin de la Société d'anatomie* (Paris). *36*, 330–57.

Bruner J. S. 1983. *Child's talk: Learning to use language.* New York: W. W. Norton.

———. 1990. *Acts of meaning.* Cambridge, Mass.: Harvard University Press.

Byrne, R. W. 1995. *The thinking ape: Evolutionary origins of intelligence.* New York: Oxford University Press.

———, and Russon, A. 1998. Learning by imitation: A hierarchical approach. *Behavioral and brain sciences. 21*, 667–722.

———, and Whiten, A. 1988. *Machiavellian intelligence: Social expertise and the evolution of intellect in monkeys, apes, and humans.* New York: Oxford University Press.

———, and ———. 1997. *Machiavellian intelligence II: Extensions and evaluations.* New York: Cambridge University Press.

Calvin, W. H. 1983. A stone's throw and its launch window: Timing precision and its implications for language and hominid brains. *Journal of theoretical biology. 104*, 121–35.

———. The brain as a Darwin machine. *Nature. 330,* 33–34.
———. 1996. *How brains think: Evolving intelligence, then and now.* New York: Basic Books.
Carlyle, T. 1857. *Critical and miscellaneous essays.* London: Chapman and Hall.
Carroll, L. 1865 (1974). *Alice's adventures in Wonderland.* London: Bodley Head.
Chalmers, D. J. 1996. *The conscious mind: In search of a fundamental theory.* New York: Oxford University Press.
Chomsky, N. 1959. Review of *Verbal behavior* by B. F. Skinner. *Language. 35,* 26–58.
———. 1965. *Aspects of the theory of syntax.* Cambridge, Mass.: MIT Press.
———. 1984. *Modular approaches to the study of the mind.* San Diego: San Diego State University Press.
———. 1995. *Language and thought.* Wakefield, R.I.: Moyer Bell.
Churchland, P. S. 1986. *Neurophilosophy: Toward a unified science of the mind-brain.* Cambridge, Mass.: MIT Press.
Clark, A. 1997. *Being there: Putting brain, body, and world together again.* Cambridge, Mass.: MIT Press.
Coltheart, M., Patterson, K., and Marshall, J. (eds.). 1980. *Deep dyslexia.* London: Routledge.
Condillac, E. Bonnot de. 1930. *Condillac's Treatise on the sensations.* Los Angeles: University of Southern California School of Philosophy.
Corballis, M. C. 1991. *The lopsided ape.* New York: Oxford University Press.
———. 1999. Phylogeny from apes to humans. In M. C. Corballis and S. E. G. Lea (eds.), *The descent of mind: Psychological perspectives on hominid evolution.* New York: Oxford University Press. 40–70.
———, and Lea, S. E. G. (eds.). 1999. *The descent of mind: Psychological perspectives on hominid evolution.* New York: Oxford University Press.
Coull, J. T., Büchel, C., Friston, K. J., and Frith, C. D. 1999. Noradrenergically mediated plasticity in a human attentional neuronal network. *Neuroimage. 10,* 705–15.
Crick, F. 1994. *The astonishing hypothesis: The scientific search for the soul.* New York: Charles Scribner & Sons.
———, and Koch, C. 1995. Are we aware of neural activity in primary visual cortex? *Nature. 375,* 121–23.
Croce, B. 1992. *The aesthetic as the science of expression and of the linguistic in general.* New York: Cambridge University Press.
Damasio, A. R. 1994. *Descartes' error: Emotion, reason, and the human brain.* New York: G. P. Putnam.
———. 1999. *The feeling of what happens: Body and emotion in the making of consciousness.* New York: Harcourt Brace & Company.
Darwin, C. 1859. *On the origin of species by means of natural selection.* London: John Murray.
———. 1871. *The descent of man.* London: John Murray.
Dawkins, R. 1982. *The extended phenotype: The gene as the unit of selection.* San Francisco: Freeman.

———. 1987. *The blind watchmaker: Why the evidence of evolution reveals a universe without design.* New York: W. W. Norton.

———. 1990. *The selfish gene.* New York: Oxford University Press.

Deacon, T. W. 1997. *The symbolic species: The co-evolution of language and the brain.* New York: W. W. Norton.

Dennett, D.C. 1991. *Consciousness explained.* Boston: Little, Brown.

———. 1992. The role of language in intelligence. *Darwin lecture*, Cambridge University, March 6, 1992. Boston: Center for Cognitive Studies, Tufts University. CCS-92-3.

———. 1995. *Darwin's dangerous idea: Evolution and the meanings of life.* New York: Simon & Schuster.

Denny, J. P. 1991. Rational thought in oral culture and literate decontextualization. In D. Olson and N. Torrance (eds.), *Literacy and orality.* New York: Cambridge University Press. 66–89.

D'Errico, F. 1991. Microscopic and statistical criteria for the identification of prehistoric systems of notation. *Rock art research. 8*, 83–93.

———. 1995. A new model and its implications for the origin of writing: The La Marche antler revisited. *Cambridge archaeological journal. 5*, 163–206.

Derrida, J. 1987. *The archeology of the frivolous: Reading Condillac.* Lincoln: University of Nebraska Press.

Descartes, R. 1993. *Discourse on method and meditations on first philosophy.* Indianapolis: Hackett Publishing.

Dodwell, P. 2000. *Brave new mind: A thoughtful inquiry into the nature and meaning of mental life.* New York: Oxford University Press.

Donald, M. W. 1970. Direct-current potentials in the human brain recorded during timed cognitive performance. *Nature. 227*, 1057–58.

———. 1983. Neural selectivity in auditory attention: Sketch of a theory. In W. Ritter and A. Gaillard (eds.), *Tutorials in ERP research: Endogenous components.* Amsterdam: Elsevier North Holland. 37–77.

———. 1991. *Origins of the modern mind: Three stages in the evolution of culture and cognition.* Cambridge, Mass.: Harvard University Press.

———. 1993a. Human cognitive evolution: What we were, what we are becoming. *Social research. 60*, 143–70.

———. 1993b. Précis of *Origins of the modern mind* with multiple reviews and author's response. *Behavioral and brain sciences. 16*, 737–91.

———. 1993c. Hominid enculturation and cognitive evolution. *Archaeological review from Cambridge. 12*, 5–24.

———. 1995. The neurobiology of human consciousness: An evolutionary approach. *Neuropsychologia. 33*, 1087–1102.

———. 1996. The role of vocalization, memory retrieval and external symbols in cognitive evolution. *Behavioral and brain sciences. 19*, 155–64.

———. 1997. The mind considered from a historical perspective: Human cognitive phylogenesis and the possibility of continuing cognitive evolution. In D. Johnson and C. Ermeling (eds.), *The future of the cognitive revolution.* New York: Oxford University Press. 355–65.

———. 1998a. Mimesis and the Executive Suite: Missing links in language evolution. In J. R. Hurford, M. Studdert-Kennedy, and C. Knight (eds.), *Approaches to the evolution of language: Social and cognitive bases*. New York: Cambridge University Press. 44–67.

———. 1998b. Hominid enculturation and cognitive evolution. In C. Renfrew and C. Scarre (eds.), *Cognition and material culture: The archaeology of symbolic storage*. Cambridge, U.K.: McDonald Institute for Archaeological Research. 7–17.

———. 1998c. Material culture and cognition: Concluding thoughts. In C. Renfrew and C. Scarre (eds.), *Cognition and material culture: The archaeology of symbolic storage*. Cambridge, U.K.: McDonald Institute for Archaeological Research. 181–87.

———. 1999. Preconditions for the evolution of protolanguages. In M. C. Corballis and S. E. G. Lea (eds.), *The descent of mind: Psychological perspectives in cognitive evolution*. New York: Oxford University Press. 138–54.

———, and Goff, W. R. 1971. Attention-related increases in cortical responsivity dissociated from the contingent negative variation. *Science. 172*, 1163–66.

———, and Young, M. 1982. A time-course analysis of attentional tuning of the auditory evoked response. *Experimental brain research. 46*, 357–67.

Dunbar, R. 1996. *Grooming, gossip and the evolution of language*. Boston: Faber & Faber.

———. 1998. Theory of mind and the evolution of language. In J. R. Hurford, M. Studdert-Kennedy, and C. Knight (eds.), *Approaches to the evolution of language: Social and cognitive bases*. New York: Cambridge University Press. 92–110.

Durham, W. H. 1978. Toward a coevolutionary theory of human biology and culture. In A. L. Caplan (ed.), *The sociobiology debate*. New York: Harper & Row.

Eccles, J. C., Ito, M., and Szentágothai, J. 1967. *The cerebellum as a neuronal machine*. Berlin: Springer-Verlag.

Edelman, G. M. 1987. *Neural Darwinism: The theory of neuronal group selection*. New York: Basic Books.

———. 1989. *The remembered present: A biological theory of consciousness*. New York: Basic Books.

———. 1992. *Bright air, brilliant fire: On the matter of the mind*. New York: Basic Books.

———, and Mountcastle, V. B. 1978. *The mindful brain: Cortical organization and the group-selective theory of higher brain function*. Cambridge, Mass.: MIT Press.

———, and Tononi, G. 2000. *A universe of consciousness: How matter becomes imagination*. New York: Basic Books.

Egan, K. 1997. *The educated mind: How cognitive tools shape our understanding*. Chicago: University of Chicago Press.

Eldredge, N. 1995. *Reinventing Darwin: The great debate at the high table of evolutionary theory*. New York: John Wiley & Sons.

Elman, J. L. (ed.). 1996. *Rethinking innateness: A connectionist perspective on development*. Cambridge, Mass.: MIT Press.

Engel, A. K., Fries, P., König, P., Brecht, M., and Singer, W. 1999. Temporal binding, binocular rivalry, and consciousness. *Consciousness and cognition. 8*, 128–51.

Engle, R. W., Kane, M. J., and Tukolsky, S. W. 1999. Individual differences in working memory capacity, and what they tell us about controlled attention, general fluid intelligence, and the function of the prefrontal cortex. In A. Miyake and P. Shah (eds.), *Models of working memory: Mechanisms of active maintenance and executive control.* London: Cambridge University Press.

Ericsson, K. A., and Kintsch, W. 1995. Long-term working memory. *Psychological review. 102*, 211–45.

Eriksen, C. W. 1960. Discrimination and learning without awareness: A methodological survey and evaluation. *Psychological review. 67*, 279–301.

Falk, D. 1992. *Braindance: New discoveries about human brain evolution.* New York: Henry Holt.

Flanagan, O. 1992. *Consciousness reconsidered.* Cambridge, Mass.: MIT Press.

Flaubert, G. 1857 (1957). *Madame Bovary.* F. Steegmuller (trans.). New York: Random House.

Flavell, J. H., and Wohlwill, J. F. 1969. Formal and functional aspects of cognitive development. In G. Dawson and K. W. Fischer (eds.), *Studies in cognitive development: Studies in honour of J. Piaget.* New York: Oxford University Press.

Fodor, J. A. 1983. *Modularity of mind: An essay on faculty psychology.* Cambridge, Mass.: MIT Press.

Foelix, R. F. 1996. *Biology of spiders.* New York: Oxford University Press.

Forman, G., and Pufell, P. B. (eds.). 1988. *Constructivism in the computer age.* Hillsdale, N.J.: Lawrence Erlbaum Associates.

Fouts, R., and Mills, S. T. 1997. *Next of kin: What chimpanzees have taught me about who we are.* New York: William Morrow.

Frisch, K. von. 1967. *The dance language and orientation of bees.* Cambridge, Mass.: Belknap Press of Harvard University Press.

Gabora, L. 1998. Autocatalytic closure in a cognitive system: A tentative scenario for the origin of culture. *Psycoloquy.* 9:67, 1–26.

Gadamer, H. G. 1976. *Philosophical hermeneutics.* Berkeley: University of California Press.

Gallese, V. 1999. Agency and the self model. *Consciousness and cognition. 8*, 387–89.

Gallup, G. G. 1970. Chimpanzees: Self recognition. *Science. 167*, 86–87.

———. 1982. Self awareness and the emergence of mind in primates. *American journal of primatology. 2*, 237–48.

Gamble, C. 1994. *Timewalkers: The prehistory of global colonization.* Cambridge, Mass.: Harvard University Press.

Gannon, P. J., Holloway, R. L., and Broadfield, A. R. B. 1997. Asymmetry of chimpanzee planum temporale: Humanlike pattern of Wernicke's brain language area homolog. *Science. 279*, 220–22.

Gardner, H. C. 1985. *The mind's new science.* New York: Basic Books.

Gardner, R. A., and Gardner, B. T. 1969. Teaching Sign language to a chimpanzee. *Science. 165*, 664–72.
Gazzaniga, M. S. 1998. *The mind's past.* Berkeley: University of California Press.
———, Ivry, R. B., and Mangun, G. R. *Cognitive neuroscience.* New York: W. W. Norton.
Georgieff, N., and Jeannerod, M. 1998. Beyond consciousness of external reality: A "who" system for consciousness of action and self-consciousness. *Consciousness and cognition. 7*, 465–77.
Geschwind, N. 1965. Disconnection syndromes in animals and man. *Brain. 88*, 237–94, 585–644.
———. 1970. The organization of language and the brain. *Science. 170*, 940–44.
Gibson, J. J. 1950. *The perception of the visual world.* Boston: Houghton Mifflin.
———. 1979. *The ecological approach to visual perception.* Boston: Houghton Mifflin.
Gibson, W. 1975. *The miracle worker.* New York: Bantam Books.
Gleick, J. 1992. *Genius.* New York: Pantheon.
Gold, I. 1999. Does 40-Hz oscillation play a role in visual consciousness? *Consciousness and cognition. 8*, 186–95.
Goldin-Meadow, S., McNeill, D., and Singleton, J. 1996. Silence is liberating: Removing the handcuffs on grammatical expression in the manual modality. *Psychological review. 103*, 34–55.
———, and Mylander, C. 1998. Spontaneous sign systems created by deaf children in two cultures. *Nature. 391*, 279–81.
———, and McNeill, D. 1999. The role of gesture and mimetic representation in making language the province of speech. In M. C. Corballis and S. E. G. Lea (eds.), *The descent of mind: Psychological perspectives on hominid evolution.* New York: Oxford University Press. 155–72.
Goldman-Rakic, P. S. 1992. Working memory and the mind. *Scientific American.* Sept. 1992, 111–17.
Goldstein, K. 1948. *Language and language disturbance.* New York: Grune & Stratton.
Gould, S. J., and Vrba, E. 1981. Exaptation: A missing term in the science of form. *Paleobiology. 8*, 4–15.
Greene, G. 1950. *The third man.* New York: Viking.
Greenfield, P. M., and Smith, J. 1976. *The Structure of communication in early language development.* New York: Academic Press.
Griffin, D. R. 1992. *Animal minds.* Chicago: University of Chicago Press.
Haldane, J. B. S. 1927. *Possible worlds and other essays.* London: Chatto & Windus.
Hamilton, W. 1859. *Lectures on metaphysics and logic.* Boston: Gould.
Hanlon, R. T. 1996. *Cephalopod behavior.* New York: Cambridge University Press.
Harris, P. L. 1998. Mimesis, imagination and role-play. In C. Renfrew and C. Scarre (eds.), *Cognition and material culture.* Cambridge, U.K.: McDonald Institute for Archaeological Research.

Hauser, M. 1996. *The evolution of communication.* Cambridge: MIT Press.

———. 2000. *Wild minds.* New York: Henry Holt.

Hebb, D. O. 1949. *The organization of behavior: A neuropsychological theory.* New York: John Wiley & Sons.

———. 1961. Distinctive features of learning in the higher animal. In J. F. Delafresnaye (ed.), *Brain mechanisms and learning.* Oxford: Blackwell.

———. 1963. The semiautonomous process: Its nature and nurture. *American psychologist. 18,* 16–27.

Heilbron, J. L. 1999. *The sun in the church.* Cambridge, Mass.: Harvard University Press.

Heller, J. 1970. *Catch-22.* New York: Dell.

Helprin, M. 1991. *A soldier of the great war.* New York: Harcourt Brace Jovanovich.

Heron, W. 1967. The pathology of boredom. In J. L. McGaugh, N. M. Weinberger, and R. E. Whalen (eds.), *Psychobiology.* San Fransisco: W. H. Freeman.

Hillyard, S. A., Hink, R. F., Schwent, V. L., and Picton, T. W. 1973. Electrical signs of selective attention in the human brain. *Science. 182,* 177–80.

Humphrey, K. 1992. *A history of the mind.* London: Chatto & Windus.

Humphrey, N. 1983. *Consciousness regained: Chapters in the development of mind.* New York: Oxford University Press.

Hurford, J. R., Studdert-Kennedy, M., and Knight, C. (eds.). 1998. *Approaches to the evolution of language: Social and cognitive bases.* New York: Cambridge University Press.

Hutchins, E. 1995. *Cognition in the wild.* Cambridge, Mass: MIT Press.

Huxley, J. 1953. *Evolution in action.* New York: Harper & Brothers.

Imamizu, H., Miyauchi, S., Tamada, T., Sasaki, Y., Takino, R., Putz, B., Yoshioka, T., and Kawato, M. 2000. Human cerebellar activity reflecting an acquired internal model of a new tool. *Nature. 403,* 192–95.

Ingold, T. 1986. *Evolution and social life.* New York: Cambridge University Press.

Jablonski, N. G., and Aiello, L. C. (eds.). 1998. *The origin and diversification of language.* San Francisco: Memoirs of the California Academy of Sciences.

Jackendoff, R. 1985. *Semantics and cognition.* Cambridge, Mass.: MIT Press.

———. 1994. *Patterns in the mind.* New York: Basic Books.

James, H. 1886 (1994). *The Bostonians.* London: J. M. Dent.

James, W. 1890. *The principles of psychology.* New York: Henry Holt.

Jasper, H. H. (ed.). 1958. *Reticular formation of the brain.* Boston: Little, Brown.

Jaynes, J. 1976. *The origin of consciousness in the breakdown of the bicameral mind.* Boston: Houghton Mifflin.

Joyce, J. 1937 (1958). *Ulysses.* London: Bodley Head Ltd.

Julesz, B. 1995. *Dialogues on perception.* Cambridge, Mass.: MIT Press.

Jusczyk, P. W. 1997. *The discovery of spoken language.* Cambridge, Mass.: MIT Press.

Karmiloff-Smith, A. 1992. *Beyond modularity: A developmental perspective on cognitive science.* Cambridge, Mass.: MIT Press.

Kawato, M. 1999. Internal models for motor control and trajectory planning. *Current opinion in neurobiology. 9*, 718–27.

Keller, H. A. 1899. Chronological statement of studies. *Helen Keller souvenir commemorating the Harvard final examination for admission to Radcliffe College, June 29–30, 1899*. Washington: Volta Bureau. 60–65.

———. 1902. *The story of my life*. New York: Grosset & Dunlap.

———. 1908. *The world I live in*. New York: Century.

———. 1927. *My religion*. New York: Swedenborg Foundation & Twayne.

Keverne, E. B., Martel, F. L., and Nevison, C. M. 1996. Primate brain evolution: Genetic and functional considerations. *Proceedings of the National Academy of Sciences. 96*, 689–96.

Kintsch, W. 1998. *Comprehension: A paradigm for cognition*. Boulder, Colo.: Cambridge University Press.

Kirby, S. 1998. Fitness and the selective adaptation of language. In J. R. Hurford, M. Studdert-Kennedy, and C. Knight (eds.), *Approaches to the evolution of language: Social and cognitive bases*. New York: Cambridge University Press. 359–83.

Kirk, R. 1994. *Raw feeling: A philosophical account of the essence of consciousness*. New York: Oxford University Press.

Kitajima, M., and Polson, P. 1997. A comprehension-based model of exploration. *Human-computer interaction. 12*, 345–89.

Knight, C. 1998. Ritual/speech coevolution: A solution to the problem of deception. In J. R. Hurford, M. Studdert-Kennedy, and C. Knight (eds.), *Approaches to the evolution of language: Social and cognitive bases*. New York: Cambridge University Press. 68–91.

Kolb, B., and Whishaw, I. Q. 1990. *Fundamentals of human neuropsychology*. 3d Ed. New York: W. H. Freeman.

Kolers, P. A., and von Grünau, M. 1976. Shape and color in apparent motion. *Vision research. 16*, 329–35.

Koriat, A. 2000. The feeling of knowing: Some metatheoretical implications for consciousness and control. *Consciousness and cognition. 9*, 149–71.

Kosslyn, S. 1988. Aspects of a cognitive neuroscience of imagery. *Science. 240*, 1621–26.

———. 1994. *Image and brain: The resolution of the imagery debate*. Cambridge, Mass.: MIT Press.

Kuper, A. 1994. *The chosen primate: Human nature and cultural diversity*. Cambridge, Mass.: Harvard University Press.

———. 1999. *Culture: The anthropologist's account*. Cambridge, Mass.: Harvard University Press.

Kutas, M., McCarthy, G., and Donchin, E. 1977. Augmenting mental chronometry: The P300 as a measure of stimulus evaluation time. *Science. 197*, 792–95.

LaBar, K. S., Gitelman, D. R., Parrish, T. B., and Mesulam, M.-M. 1999.

Neuroanatomic overlap of working memory and spatial attention networks: A functional MRI comparison within subjects. *Neuroimage. 10*, 695–704.

LaBerge, D. 1995. *Attentional processing: The brain's art of mindfulness.* Cambridge, Mass.: Harvard University Press.

———. 1997. Attention, awareness, and the triangular circuit. *Consciousness and cognition. 6*, 149–81.

Lakoff, G., and Johnson, M. 1980. *Metaphors we live by.* Chicago: University of Chicago Press.

———, and ———. 1999. *Philosophy in the flesh: The embodied mind and its challenge to Western thought.* New York: Basic Books.

Lane, H. (ed.). 1984. *The deaf experience: Classics in language and education.* Cambridge, Mass.: Harvard University Press.

Langer, J., and Killen, M. (eds.). 1998. *Piaget, evolution and development.* Mahwah, N.J.: Lawrence Erlbaum Associates.

Langer, S. K. 1967. *Mind: An essay on human feeling.* Baltimore: Johns Hopkins University Press.

Lash, J. P. 1980. *Helen and teacher: The story of Helen Keller and Anne Sullivan Macy.* New York: Delacorte Press/Seymour Lawrence.

Le Carré, J. 1975. *The spy who came in from the cold.* Toronto: Bantam.

Lecours, A. R., and Joanette, Y. 1980. Linguistic and other aspects of paroxysmal aphasia. *Brain and language. 10*, 1–23.

Lehmann, H., Ban, T., and Donald, M. 1965. Rating the rater: An experimental approach to the methodological problem of interrater agreement. *Archives of general psychiatry. 13*, 67–75.

Leiber, J. 1996. Helen Keller as cognitive scientist. *Philosophical psychology. 9*, 419–40.

Leibniz, G. W. 1714 (1951). The Monadology. In P. Weiner (ed.), *Leibnitz: Selections.* New York: Charles Scribner & Sons.

Leiner, H. C., Leiner, A. L., and Dow, R. S. 1993. Cognitive and language functions of the human cerebellum. *Trends in the neurosciences. 16*, 444–47.

Lemon, R. N., Armand, J., Olivier, E., and Edgley, S. A. 1999. Skilled action and the development of the corticospinal tract in primates. In K. Connolly and H. Forssberg (eds.), *Neurophysiology and neuropsychology of motor development.* New York: Cambridge University Press.

Levelt, W. J. M. 1989. *Speaking: From intention to articulation.* Cambridge, Mass.: MIT Press.

Libet, B. 1985a. Subjective antedating of a sensory experience and mind-brain theories. *Journal of theoretical biology. 114*, 563–70.

———. 1985b. Unconscious cerebral initiative and the role of conscious will in voluntary action. *Behavioral and brain sciences. 8*, 529–66.

———. 1993. *Neurophysiology of consciousness: Selected papers and new essays of Benjamin Libet.* Boston: Birkhäuser.

Lieberman, P. 1984. *The biology and evolution of language.* Cambridge, Mass.: Harvard University Press.

———. 1991. *Uniquely human: The evolution of speech, thought, and selfless behavior.* Cambridge, Mass.: Harvard University Press.

Lindsley, D. B. 1961. The reticular activation system and perceptual integration. In D. E. Sheer (ed.), *Electrical stimulation of the brain.* Austin: University of Texas Press. 331–49.

———, Schreiner, L. H., Knowles, W. B., and Magoun, H. W. 1950. Behavioral and EEG changes following chronic brain stem lesions in the cat. *Electroencephalography and clinical neurophysiology. 2,* 483–98.

Luria, A. R. 1966. *Human brain and psychological processes.* New York: Harper & Row.

———. 1987. *The man with a shattered world: The history of a brain wound.* Cambridge, Mass.: Harvard University Press.

———, and Vygotsky, L. S. 1992. *Ape, primitive man, and child: Essays in the history of behavior.* New York: Harvester Wheatsheaf.

Mach, E. 1959. *The analysis of sensations and the relation of the physical to the psychical.* New York: Dover Publications.

Mack, A., and Rock, I. 1998. *Inattentional blindness.* Cambridge, Mass.: MIT Press.

MacNeilage, P. F. 1998a. The frame-content theory of the evolution of speech production. *Behavioral and brain sciences. 21,* 499–546.

———. 1998b. Evolution of the mechanisms of language output: Comparative neurobiology of vocal and manual communication. In J. R. Hurford, M. Studdert-Kennedy, and C. Knight (eds.), *Approaches to the evolution of language: Social and cognitive bases.* New York: Cambridge University Press. 222–41.

———. 1999. Whatever happened to articulate speech? In M. C. Corballis and S. E. G. Lea (eds.), *The descent of mind: Psychological perspectives on hominid evolution.* New York: Oxford University Press. 116–37.

Magoun, H. W. 1965. *The waking brain.* Springfield, Ill.: Charles C. Thomas.

Mandler, G. 1959. *The language of psychology.* New York: John Wiley.

———. 1985. *Cognitive psychology: An essay in cognitive science.* Hillsdale, N.J.: Lawrence Erlbaum Associates.

———. 1997. *Human nature explored: Psychology, evolution, society.* New York: Oxford University Press.

Marcel, A. J. 1988. Phenomenal experience and functionalism. In A. J. Marcel and E. Bisiach (eds.), *Consciousness in contemporary science.* Oxford: Clarendon Press.

Marcus Aurelius. 170 (1964). *Meditations.* New York: Penguin.

Marr, D. 1982. *Vision: A computational investigation into the human representation and processing of visual information.* San Francisco: W. H. Freeman.

Marshack, A. 1972. *The roots of civilization: The cognitive beginnings of man's first art, symbol, and notation.* New York: McGraw-Hill.

Mayr, E. 1974. Behavioral programs and evolutionary strategy. *American scientist. 62,* 650–59.

———. 1982. *The growth of biological thought: Diversity, evolution, and inheritance.* Cambridge, Mass.: Belknap Press of Harvard University Press.

———. 1988. *Toward a new philosophy of biology: Observations of an evolutionist.* Cambridge, Mass.: Belknap Press of Harvard University Press.

———. 1991. *One long argument: Charles Darwin and the genesis of modern evolutionary thought.* Cambridge, Mass.: Harvard University Press.

McCarthy, R., and Warrington, E. 1990. *Cognitive neuropsychology.* San Diego: Academic Press.

McGinn, C. 1991. *The problem of consciousness: Essays toward a resolution.* Cambridge, Mass.: Blackwell.

McGurk, H., and MacDonald, J. 1976. Hearing lips and seeing voices. *Nature. 264,* 746–48.

McIntosh, A. R., Grady, C. L., Ungerleider, L. G., Haxby, J. V., Rapoport, S. I., and Horwitz, B. 1994. Network analysis of cortical visual pathways mapped with PET. *Journal of neuroscience. 14,* 655–66.

McNeill, D. 1992. *Hand and mind: What gestures reveal about thought.* Chicago: University of Chicago Press.

Meltzoff, A. N. and Moore, M. K. 1997. Explaining facial imitation: A theoretical model. *Early Development and Parenting. 6,* 179–92.

Merzenich, M. M., Nelson, R. J., Kaas, J. H., Stryker, M. P., Jenkins, W. M., Zook, J. M., Cynader, M. S., and Schoppana, A. 1987. Variability in hand surface representations in areas 3b and 1 in adult owl and squirrel monkeys. *Journal of comparative neurology. 258,* 281–96.

Miller, G. A. 1956. The magical number seven, plus or minus two: Some limits on our capacity to process information. *Psychological review. 63,* 81–97.

Milner, A. D., and Goodale, M. A. 1992. Separate visual pathways for perception and action. *Trends in the neurosciences. 15,* 20–25.

———, and ———. 1995. *The visual brain in action.* New York: Oxford University Press.

Milner, P. 1957. The cell assembly: Mark II. *Psychological review. 64,* 242–52.

Miltner, W. H. R., Braun, C., Arnold, M., Witte, H., and Taub, E., 1999. Coherence of gamma-band EEG activity as a basis for associative learning. *Nature. 397,* 434–36.

Milton, K. 1980. *The foraging strategy of howler monkeys: A study in primate economics.* New York: Columbia University Press.

Minsky, M. L. 1986. *The society of mind.* New York: Simon and Schuster.

Mithen, S. 1996. *The prehistory of the mind: A search for the origins of art, religion and science.* London: Thames and Hudson.

Moruzzi, G., and Magoun, H. W. 1949. Brainstem reticular formation and activation of the EEG. *Electroencephalography and clinical neurophysiology. 1,* 455–73.

Mountcastle, V. B. 1998. *Perceptual neuroscience: The cerebral cortex.* Cambridge, Mass.: Harvard University Press.

Näätänen, R. 1992. *Attention and brain function.* Hillsdale, N.J.: Lawrence Erlbaum Associates.

Nauta, W. J. H. 1993. *Neuroanatomy: Selected papers of Walle J. H. Nauta.* Boston: Birkhäuser.

———, and Feirtag, M. 1986. *Fundamental neuroanatomy*. New York: W. H. Freeman.

Neisser, U. 1982. *Memory observed: Remembering in natural contexts*. New York: W. H. Freeman.

Nelson, K. 1986. *Event knowledge: Structure and function in development*. Hillsdale, N.J.: Lawrence Erlbaum Associates.

———. 1993. Psychological and social origins of autobiographical memory. *Psychological science. 4*, 7–14.

———. 1996. *Language in cognitive development: Emergence of the mediated mind*. New York: Cambridge University Press.

———, and Shaw, L. K. 2000. Developing a socially shared symbolic system. In J. Byrnes and E. Amsel (eds.), *Language, literacy, cognitive development*. New York: Erlbaum. 1–52.

Neville, H. J., and Bavelier, D. 1999. Specificity and plasticity in neurocognitive development in humans. In M. Gazzaniga (ed.), *The cognitive neurosciences*. Cambridge, Mass.: MIT Press. 83–98.

Newman, J. 1995. Review: Thalamic contributions to attention and consciousness. *Consciousness and cognition. 4*, 172–93.

———, and Baars, B. J. 1993. A neural attentional model for access to consciousness: A global workspace perspective. *Concepts in neuroscience. 4*, 255–90.

———, and Grace, A. A. 1999. Binding across time: The selective gating of frontal and hippocampal systems modulating working memory and attentional states. *Consciousness and cognition. 8*, 196–212.

Newmeyer, F. J. 1998. On the supposed "counter functionality" of Universal Grammar: Some evolutionary implications. In J. R. Hurford, M. Studdert-Kennedy, and C. Knight (eds.), *Approaches to the evolution of language: Social and cognitive bases*. New York: Cambridge University Press. 305–19.

Noble, W., and Davidson, I. 1996. *Human evolution, language and mind: A psychological and archaeological inquiry*. New York: Cambridge University Press.

Norman, D. A., and Shallice, T. 1986. Attention to action: Willed and automatic control of behavior. In R. J. Davidson, G. E. Schwartz, and D. Shapiro (eds.), *Consciousness and self-regulation*. New York: Plenum Press.

Norris, C. 1993. *The truth about Postmodernism*. Oxford: Basil Blackwell.

Nottebohm, F. 1984. Learning, forgetting and brain repair. In N. Geschwind and A. M. Galaburda (eds.), *Cerebral dominance: The biological foundations*. Cambridge, Mass.: Harvard University Press.

O'Brien, G., and Opie, J. 1999. A connectionist theory of phenomenal experience. *The behavioral and brain sciences. 22*, 127–96.

Olson, D. 1991. Literacy as metalinguistic activity. In D. Olson and N. Torrance (eds.), *Literacy and orality*. New York: Cambridge University Press. 251–70.

———. 1994. *The world on paper: The conceptual and cognitive implications of writing and reading*. Cambridge, U.K.: Cambridge University Press.

Olson, J. R. and Olson, G. M. 1991. User-centered design of collaboration technology. *Journal of organizational computing. 1*, 61–83.

Parasuraman, R. (ed.). 1998. *The attentive brain*. Cambridge, Mass.: MIT Press.

Passingham, R. 1982. *The human primate*. New York: W. H. Freeman.

Penrose, R. 1990. *The emperor's new mind: Concerning computers, minds, and the laws of physics*. New York: Vintage.

———. 1995. *Shadows of the mind: A search for the missing source of consciousness*. London: Vintage.

Pepperberg, I. M. 1987. Acquisition of the same/different concept by African grey parrot *Psittaceus erithacus*. *Animal learning and behavior*. *15*, 423–32.

Peterson, J. 1999. *Maps of meaning: The architecture of belief*. New York: Routledge.

Piaget, J. 1976. *The grasp of consciousness: Action and concept in the young child*. Cambridge, Mass.: Harvard University Press.

Picton, T. W., and Hillyard, S. A. 1974. Human auditory evoked potentials II: Effects of attention. *Electroencephalography and clinical neurophysiology*. *36*, 191–99.

Pinker, S. 1994. *The language instinct*. New York: William Morrow.

———. 1997. *How the mind works*. New York: W. W. Norton.

———, and Bloom, P. 1990. Natural language and natural selection. *Behavioral and brain sciences*. *13*, 707–84.

Plotkin, H. 1993. *Darwin machines and the nature of knowledge*. Cambridge, Mass.: Harvard University Press.

———. 1998. *Evolution in mind: An introduction to evolutionary psychology*. Cambridge, Mass.: Harvard University Press.

Popper. K. R., and Eccles, J. C. 1977. *The self and its brain*. New York: Springer International.

Posner, M. I., and Raichle, M. E. 1994. *Images of mind*. New York: Scientific American Library.

———, DiGirolamo, G. J., and Fernandes-Duque, D. 1997. Brain mechanisms of cognitive skills. *Consciousness and cognition*. *6*, 267–90.

Potts, R. 1997. *Humanity's descent: The consequences of ecological instability*. New York: Avon Books.

Premack, D. 1976. *Intelligence in ape and man*. Hillsdale, N.J.: Lawrence Erlbaum Associates.

———. 1987. *Gavagai*. Cambridge, Mass.: MIT Press.

———. 1993. Prolegomenon to evolution of cognition. In T. A. Poggio and D. A. Glaser (eds.), *Exploring brain functions: Models in neuroscience*. New York: John Wiley.

Pribram, K. H. 1971. *Languages of the brain: Experimental paradoxes and principles in neuropsychology*. Englewood Cliffs, N.J.: Prentice-Hall.

———. 1999. The self as me and I. *Consciousness and cognition*. *8*, 385–86.

Quartz, S. R., and Sejnowski, T. J. 1994. Beyond modularity: Neural evidence for constructivist principles in development. *Behavioral and brain sciences*. *17*, 725–26.

Rapoport, S. I. 1999. How did the human brain evolve? A proposal based on new evidence from *in vivo* brain imaging during attention and ideation. *Brain research bulletin*. *50*, 149–65.

Reber, A. 1996. *Implicit learning and tacit knowledge: An essay on the cognitive unconscious.* New York: Oxford University Press.
Renfrew, C. 1987. *Archaeology and language.* London: Jonathan Cape.
———, and Scarre, C. (eds.). 1998. *Cognition and material culture: The archaeology of symbolic storage.* Cambridge, U.K.: McDonald Institute for Archaeological Research.
Revonsuo, A., and Newman, J. 1999. Binding and consciousness. *Consciousness and cognition. 8,* 123–27.
Rey, G. 1997. *Contemporary philosophy of mind: A contentiously classical approach.* Cambridge, Mass.: Blackwell.
Rizzolatti, G., Fadiga, L., Gallese, V., and Fogassi, L. 1996. Premotor cortex and the recognition of motor actions. *Cognitive brain research. 3,* 131–41.
———, and Arbib, M. A. 1998. Language within our grasp. *Trends in neurosciences. 21,* 188–94.
Robb. G. 1994. *Balzac: A biography.* New York: W. W. Norton.
Rodriguez, E., George, N., Lachaux, J., Martinerie, J. Renault, B., and Varela, F. 1999. Perception's shadow: Long-distance synchronization of human brain activity. *Nature. 397,* 430–33.
Rorty, R. 1989. *Contingency, irony, and solidarity.* New York: Cambridge University Press.
Rubin, D. C. 1995. *Memory in oral traditions: The cognitive psychology of epic, ballads, and counting-out rhymes.* New York: Oxford University Press.
Rumbaugh, D. 1974. *Language learning by a chimpanzee: The Lana project.* New York: Academic Press.
Sacks, O. W. 1970. *The man who mistook his wife for a hat and other clinical tales.* New York: HarperCollins.
———. 1995. *An anthropologist on Mars: Seven paradoxical tales.* New York: Knopf.
Saussure, F. de. 1983. *Course in general linguistics.* C. Bally, A. Sechehaye, and A. Riedlinger (eds.), R. Harris (trans.). London: G. Duckworth.
Sauvé, K. 1999. Gamma-band synchronous oscillations: Recent evidence regarding their functional significance. *Consciousness and cognition. 8,* 213–24.
Savage-Rumbaugh, E. S. 1986. *Ape language: From conditioned response to symbol.* New York: Columbia University Press.
———. 1991. Multi-tasking: The *Pan*-human rubicon. *Neurosciences. 3,* 417–22.
———, and Lewin, R. 1994. *Kanzi: The ape at the brink of the human mind.* New York: John Wiley.
———, Shanker, S. G., and Taylor, T. J. 1998. *Apes, language, and the human mind.* New York: Oxford University Press.
———, Murphy, J., Sevcik, R. A., Brakke, K. E., Williams, S. L., and Rumbaugh, D. M. 1993. *Language comprehension in ape and child.* Chicago: Monographs of the Society for Research in Child Development. *58.*
Schacter, D. L., and Scarry, E. (eds.). 2000. *Memory, brain, and belief.* Cambridge, Mass.: Harvard University Press.
Schaller, S. 1991. *A man without words.* New York: Summit Books.
Searle, J. R. 1997. *The mystery of consciousness.* London: Granta Books.

Shahn, B. 1957. *The shape of content.* Cambridge, Mass.: Harvard University Press.

Shallice, T. 1988. *From neuropsychology to mental structure.* New York: Cambridge University Press.

Simon, H. 1969. *The sciences of the artificial.* Cambridge, Mass.: MIT Press.

Singer, W. 1994. The organization of sensory motor representations in the neocortex: A hypothesis based on temporal coding. In C. Umilta and M. Moscovitch (eds.), *Attention and performance XV: Conscious and nonconscious information processing.* Cambridge, Mass.: MIT Press. 77–107.

Sinnott, E. W. 1961. *Cell and psyche: The biology of purpose.* New York: Harper & Brothers.

Skinner, B. F. 1959. *Verbal behavior.* New York: Appleton-Century-Crofts.

Smith, J. D. 1992. Knowing thyself, knowing the other: They're not the same. *Behavioral and brain sciences. 15,* 166–67.

Sober, E., and Wilson, D. S. 1998. *Unto others: The evolution of psychology and unselfish behavior.* Cambridge, Mass.: Harvard University Press.

Sokolov, A. N. 1977. *Inner speech and thought.* New York: Plenum Press.

Spencer, H. 1855. *The principles of psychology.* London: Longman, Brown, Green & Longmans.

Stendhal (Henri-Marie Beyle). 1830 (1970). *The red and the black.* New York: Penguin.

Studdert-Kennedy, M. 1998. The particulate origins of language generativity, from syllable to gesture. In J. R. Hurford, M. Studdert-Kennedy, and C. Knight (eds.), *Approaches to the evolution of language: Social and cognitive bases.* New York: Cambridge University Press. 202–21.

Stuss, D., and Benson, F. 1986. *The frontal lobes.* New York: Raven Press.

Suchman, L. 1987. *Plans and situated actions: The problem of human-machine communication.* New York: Cambridge University Press.

Suddendorf, T. 1999. The rise of the metamind. In M. C. Corballis and S. E. G. Lea (eds.), *The descent of mind: Psychological perspectives on hominid evolution.* New York: Oxford University Press. 218–60.

Sullivan, A. M. 1891. *Report to the Perkins Institution.*

———. 1899. Instruction of Helen Keller. *Helen Keller souvenir commemorating the Harvard final examination for admission to Radcliffe College, June 29–30, 1899.* Washington: Volta Bureau. 12–23.

Szentágothai, J. 1969. Architecture of the cerebral cortex. In H. H. Jasper, A. A. Ward, Jr., and A. Pope (eds.), *Basic mechanisms of the epilepsies.* Boston: Little, Brown. 13–28.

———, Hámori, J., and Vizi, E. S. 1977. *Neuron concept today: Symposium held in Tihany, Hungary, August 26–28, 1976.* Budapest: Akadémiai Kiadó.

Tamada, T., Miyauchi, S., Imamizu, H., Yoshioka, T., and Kawato, M. Cerebro-cerebellar functional connectivity revealed by the laterality index in tool-use learning. *Neuroreport. 10,* 325–31.

Terrace, H. S. 1986. *Nim.* New York: Columbia University Press.

———, and Bever, T. G. 1980. What might be learned from studying language in the chimpanzee? The importance of symbolizing oneself. In T. Sebeok and J. Sebeok (eds.), *Speaking of apes*. New York: Plenum.

Thatch, W. T. 1998. What is the role of the cerebellum in motor learning and cognition? *Trends in the cognitive sciences. 2*, 331–37.

Tilney, Frederick. 1929. A comparative sensory analysis of Helen Keller and Laura Bridgman. *Archives of neurology and psychiatry. 21*, 1228–69.

Tomasello, M. 1999. *The cultural origins of human cognition*. Cambridge, Mass.: Harvard University Press.

Tooby, J., and Cosmides, L. 1992. Psychological foundations of culture. In J. Barkow, L. Cosmides, and J. Tooby (eds.), *The adapted mind: Evolutionary psychology and the generation of culture*. New York: Oxford University Press, pp. 19–136.

Treisman, A., and Gelade, G. 1980. A feature integration theory of attention. *Cognitive psychology. 12*, 97–136.

Trevarthen, C. 1979. Instincts for human understanding and for cultural cooperation: Their development in infancy. In M. von Cranach, K. Foppa, W. Lepenies, and D. Ploof (eds.), *Human ethology: Claims and limits of a new discipline*. Cambridge, UK: Cambridge University Press.

———. 1980. The foundations of intersubjectivity: Development of interpersonal and cooperative understanding in infants. In B. de Boysson-Bardies, S. de Schonen, P. Juszyck, P. MacNeilage, and J. Morton (eds.), *Developmental neurocognition: Speech and face processing in the first year of life*. Dordrecht, Netherlands: *Kluwer*.

Tulving, E. 1983. *Elements of episodic memory*. Oxford: Clarendon Press.

———. 1984. Précis of *Elements of episodic memory*. *The behavioral and brain sciences. 7*, 223–68.

Turner, M. 1996. *The literary mind: The origins of thought and language*. New York: Oxford University Press.

Varela, F. J. and Shear, J. 1999. *The view from within*. Thorverton, UK: Imprint Academic Press.

———, Thompson, E., and Rosch, E. 1995. *The embodied mind: Cognitive science and human experience*. Cambridge, Mass.: MIT Press.

Vogeley, K., Kurthen, M., Falkai, P., and Maier, W. 1999. Essential functions of the human self model are implemented in the prefrontal cortex. *Consciousness and cognition. 8*, 343–63.

Vrba, E. S. 1998. Multiphasic growth models and the evolution of prolonged growth exemplified by human brain evolution. *Journal of theoretical biology. 190*, 227–39.

Vygotsky, L. S. 1978. *Mind in society: The development of higher psychological processes*. Cambridge, Mass.: Harvard University Press.

———. 1986. *Thought and language*. Cambridge, Mass.: MIT Press.

Waal, F. B. M. de. 1989. *Peacemaking among primates*. Cambridge, Mass.: Harvard University Press.

———. 1996. *Good natured: The origins of right and wrong in humans and other animals*. Cambridge, Mass.: Harvard University Press.

———. 1997. *Bonobo: The forgotten ape*. Berkeley: University of California Press.

Weiskrantz, L. 1997. *Consciousness lost and found: A neurological exploration*. Oxford: Oxford University Press.

Wells, H. G. 1992. *The invisible man*. New York: Dover Publications.

Wernicke, C. 1874. *Der aphasische Symptomenkomplex*. Breslau: Cohn and Weigart. *Boston studies in philosophy and science. 4*, 34–97.

Wills, C. 1993. *The runaway brain: The evolution of human uniqueness*. New York: Basic Books.

Willshaw, D. J., Buneman, O. P., and Longuet-Higgins, H. C. 1954. Non-holographic associative memory. *Nature. 222*, 960–62.

Wilson, E. O. 1971. *The insect societies*. Cambridge, Mass.: Belknap Press of Harvard University Press.

———. 1978. *On human nature*. Cambridge, Mass.: Harvard University Press.

———. 1992. *The diversity of life*. Cambridge, Mass.: Belknap Press of Harvard University Press.

Wittgenstein, L. 1963. *Philosophical investigations*. Oxford: Basil Blackwell.

———. 1980. *Remarks on the philosophy of psychology.* Chicago: University of Chicago Press.

Wolpert, D. M., Miall, R. C., and Kawato, M. 1998. Internal models in the cerebellum. *Trends in the cognitive sciences. 2*, 338–47.

Woolf, N. J. 1997. A possible role for cholinergic neurons of the basal forebrain and pontomesencephalon in consciousness. *Consciousness and cognition. 6*, 574–96.

———. 1999. Dendritic encoding: An alternative to temporal synaptic coding of conscious experience. *Consciousness and cognition. 8*, 447–54.

Wrangham, R. W., McGrew, W. C., Waal, F. B. M. de, and Heltne, P. G. (eds.). 1994. *Chimpanzee cultures*. Cambridge, Mass.: Harvard University Press.

Young, J. Z. 1960. *A model of the brain*. Oxford: Clarendon Press.

Zelazo, P. D. 2000. Self-reflection and the development of consciously controlled processing. In P. Mitchell and K. J. Riggs (eds.), *Children's reasoning and the mind.* Hove, UK: Psychology Press. 169–89.

Zhang, J., and Norman, D. 1995. A representational analysis of numeration systems. *Cognition. 57*, 271–95.

Acknowledgments

When I started to assemble this book, I had only a few vague ideas about the connection between consciousness and human evolution. Those were summarized briefly in my first book, but the precise role of consciousness in human evolution escaped me completely at that time. Moreover, the literature on consciousness was exploding, and I abandoned this project for something more reassuringly predictable. My interest in the project was rekindled when I received an invitation from French neurobiologist Jean Delacour to write a paper on consciousness for the journal *Neuropsychologia,* which I accepted. The exercise convinced me that it might be worthwhile to pursue these ideas further, and this book is the result. It may be just a small brick in a massive theoretical edifice whose completion is not on the horizon, but I can offer the reader at least one notion that is new: Consciousness played a causal role in the evolution of the human mind. We are a supremely conscious species; this was the key to our success in the past, and it will continue to be in the future.

The completion of this book has been helped by a number of institutions and people. First, I must thank my editor at W. W. Norton, Angela von der Lippe, for her wise guidance, and Stefanie Diaz, also at Norton, for her help in shepherding this volume through various stages of production. The Social Science and Humanities Research Council of Canada provided me with vital operating grant support during this project, and I wish also to thank the Killam Trust, the Canada Council, and the Faculty of Arts and Science of Queen's University, for their financial support. The Departments of Psychology of Harvard University and University College, London, provided me with space and secretarial support while on leave. I gratefully acknowledge the efforts of my hosts (and good friends) at those two institutions, professors Sheldon White and Henry Plotkin. I also acknowledge the enthusiastic cooperation I received from Ken Stuckey of the Keller Library, Perkins School for the Blind.

On the academic side, I should especially thank Peter Dodwell and Katherine Nelson for their extensive comments on the penultimate version of this manuscript. Harlan Lane, Donald Brunet, and Oliver Sacks provided me with feeback on specific passages. Too many other people have helped me, sometimes inadvertently, with ideas, feedback, anecdotes, and examples, to list here, but I would like to thank some of my colleagues at Queen's, particularly Barrie Frost, Hans Dringenberg, Kang Lee, Darwin Muir, and Cella Olmstead, for regularly vetting ideas and trading knowledge. I also thank Alex Kirlik, of the Georgia Institute of Technology, for graciously sending me one of the opening quotations for Chapter 8.

Shelagh Freedman served as my principal library researcher for the later stages of this project, and kept my ever-chaotic and constantly growing filing system from disintegrating completely. Several other research assistants also helped me during the earlier stages, including Peter Donald, Michelle Eskritt, Jessica Goldberg, Brenda Orser, and Samara Warren. Monica Hurt ably produced the figures. Julian Donald was a crucial player in the final production; he edited endnotes and references, helped generate the index, did library searches, and entered editorial changes to the penultimate manuscript.

Finally, I cannot thank my wife, Thais, enough, for agreeing to apply her enormous editorial talents to this manuscript, and for finding the perfect line (from Shakespeare's *Cymbeline,* I, 6, 16–17) for the title. Without her support, dedication, and constant companionship, this project could not have been brought to completion.

KINGSTON, ONTARIO,
DECEMBER 2000

Index